Non-Instantaneous Impulsive Differential Equations

Basic theory and computation

Non-Instantaneous Impulsive Differential Equations

Basic theory and computation

JinRong Wang
Guizhou University, People's Republic of China

Michal Fečkan
Comenius University in Bratislava

IOP Publishing, Bristol, UK

ISBN 978-0-7503-1704-7 (ebook)
ISBN 978-0-7503-1702-3 (print)
ISBN 978-0-7503-1703-0 (mobi)

DOI 10.1088/2053-2563/aada21

Version: 20181101

IOP Expanding Physics
ISSN 2053-2563 (online)
ISSN 2054-7315 (print)

British Library Cataloguing-in-Publication Data: A catalogue record for this book is available from the British Library.

Published by IOP Publishing, wholly owned by The Institute of Physics, London

IOP Publishing, Temple Circus, Temple Way, Bristol, BS1 6HG, UK

US Office: IOP Publishing, Inc., 190 North Independence Mall West, Suite 601, Philadelphia, PA 19106, USA

To our beloved families.

Contents

Preface

Many real life processes and phenomena are characterized by rapid changes in their state. There are two main types of such changes. In the first type the changes take place over a relatively short time compared to the overall duration of the whole process. Mathematical models in such cases are created using impulsive equations. In the second type the changes are not negligibly short in duration and these changes begin impulsively at some points and remain active over certain time intervals. Mathematical models of these situations use non-instantaneous impulsive differential equations. These equations give rise to a new hybrid dynamical system which contains a continuous-time dynamical system, a discrete-time dynamical system and an algebraic system. Non-instantaneous impulsive differential equations provide an excellent tool for the description of the injection of drugs in the bloodstream, and their consequent absorption in the body as a random, periodic, gradual and continuous process. Other examples of such processes can be found in physics, biology, population dynamics, ecology and economics.

This monograph is devoted to non-instantaneous impulsive differential equations. The concepts, existence and asymptotical behaviors of solutions, such as orbital Hausdorff continuous dependence, random effects and Ulam's stability, are presented. Periodic and almost periodic solutions are also obtained. The broad variety of concepts and results with rigorous proofs that are provided, as well as the many examples, make this monograph unique. The material herein is based on research carried out by the authors over the past four years. The monograph is self-contained and unified in presentation, and provides the necessary background material required to go further into the subject and explore the rich research literature. Each chapter concludes with a section devoted to notes and bibliographical remarks, and all abstract results are illustrated by examples.

This monograph is useful for researchers and graduate students studying evolution equations and other nonlinear problems with non-instantaneous impulsive effects, as well as for research, seminars and advanced graduate courses in pure and applied mathematics and related disciplines.

JinRong Wang and Michal Fečkan
Guizhou University, Guiyang, Guizhou 550025, People's Republic of China
Faculty of Mathematics, Physics and Informatics,
Comenius University in Bratislava, 842 48 Bratislava, Slovakia

Acknowledgement

We would like to thank Professors M Benchohra, K Dishlieva, A G Ibrahim, D O'Regan, J J Trujillo and Y Zhou for their support. We also wish to express appreciation to our colleagues Dr W Wei and Q Chen, and graduate students D Yang, M M Li, S D Liu and Z L You for their help. Finally, we are grateful for the editorial assistance of IOP Publishing.

We acknowledge with gratitude the support of the National Natural Science Foundation of China (11661016), the Slovak Research and Development Agency under contract No. APVV-14-0378, and the Slovak Grant Agency VEGA No. 1/0078/17.

Author Biographies

JinRong Wang

JinRong Wang has been a Professor of the Department of Mathematics at the School of Mathematics and Statistics at Guizhou University, People's Republic of China since 2011. He received his master's degree from Guizhou University in 2006 and PhD from the same university in 2009. He is interested in impulsive differential equations, fractional differential equations, delay differential equations and iterative learning controls. He is the 2015-2018 Highly Cited Researcher in Mathematics.

Michal Fečkan

Michal Fečkan has been a Professor of Mathematics at the Department of Mathematical Analysis and Numerical Mathematics in the Faculty of Mathematics, Physics and Informatics at the Comenius University in Bratislava, Slovak Republic since 2003. He received the master's degree (mathematics) from Comenius University in Bratislava in 1985 and PhD (mathematics) from the Mathematical Institute of Slovak Academy of Sciences in Bratislava, Slovak Republic in 1993. He is interested in nonlinear functional analysis, bifurcation theory, dynamical systems and fractional calculus with applications to mechanics, vibrations and economics.

Chapter 1

Linear and perturbed equations

In this chapter we establish a unified framework to investigate the stability and Lyapunov regularity of linear and perturbed equations. In section 1.1 we discuss stability in relation to linear and perturbed equations and give sufficient conditions to guarantee that the desired equations are asymptotically and exponentially stable. Section 1.2.2 is devoted to studying Lyapunov regularity, a new version of Perron's theorem, criteria for nonuniform exponential behavior, and lower and upper bounds for regularity coefficients.

1.1 Stability analysis

1.1.1 Introduction

Impulsive differential equations are widely used in mechanical engineering, medical science, the life sciences and many other fields. Many researchers have made great contributions to this topic. In general, two types of impulsive effects are described by differential equations. The first type is differential equations with instantaneous impulses [1], i.e. the duration of these changes is relatively short compared to the overall duration of the whole process. The second type is differential equations with non-instantaneous impulses [2], i.e. the impulsive action starts at an arbitrary fixed point and remains active over a finite time interval.

Inspired by the contribution from Myshkis and Samoilenko [1] who provided differential equations with instantaneous impulses, Bainov and Simeonov [3–5], Lakshmikantham *et al* [6], Samoilenko *et al* [7] and Benchohra *et al* [8] presented the fundamental theory for instantaneous impulsive differential equations. There are also some recent works on qualitative analysis for instantaneous impulsive differential equations, for example, see [9–14]. However, the development of the basic theory for non-instantaneous impulsive differential equations has been very slow. Hernández and O'Regan [2] originally introduce a new class of differential equations with non-instantaneous impulses based on the context of injecting therapeutic drugs into a person. In their works, a mild solution was provided by using strongly continuous semigroup and fixed point methods. Next, Pierri and O'Regan [15] derived a new existence result for the same problem by using analytic semigroup theory and fixed point methods in fractional power space. Motivated by [2], Wang and Fečkan [16, 17] extended the study to a new generalized class of differential equations with non-instantaneous impulses which depend on the previous state. For more recent works on this topic, see O'Regan *et al* [19–22], Wang *et al* [23–25], Abbas and Benchohra [28], Gautam and Dabas [29], and Yan and Lu [30].

Essentially, linear non-instantaneous impulsive problems correspond to a more general evolution process than linear instantaneous impulsive differential equations that change instantaneously at certain times. The basic properties of the standard impulsive Cauchy matrix

$$X(t, t_0) = e^{A(t-t_i)} \prod_{t_0<t_j<t_i} (\mathrm{Id} + B)e^{A(t_j-t_{j-1})},$$
$$t_i < t \leqslant t_{i+1}$$

where 'Id' denotes the identity matrix, and the exponential stability of linear instantaneous impulsive differential equations

$$\begin{cases} x'(t) = Ax(t), & t \neq t_i, \ i = 1, 2, \dots, \\ x(t_i^+) - x(t_i^-) = Bx(t_i^-), & i = 1, 2, \dots. \end{cases}$$

was established in [7], where $x(t) \in \mathbb{R}^n$, A, B are constant $n \times n$ matrices with $AB = BA$, and t_i acts as an impulsive point satisfying $t_0 = 0 < t_1 < t_2 \cdots < t_i < t_{i+1} \cdots$, $t_i \to \infty$. $x(t_i^+)$ and $x(t_i^-)$ represent the right and left limits of $x(t)$ at $t = t_i$, respectively. In addition, we set $x(t_i^-) = x(t_i)$. However, to the best of our knowledge, there is no literature investigating the associated issues for linear non-instantaneous impulsive problems.

We develop the approach of [7] and investigate three basic problems:

1. The asymptotic stability of linear non-instantaneous impulsive differential equations of the type

$$\begin{cases} x'(t) = Ax(t), & t \in [s_i, t_{i+1}], \ i = 0, 1, 2, \dots, \\ x(t_i^+) = Bx(t_i^-), & i = 1, 2, \dots, \\ x(t) = Bx(t_i^-), & t \in (t_i, s_i], \ i = 1, 2, \dots, \\ x(s_i^+) = x(s_i^-), & i = 1, 2, \dots, \end{cases} \tag{1.1}$$

where s_i acts as a junction point satisfying $t_0 = s_0 = 0 < t_1 < s_1 < t_2 \cdots < t_i < s_i < t_{i+1} \cdots$, $t_i \to \infty$.

2. The asymptotic stability of linear non-instantaneous impulsive differential equations of the type

$$\begin{cases} x'(t) = Ax(t) + P(t)x(t), & t \in [s_i, t_{i+1}], \ i = 0, 1, 2, \dots, \\ x(t_i^+) = Bx(t_i^-) + I_i x(t_i^-), & i = 1, 2, \dots \\ x(t) = Bx(t_i^-) + I_i x(t_i^-), & t \in (t_i, s_i], \ i = 1, 2, \dots, \\ x(s_i^+) = x(s_i^-), & i = 1, 2, \dots, \end{cases} \tag{1.2}$$

where $P(t)$ is a continuous $n \times n$ matrix for $t \geqslant 0$ and I_i are constant $n \times n$ matrices, $i = 1, 2, \ldots$.

3. The existence and uniqueness of solutions and the Ulam–Hyers–Rassias stability of nonlinear non-instantaneous impulsive equations of the type

$$\begin{cases} y'(t) = Ay(t) + g(t, y(t)), \ t \in [s_i, t_{i+1}], \ i = 0, 1, 2, \ldots, \\ y(t_i^+) = By(t_i^-) + b_i, \ i = 1, 2, \ldots, \\ y(t) = By(t_i^-) + b_i, \ t \in (t_i, s_i], \ i = 1, 2, \ldots, \\ y(s_i^+) = y(s_i^-), \ i = 1, 2, \ldots, \end{cases} \tag{1.3}$$

where $g \in C(J \times \mathbb{R}^n, \mathbb{R}^n)$ and $J = [0, \infty)$.

In section 1.1.2 we derive the formula for a solution involving a non-instantaneous impulsive Cauchy matrix for (1.1), and discuss the asymptotic stability and instability for (1.1) and provide sufficient conditions. In section 1.1.3 we give sufficient conditions to guarantee asymptotic stability of the solution of (1.2). In section 1.1.4 we study the existence and uniqueness of the solutions and Ulam–Hyers–Rassias stability of (1.3).

1.1.2 Solutions of linear non-instantaneous impulsive equations

We first present the formula of the solution of (1.1) for commutative matrices A and B via a non-instantaneous impulsive Cauchy matrix.

Let $i(t, 0)$ denote the number of impulsive points which belong to $(0, t)$ and $z^+ = \max\{0, z\}$ for $z \in \mathbb{R}$.

The following result is fundamental although its proof is straightforward (see [7, chapter 2.1] for a similar discussion).

Theorem 1.1.1. *The solution $x(t, s, x_0)$ of (1.1) with initial value condition $x(s) = x_0$ has the form*

$$x(t, s, x_0) = W(t, s)x_0, \quad \forall\ 0 \leqslant s \leqslant t,$$

where $W(\cdot, \cdot)$ is a non-instantaneous impulsive Cauchy matrix given by

$$W(t, s) = B^{i(t,\,0)-i(s,\,0)} e^{A\left[(t-s_{i(t,\,0)})^+ - (s-s_{i(s,\,0)})^+ + \sum_{k=i(s,0)}^{i(t,0)-1} (t_{k+1}-s_k)\right]} \tag{1.4}$$

and we set $\sum_{k=i(s,0)}^{i(t,0)-1} = 0$ for $i(s, 0) = i(t, 0)$. In particular,

$$W(t, 0) = B^{i(t,\,0)} e^{A\left[(t-s_{i(t,\,0)})^+ + \sum_{k=0}^{i(t,0)-1} (t_{k+1}-s_k)\right]}$$

and

$$x(t, x_0) := x(t, 0, x_0) = B^{i(t,\,0)} e^{A\left[(t-s_{i(t,\,0)})^+ + \sum_{k=0}^{i(t,0)-1} (t_{k+1}-s_k)\right]} x_0. \tag{1.5}$$

Definition 1.1.2. *The solution $x(t, x_0)$ is locally asymptotically stable if there exists $\delta > 0$ such that for any $y_0 \in \mathbb{R}^n$ with $\|x_0 - y_0\| < \delta$, the following holds:*

$$\lim_{t\to\infty} \|x(t, x_0) - x(t, y_0)\| = 0.$$

If δ can be arbitrary then $x(t, x_0)$ is globally asymptotically stable.

Remark 1.1.3. *Local and global asymptotic stabilities are coincident for (1.1). Moreover, it is enough to study the stability of the zero solution $x(t, 0) = 0$. Hence we say that (1.1) is asymptotically stable if its zero solution is so. Moreover, from (1.5), it is sufficient to estimate $W(t, 0)$.*

Now, we introduce the concept of the exponential stability of (1.1).

Definition 1.1.4. *System (1.1) is exponentially stable if there exist constants $K \geqslant 1$ and $\gamma > 0$ such that*

$$\|W(t, s)\| \leqslant Ke^{-\gamma(t-s)}, \quad 0 \leqslant s \leqslant t.$$

To estimate $\|W(t, 0)\|$ in terms of the eigenvalues of the involved matrices for the study of the stability of (1.1), the following well-known lemmas are applied [34].

Lemma 1.1.5. *Let $|\cdot|$ be a norm on $\mathbb{R}^n$ and A be an $n \times n$ matrix. Set $\alpha(A) = \max\{\Re\lambda | \lambda \in \sigma(A)\}$ and $\beta(A) = \min\{\Re\lambda | \lambda \in \sigma(A)\}$. Then for any $\varepsilon > 0$ there are $K_\varepsilon \geqslant 1$ and $0 < k_\varepsilon \leqslant 1$ such that*

$$k_\varepsilon e^{(\beta(A)-\varepsilon)t} \leqslant \|e^{At}\| \leqslant K_\varepsilon e^{(\alpha(A)+\varepsilon)t}$$

for any $t \geqslant 0$. Here $\|\cdot\|$ is the matrix norm generated by $|\cdot|$, i.e. $\|A\| = \max_{|x|=1}|Ax|$ and $\sigma(A)$ is the spectrum of A.

Lemma 1.1.6. *Let $|\cdot|$ be a norm on $\mathbb{R}^n$ and B be an $n \times n$ matrix. For any $\varepsilon > 0$ there is $K_\varepsilon \geqslant 1$ such that $\|B^k\| \leqslant K_\varepsilon(\rho(B) + \varepsilon)^k$ for any non-negative integer k, where $\rho(B)$ is the spectral radius of B.*

Then we have the following lemma.

Lemma 1.1.7. *For any $\varepsilon > 0$ there exists $K_\varepsilon \geqslant 1$ such that*

$$\|W(t, 0)\| \leqslant K_\varepsilon e^{(\alpha(A)+\varepsilon)\left[(t-s_{i(t,0)})^+ + \sum_{k=0}^{i(t,0)-1}(t_{k+1}-s_k)\right]}(\rho(B) + \varepsilon)^{i(t,0)}.$$

Proof. From lemmas 1.1.5 and 1.1.6, for any $\varepsilon > 0$ there is $\bar{K}_\varepsilon \geqslant 1$ such that

$$\|B^{i(t,0)}\| \leqslant K_\varepsilon(\rho(B) + \varepsilon)^{i(t,0)},$$

$$\left\| e^{A\left[(t-s_{i(t,0)})^+ + \sum_{k=0}^{i(t,0)}(t_{k+1}-s_k)\right]} \right\| \leqslant \bar{K}_\varepsilon e^{(\alpha(A)+\varepsilon)\left[(t-s_{i(t,0)})^+ + \sum_{k=0}^{i(t,0)-1}(t_{k+1}-s_k)\right]}.$$

Thus,

$$\| W(t, 0)\| \leqslant K_\varepsilon e^{(\alpha(A)+\varepsilon)\left[(t-s_{i(t,0)})^+ + \sum_{k=0}^{i(t,0)-1}(t_{k+1}-s_k)\right]}(\rho(B) + \varepsilon)^{i(t,0)},\ \ K_\varepsilon = \bar{K}_\varepsilon^2.$$

The proof is complete. □

Suppose that the distance between the impulsive points t_i and junction points s_i satisfies

$$0 < \tilde{\theta}_1 \leqslant t_{k+1} - s_k \leqslant \tilde{\theta}_2,\ k = 0, 1, 2, 3, \ldots. \tag{1.6}$$

Define

$$\tilde{\theta} = \begin{cases} \tilde{\theta}_1, & \alpha(A) < 0, \\ \tilde{\theta}_2, & \alpha(A) \geqslant 0. \end{cases}$$

Now we give sufficient conditions to guarantee asymptotic stability of (1.1), and extend the results in [7, chapter 3.1].

Theorem 1.1.8. *Assume that (1.6) holds. If the following inequality holds*

$$\Xi := \alpha(A) + \frac{1}{\tilde{\theta}} \ln \rho(B) < 0, \tag{1.7}$$

then (1.1) is asymptotically stable.

Proof. For any $\varepsilon > 0$, from lemma 1.1.7, we have

$$\begin{aligned} \|W(t, 0)\| &\leqslant K_\varepsilon e^{(\alpha(A)+\varepsilon)\left[(t-s_{i(t,0)})^+ + \sum_{k=0}^{i(t,0)-1}(t_{k+1}-s_k)\right]}(\rho(B) + \varepsilon)^{i(t,0)} \\ &\leqslant K_\varepsilon e^{(\alpha(A)+\varepsilon)\left[(t-s_{i(t,0)})^+ + i(t,0)\tilde{\theta}\right]}(\rho(B) + \varepsilon)^{i(t,0)} \\ &\leqslant K_\varepsilon e^{(\alpha(A)+\varepsilon)\tilde{\theta}}\left[e^{(\alpha(A)+\varepsilon)\tilde{\theta}}(\rho(B) + \varepsilon)\right]^{i(t,0)}. \end{aligned} \tag{1.8}$$

It follows from the condition (1.7) that one can choose ε such that

$$e^{(\alpha(A)+\varepsilon)\tilde{\theta}}(\rho(B) + \varepsilon) \leqslant e^{\frac{\tilde{\theta}\Xi}{2}} < 1.$$

Then (1.8) gives

$$\|W(t,0)\| \leqslant K_\varepsilon e^{(\alpha(A)+\varepsilon)\tilde{\theta}} e^{\frac{\Xi}{2} i(t,0)} \to 0 \quad \text{for} \quad t \to \infty,$$

since $i(t, 0) \to \infty$ as $t \to \infty$. The proof is complete. □

Remark 1.1.9. *From (1.6) we obtain* $i(t, 0) \leqslant \frac{t}{\tilde{\theta}_1}$. *Hence* $\limsup_{t\to\infty} \frac{i(t,0)}{t} \leqslant \frac{1}{\tilde{\theta}_1}$. *Clearly* $\liminf_{t\to\infty} \frac{i(t,0)}{t} \geqslant 0$. *This cannot be improved in general, since taking* $t_k = e^k$ *and* $s_k = t_{k+1} - 1$ *for* $k \geqslant 1$, *we obtain* $\tilde{\theta}_1 = \tilde{\theta}_2 = \tilde{\theta} = 1$ *and* $i(t, 0) = \lceil \ln t \rceil - 1$ *for* $t > e$ *and* $t \to \lceil t \rceil$ *is the ceiling function, which denotes the smallest integer greater than or equal to t. Then* $\lim_{t\to\infty} \frac{i(t,0)}{t} = 0$. *This makes a difference with instantaneous impulses* $t_i = s_i$ *for any* $i \geqslant 1$, *since then (1.6) implies* $\liminf_{t\to\infty} \frac{i(t,0)}{t} \geqslant \frac{1}{\tilde{\theta}_2}$, *and so condition (1.7) gives the exponential stability of (1.1).*

Motivated by the previous remark, we arrive at the following results.

Theorem 1.1.10. *If one of the following conditions holds*

$$\rho(B) \geqslant 1, \quad \limsup_{t\to\infty} \frac{i(t,0)}{(t - s_{i(t,0)})^+ + \sum_{k=0}^{i(t,0)-1} (t_{k+1} - s_k)} := \tilde{p} < \infty, \tag{1.9}$$

$$\rho(B) < 1, \quad \liminf_{t\to\infty} \frac{i(t,0)}{(t - s_{i(t,0)})^+ + \sum_{k=0}^{i(t,0)-1} (t_{k+1} - s_k)} := \tilde{p} < \infty, \tag{1.10}$$

then (1.1) is asymptotically stable provided that the following inequality

$$\Upsilon = \alpha(A) + \tilde{p} \ln \rho(B) < 0 \tag{1.11}$$

is satisfied.

Proof. For any $\varepsilon > 0$, from lemma 1.1.7, there is $K_\varepsilon > 0$ such that

$$\|W(t,0)\| \leqslant K_\varepsilon e^{(\alpha(A)+\varepsilon)\left[(t - s_{i(t,0)})^+ + \sum_{k=0}^{i(t,0)-1} (t_{k+1} - s_k)\right]} (\rho(B) + \varepsilon)^{i(t,0)}.$$

From (1.9), we obtain

$$i(0, t) < (\tilde{p} + \varepsilon)\left[(t - s_{i(t,0)})^+ + \sum_{k=0}^{i(t,0)-1} (t_{k+1} - s_k)\right]$$

for any $t \gg 1$ large enough. Consequently, using $\rho(B) \geqslant 1$, we derive

$$\|W(t, 0)\| \leqslant K_\varepsilon e^{(\alpha(A)+\varepsilon+(\tilde{p}+\varepsilon)\ln(\rho(B)+\varepsilon))\left[(t-s_{i(t,0)})^{+}+\sum_{k=0}^{i(t,0)-1}(t_{k+1}-s_k)\right]}.$$

For ε sufficiently small, $\alpha(A) + \varepsilon + (\tilde{p}+\varepsilon)\ln(\rho(B) + \varepsilon) < \Upsilon/2 < 0$, so then

$$\|W(t, 0)\| \leqslant K_\varepsilon e^{\frac{\Upsilon}{2}\left[(t-s_{i(t,0)})^{+}+\sum_{k=0}^{i(t,0)-1}(t_{k+1}-s_k)\right]} \to 0 \quad \text{for} \quad t \to \infty,$$

since $i(t, 0) \to \infty$ as $t \to \infty$.

Similarly, from (1.10), we obtain

$$i(0, t) > (\tilde{p}-\varepsilon)\left[(t - s_{i(t,0)})^{+}+\sum_{k=0}^{i(t,0)-1}(t_{k+1} - s_k)\right]$$

for any $t \gg 1$ large enough. Consequently, using $\rho(B) < 1$, we assume $\rho(B) + \varepsilon < 1$ and derive

$$\|W(t, 0)\| \leqslant K_\varepsilon e^{(\alpha(A)+\varepsilon+(\tilde{p}-\varepsilon)\ln(\rho(B)+\varepsilon))\left[(t-s_{i(t,0)})^{+}+\sum_{k=0}^{i(t,0)-1}(t_{k+1}-s_k)\right]}.$$

For ε sufficiently small, $\alpha(A) + \varepsilon + (\tilde{p}-\varepsilon)\ln(\rho(B) + \varepsilon) < \Upsilon/2 < 0$, so then

$$\|W(t, 0)\| \leqslant K_\varepsilon e^{\frac{\Upsilon}{2}\left[(t-s_{i(t,0)})^{+}+\sum_{k=0}^{i(t,0)-1}(t_{k+1}-s_k)\right]} \to 0 \quad \text{for} \quad t \to \infty,$$

since $i(t, 0) \to \infty$ as $t \to \infty$. The proof is complete. □

Note that

$$\begin{aligned} W(t, 0) &= e^{A\left[(t-s_{i(t,0)})^{+}+\sum_{k=0}^{i(t,0)-1}(t_{k+1}-s_k)\right]}B^{i(t,0)} \\ &= e^{\Lambda\left[(t-s_{i(t,0)})^{+}+\sum_{k=0}^{i(t,0)-1}(t_{k+1}-s_k)\right]}e^{\ln B\left(i(t,0)-\tilde{p}\left[(t-s_{i(t,0)})^{+}+\sum_{k=0}^{i(t,0)-1}(t_{k+1}-s_k)\right]\right)}, \end{aligned} \tag{1.12}$$

because of $A \ln B = \ln BA$.

Set $\Lambda = A + \tilde{p}\ln B$. Now we improve theorem 1.1.10 by changing the condition on $\alpha(A) + \tilde{p}\ln\rho(B)$ to the view of matrix Λ directly.

Theorem 1.1.11. *Let B be nonsingular and suppose*

$$\lim_{t\to\infty} \frac{i(t, 0)}{(t - s_{i(t,0)})^{+}+\sum_{k=0}^{i(t,0)-1}(t_{k+1} - s_k)} := \tilde{p}<\infty. \tag{1.13}$$

Then it holds that:

(i) *If* $\alpha(\Lambda) < 0$ *then (1.1) is asymptotically stable.*

(ii) *If* $\alpha(\Lambda) > 0$ *then (1.1) is unstable.*

(iii) *If* $\beta(\Lambda) > 0$ *then all nonzero solutions of (1.1) tend to infinity as* $t \to \infty$.

Proof. (i) From $\alpha(\Lambda) < 0$ and lemma 1.1.5, there is $K \geqslant 1$ such that

$$\| e^{\Lambda t} \| \leqslant K e^{-\gamma t}, \quad t \geqslant 0 \tag{1.14}$$

for $\gamma = -\frac{\alpha(\Lambda)}{2}$. From (1.13), there is $t_0 > 0$ such that for any $t \geqslant t_0$, it holds that

$$\left| \frac{i(t, 0)}{(t - s_{i(t,0)})^{+} + \sum_{k=0}^{i(t,0)-1} (t_{k+1} - s_k)} - \tilde{p} \right| \leqslant \frac{\gamma}{2\|\ln B\|},$$

which implies

$$\begin{aligned} \left\| e^{\ln B\left(i(t,0) - \tilde{p}\left[(t - s_{i(t,0)})^{+} + \sum_{k=0}^{i(t,0)-1} (t_{k+1} - s_k)\right]\right)} \right\| &\leqslant e^{\left| i(t,0) - \tilde{p}\left[(t - s_{i(t,0)})^{+} + \sum_{k=0}^{i(t,0)-1} (t_{k+1} - s_k)\right]\right| \|\ln B\|} \\ &\leqslant e^{\frac{\gamma}{2}\left[(t - s_{i(t,0)})^{+} + \sum_{k=0}^{i(t,0)-1} (t_{k+1} - s_k)\right]}. \end{aligned} \tag{1.15}$$

Using (1.15) and (1.14) for (1.12), we derive

$$\| W(t, 0) \| \leqslant K e^{-\frac{\gamma}{2}\left[(t - s_{i(t,0)})^{+} + \sum_{k=0}^{i(t,0)-1} (t_{k+1} - s_k)\right]} \to 0 \quad \text{for} \quad t \to \infty, \tag{1.16}$$

since $i(t, 0) \to \infty$ as $t \to \infty$. The proof of (i) is finished.

(ii) The equality (1.12) can be rewritten as

$$e^{-\ln B\left(i(t,0) - \tilde{p}\left[(t - s_{i(t,0)})^{+} + \sum_{k=0}^{i(t,0)-1} (t_{k+1} - s_k)\right]\right)} W(t, 0) = e^{(A + \tilde{p}\ln B)\left[(t - s_{i(t,0)})^{+} + \sum_{k=0}^{i(t,0)-1} (t_{k+1} - s_k)\right]}. \tag{1.17}$$

From $\alpha(\Lambda) > 0$ and using [7, lemma 12, p 116], there exist $\bar{K} > 0$, $\lambda_0 > 0$ and $x_0 \in \mathbb{R}^n$, $\|x_0\| = 1$ such that

$$\| e^{\Lambda t} x_0 \| \geqslant \bar{K} e^{\lambda_0 t}, \quad t \geqslant 0 \tag{1.18}$$

for $\lambda_0 = \frac{\alpha(\Lambda)}{2}$. From (1.13), there is $t_1 > 0$ such that for any $t \geqslant t_1$, it holds that

$$\left\| e^{-\ln B\left(i(t,0) - \tilde{p}\left[(t - s_{i(t,0)})^{+} + \sum_{k=0}^{i(t,0)-1} (t_{k+1} - s_k)\right]\right)} \right\| \leqslant e^{\frac{\lambda_0}{2}\left[(t - s_{i(t,0)})^{+} + \sum_{k=0}^{i(t,0)-1} (t_{k+1} - s_k)\right]}. \tag{1.19}$$

Using (1.18) and (1.19) to (1.17) for $t \geqslant t_1$, we have

$$\bar{K}e^{\lambda_0\left[(t-s_{i(t,0)})^+ + \sum_{k=0}^{i(t,0)-1}(t_{k+1}-s_k)\right]} \leqslant \left\| e^{\Lambda\left[(t-s_{i(t,0)})^+ + \sum_{k=0}^{i(t,0)-1}(t_{k+1}-s_k)\right]}x_0 \right\|$$

$$\leqslant \|W(t,0)x_0\| \left\| e^{-\ln B\left(i(t,0)-\tilde{p}\left[(t-s_{i(t,0)})^+ + \sum_{k=0}^{i(t,0)-1}(t_{k+1}-s_k)\right]\right)} \right\|$$

$$\leqslant \|W(t,0)x_0\| e^{\frac{\lambda_0}{2}\left[(t-s_{i(t,0)})^+ + \sum_{k=0}^{i(t,0)-1}(t_{k+1}-s_k)\right]},$$

thus

$$\|W(t,0)x_0\| \geqslant \bar{K}e^{\frac{\lambda_0}{2}\left[(t-s_{i(t,0)})^+ + \sum_{k=0}^{i(t,0)-1}(t_{k+1}-s_k)\right]} \to \infty \quad \text{for} \quad t \to \infty,$$

since $i(t,0) \to \infty$ as $t \to \infty$.

(iii) From $\beta(\Lambda) > 0$ and lemma 1.1.5, there is $K > 0$ such that

$$\| e^{\Lambda t} \| \geqslant Ke^{\gamma t}, \quad t \geqslant 0$$

for $\gamma = \frac{\beta(\Lambda)}{2}$. The rest of the proof is the same as for case (ii), so we omit the details. The proof is complete. □

To end this section, two examples are given to illustrate our theoretical results above.

The first numerical example will show that how to use the result in theorem 1.1.8.

Example 1.1.12. *Consider (1.1) with*

$$A = \begin{pmatrix} -1 & 1 \\ 0 & -1 \end{pmatrix} \in \mathbb{R}^{2\times 2}, \quad B = \begin{pmatrix} 1 & 1 \\ 0 & 1 \end{pmatrix} \in \mathbb{R}^{2\times 2}, \quad x_0 = \begin{pmatrix} 0 \\ 1 \end{pmatrix} \in \mathbb{R}^2. \tag{1.20}$$

Then $\alpha(A) = \beta(A) = -1$, $AB = BA = -I$, $\rho(B) = 1$ and the exact solution of (1.1); (1.20) is

$$x(t, x_0) = e^{-\left[(t-s_{i(t,0)})^+ + \sum_{k=0}^{i(t,0)-1}(t_{k+1}-s_k)\right]} \begin{pmatrix} (t - s_{i(t,0)})^+ + \sum_{k=0}^{i(t,0)-1}(t_{k+1}-s_k) + i(t,0) \\ 1 \end{pmatrix}.$$

Meanwhile, let the distance between the impulsive points and junction points satisfy

$$\frac{1}{8} \leqslant t_{k+1} - s_k \leqslant \frac{1}{4}, \quad k = 0, 1, 2, \ldots$$

then theorem 1.1.8 can be applied such that (1.1);(1.20) is asymptotically stable.

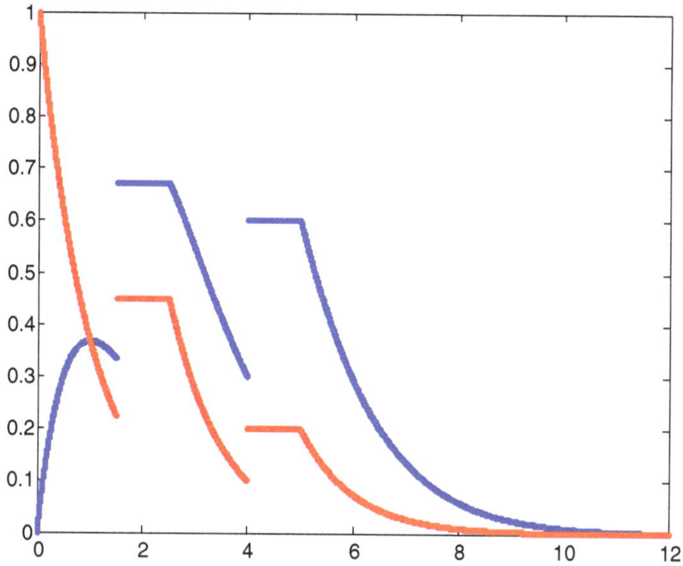

Figure 1.1. The red line denotes $x_1(t)$ and the blue line denotes $x_2(t)$ for the solution $x(t) = (x_1(t)\ x_2(t))^T$ of (1.1). Reproduced from [26]. (Copyright Springer International Publishing 2017). With permission of Springer Nature.

Next, if we choose two impulsive points and junction points, i.e. $i(t, 0) = 2$ for any t large enough 1.1.10 and 1.1.11 can be applied such that (1.1);(1.20) is asymptotically stable (see figure 1.1).

The second example is motivated by [7, section 3.1, pp 110–2]. Example 1.1.13 is a general case and reduces to the example in [7, section 3.1, pp 110–2] when s_i goes back to t_i for each i.

Example 1.1.13. *Consider the model of a linear oscillator subjected to non-instantaneous impulsive effects*

$$\begin{cases} x_1'(t) = \omega x_2(t), \quad x_2'(t) = -\omega x_1(t) - 2\lambda x_2(t), \ t \in [s_i, t_{i+1}], \\ \qquad i = 0, 1, 2, \ldots, \\ x_1(t_i^+) = b x_1(t_i^-), \quad x_2(t_i^+) = b x_2(t_i^-), \ i = 1, 2, \ldots, \\ x_1(t) = b x_1(t_i^-), \quad x_2(t) = b x_2(t_i^-), \ t \in (t_i, s_i], \ i = 1, 2, \ldots, \\ x_1(s_i^+) = x_1(s_i^-), \quad x_2(s_i^+) = x_2(s_i^-), \ i = 1, 2, \ldots, \\ x_1(0) = x_{01}, \quad x_2(0) = x_{02}, \end{cases} \tag{1.21}$$

where ω, b, λ are positive constants and $\omega > \lambda$. Denoting

$$x(t) = \begin{pmatrix} x_1(t) \\ x_2(t) \end{pmatrix} \in \mathbb{R}^2, \quad A = \begin{pmatrix} 0 & \omega \\ -\omega & -2\lambda \end{pmatrix} \in \mathbb{R}^{2\times 2}, \quad B = \begin{pmatrix} b & 0 \\ 0 & b \end{pmatrix} \in \mathbb{R}^{2\times 2}$$

and $t_0 = s_0 = 0$, *equation (1.21) has the form of (1.1). Assume that* $t_{k+1} - s_k = 1$, $k = 0, 1, 2, 3\ldots$, *then* $\tilde{\theta} = 1$. *Noting*

$$\frac{i(t,0)}{i(t,0)+1} = \frac{i(t,0)}{\sum_{k=0}^{i(t,0)}(t_{k+1} - s_k)}$$
$$\leqslant \frac{i(t,0)}{(t - s_{i(t,0)})^+ + \sum_{k=0}^{i(t,0)-1}(t_{k+1} - s_k)} \leqslant \frac{i(t,0)}{\sum_{k=0}^{i(t,0)-1}(t_{k+1} - s_k)} = 1$$

for $i(t, 0) > 1$, *we derive that*

$$\tilde{p} = \lim_{t\to\infty} \frac{i(t,0)}{(t - s_{i(t,0)})^+ + \sum_{k=0}^{i(t,0)-1}(t_{k+1} - s_k)} = 1.$$

For $\lambda_{1,2}(A) = -\lambda \pm i\sqrt{\omega^2 - \lambda^2}$, *we have* $\alpha(A) = \beta(A) = -\lambda < 0$. *Moreover,*

$$AB = BA, \ \sigma(B) = \{b\}.$$

Next,

$$\Lambda = A + \tilde{p}\ln B = \begin{pmatrix} \ln b & \omega \\ -\omega & -2\lambda + \ln b \end{pmatrix}$$

and $\lambda_{1,2}(\Lambda) = -\lambda + \ln b \pm i\sqrt{\omega^2 - \lambda^2}$. *Thus, we have the following conclusions:*

- *From theorems 1.1.8, 1.1.10 and 1.1.11, equation (1.21) is asymptotically stable, if* $\lambda > \ln b$.
- *From theorem 1.1.11, all nonzero solutions of (1.21) tend to infinity as* $t \to \infty$, *if* $\lambda < \ln b$.

1.1.3 The stability of perturbed problems

In order to discuss the stability of (1.2), it is necessary to derive the formula for its solution. Thus, we first seek the explicit formula of the solution for the following nonhomogeneous non-instantaneous impulsive perturbed problems:

$$\begin{cases} x'(t) = Ax(t) + g(t), \ t \in [s_i, t_{i+1}], \ i = 0, 1, 2, \ldots, \\ x(t_i^+) = Bx(t_i^-) + b_i, \ i = 1, 2, \ldots, \\ x(t) = Bx(t_i^-) + b_i, \ t \in (t_i, s_i], \ i = 1, 2, \ldots, \\ x(s_i^+) = x(s_i^-), \ i = 1, 2, \ldots, \end{cases} \tag{1.22}$$

where A, B, s_i and t_i are the same as in (1.1), $b_i \in \mathbb{R}^n$ and g: $J \to \mathbb{R}^n$ is continuous. By using the method of variation of constants, the solution of (1.22) with $x(0) = x_0 \in \mathbb{R}^n$ has the following form:

$$\begin{aligned} x(t, x_0) = W(t, 0)x_0 &+ \sum_{k=0}^{i(t,0)-1} \int_{s_k}^{t_{k+1}} W(t, s)g(s)ds \\ &+ \int_{s_{i(t,\,0)}}^{t} W(t, s)g(s)ds + \sum_{k=1}^{i(t,0)} W(t, s_k)b_k, \end{aligned} \tag{1.23}$$

where we note $W(t, s_k) = W(t, t_k^+)$.

From (1.23), any solution of (1.2) with $x(0) = x_0$ can be rewritten as

$$\begin{aligned} x(t, x_0) = W(t, 0)x_0 &+ \sum_{k=0}^{i(t,0)-1} \int_{s_k}^{t_{k+1}} W(t, s)P(s)x(s, x_0)ds \\ &+ \int_{s_{i(t,\,0)}}^{t} W(t, s)P(s)x(s, x_0)ds \\ &+ \sum_{k=1}^{i(t,0)} W(t, s_k)I_k x(t_k^-, x_0). \end{aligned} \tag{1.24}$$

We also need the following impulsive Grönwall inequality.

Lemma 1.1.14. *(See [7, lemma 1, p 12].) Let a non-negative piecewise continuous function $u(t)$ satisfy for $t \geqslant 0$ the inequality*

$$u(t) \leqslant C + \int_0^t mu(s)ds + \sum_{0<t_k<t} \beta_k u(t_k^-),$$

where $C,\ m,\ \beta_k > 0$. Then, for $t \geqslant 0$, the following inequality is valid:

$$u(t) \leqslant C \prod_{0<t_k<t} (1 + \beta_k)e^{mt}.$$

The following theorem illustrates that the stability of (1.1) would imply the stability of (1.2).

Theorem 1.1.15. *Suppose (1.1) is exponentially stable with constants $K \geqslant 1$ and $\gamma > 0$. If*

$$\limsup_{t\to\infty} \frac{i(t, 0)}{t} = \bar{p} < \infty, \quad \limsup_{t\to\infty} \|P(t)\| < c, \quad \limsup_{i\to\infty} \|I_i\| < c$$

for $\bar{p}$ and c satisfying

$$\Phi = \gamma - Kc - \bar{p}\ln(1 + Kc) > 0.$$

Then (1.2) is exponentially stable.

Proof. Since (1.1) is exponentially stable, there are positive numbers K and γ such that

$$\|W(t, s)\| \leqslant Ke^{-\gamma(t-s)}$$

for $0 \leqslant s \leqslant t$. Let $i_0 \in \mathbb{N}$ be so large that $\|P(t)\| \leqslant c$ and $\|I_i\| \leqslant c$ for any $t > t_{i_0}$ and $i > i_0$. Then for $t \in [0, t_{i_0+1}]$, formula (1.24) implies

$$\|x(t, x_0)\| \leqslant K\|x_0\| + KP_0 \int_0^t \|x(s, x_0)\|ds + KI_0 \sum_{k=1}^{i(t,0)} \|x(t_k^-, x_0)\|$$

for $P_0 = \max_{t\in[0,\, t_{i_0+1}]}\|P(t)\|$ and $I_0 = \max_{k\in\{1, 2, \ldots, i_0\}}\|I_k\|$. So lemma 1.1.14 gives

$$\|x(t, x_0)\| \leqslant K\|x_0\|(1 + I_0)^{i(t,\, 0)}e^{KP_0t} \leqslant K\|x_0\|(1 + I_0)^{i_0}e^{KP_0t_{i_0}} = \bar{K}_0\|x_0\|$$

for $t \in [0, t_{i_0+1}]$, since $i(t, 0) \leqslant i_0$. Next, taking the norm for both sides of (1.24) and multiplying by $e^{\gamma\, t}$, we obtain

$$\begin{aligned} e^{\gamma t}\|x(t, x_0)\| \leqslant{}& (K + K_0)\|x_0\| + Kc \sum_{k=i_0}^{i(t,0)-1} \int_{s_k}^{t_{k+1}} e^{\gamma s}\|x(s, x_0)\|ds \\ &+ Kc \int_{s_{i(t,\,0)}}^{t} e^{\gamma s}\|x(s, x_0)\|ds \\ &+ Kc \sum_{k=i_0}^{i(t,0)} e^{\gamma s_k}\|x(t_k^-, x_0)\| \leqslant \tilde{K}_0\|x_0\| \\ &+ Kc \int_0^t e^{\gamma s}\|x(s, x_0)\|ds + Kc \sum_{k=1}^{i(t,0)} e^{\gamma s_k}\|x(t_k^-, x_0)\| \end{aligned}$$

for $t > t_{i_0+1}$ and $\tilde{K}_0 = K + \bar{K}_0$ with

$$K_0 = K\bar{K}_0 \sum_{k=0}^{i_0-1} \int_{s_k}^{t_{k+1}} e^{\gamma s}\|P(s)\|ds + K\bar{K}_0 \sum_{k=1}^{i_0-1} e^{\gamma s_k}\|I_k\|,$$

since $i(t, 0) \geqslant i_0 + 1$. Summarizing, for any $t \geqslant 0$, we have

$$e^{\gamma t}\|x(t, x_0)\| \leqslant K_1\|x_0\| + Kc \int_0^t e^{\gamma s}\|x(s, x_0)\|ds + Kc \sum_{k=1}^{i(t,0)} e^{\gamma s_k}\|x(t_k^-, x_0)\|$$

for $K_1 = \max\{\tilde{K}_0, e^{\gamma t_{i_0}}\bar{K}_0\}$. By using lemma 1.1.14, we have

$$e^{\gamma t}\|x(t, x_0)\| \leqslant K_1(1 + Kc)^{i(t,\, 0)}e^{Kct}\|x_0\|,$$

which implies

$$\|x(t, x_0)\| \leqslant K_1 e^{-(\gamma - Kc - \frac{i(t,\, 0)}{t}\ln(1+Kc))t}\|x_0\|$$

for $t \geqslant 0$. Since

$$\liminf_{t\to\infty}\left(\gamma - Kc - \frac{i(t, 0)}{t}\ln(1 + Kc)\right) = \gamma - Kc - \bar{p}\ln(1 + Kc) = \Phi > 0$$

there is $t_1 > 0$ such that

$$\gamma - Kc - \frac{i(t, 0)}{t}\ln(1 + Kc) \geqslant \frac{\Phi}{2}$$

for $t \geqslant t_1$. Consequently, we arrive at

$$\|x(t, x_0)\| \leqslant K_1 e^{-\frac{\Phi}{2}t}\|x_0\|$$

for $t \geqslant t_1$. Then the proof is complete. □

1.1.4 Existence, uniqueness and Ulam–Hyers–Rassias stability

Consider the Banach space $PC(J, \mathbb{R}^n) := \{x: J \to \mathbb{R}^n: x \in C((t_k, t_{k+1}], \mathbb{R}^n)$ for $k \in \mathbb{N}$, and there exist $x(t_k^-)$ and $x(t_k^+)$, $k \in \mathbb{N}$ with $x(t_k^-) = x(t_k)\}$, endowed either with the Chebyshev PC-norm $\|x\|_{PC} := \sup\{\|x(t)\|: t \in J\}$ or with the Bielecki PCB-norm $\|x\|_{PCB} := \sup\{\|x(t)\|e^{-\omega t}: t \in J\}$ for some $\omega \in \mathbb{R}$.

From (1.23), the solution of (1.3) with $y(0) = y_0 \in \mathbb{R}^n$ can be expressed in the following form:

$$\begin{aligned} y(t, y_0) = W(t, 0)y_0 &+ \sum_{k=0}^{i(t,0)-1}\int_{s_k}^{t_{k+1}} W(t, s)g(s, y(s, y_0))ds \\ &+ \int_{s_{i(t,0)}}^{t} W(t, s)g(s, y(s, y_0))ds + \sum_{k=1}^{i(t,0)} W(t, s_k)b_k. \end{aligned} \tag{1.25}$$

We introduce the following conditions:

(H1) Assume that (1.1) is exponentially growing, i.e. there exist $K \geqslant 1$ and $\gamma > 0$ such that $\|W(t, s)\| \leqslant Ke^{\gamma(t-s)}$ for $0 \leqslant s \leqslant t$.

(H2) $g \in C(J \times \mathbb{R}^n, \mathbb{R}^n)$ and there exists a positive constant L_g such that

$$\|g(t, y_1) - g(t, y_2)\| \leqslant L_g\|y_1 - y_2\| \tag{1.26}$$

for each $t \in J$ and all $y_1, y_2 \in \mathbb{R}^n$. Moreover, $\|g\|_\infty := \sup_{t\in J}\|g(t, 0)\| < \infty$ and $\|b\|_\infty := \sup_{i\in\mathbb{N}}\|b_i\| < \infty$.

(H3) Suppose that there is $\omega_0 > 0$ such that

$$L := \sup_{t\in J} e^{-\omega_0 t}\sum_{k=1}^{i(t,0)} e^{-\gamma(t-s_k)^+} < \infty.$$

Now we are ready to present the existence and uniqueness result of the solutions.

Theorem 1.1.16. *Suppose that $[H_1]$–$[H_3]$ are satisfied. Then (1.3) with $y(0) = y_0 \in \mathbb{R}^n$ has a unique solution $y \in PCB(J, \mathbb{R}^n)$ for*

$$\omega = \max\left\{\gamma + 2KL_g, \gamma + \omega_0\right\}. \tag{1.27}$$

Proof. Supposing $\omega > \gamma$, we consider a mapping G: $PC(J, \mathbb{R}^n) \to PC(J, \mathbb{R}^n)$ defined by

$$(Gy)(t) = W(t, 0)y_0 + \sum_{k=0}^{i(t,0)-1} \int_{s_k}^{t_{k+1}} W(t, s)g(s, y(s))ds + \int_{s_{i(t,0)}}^{t} W(t, s)g(s, y(s))ds + \sum_{k=1}^{i(t,0)} W(t, s_k)b_k.$$

For any $y, \bar{y} \in PC(J, \mathbb{R}^n)$ and $t \in (t_{i(t,0)}, t_{i(t,0)+1}]$, we have

$$\begin{aligned}
&\|(Gy)(t) - (G\bar{y})(t)\| \\
&\leqslant \sum_{k=0}^{i(t,0)-1} \int_{s_k}^{t_{k+1}} \|W(t, s)(g(s, y(s)) - g(s, \bar{y}(s)))\|ds \\
&\quad + \int_{s_{i(t,0)}}^{t} \|W(t, s)(g(s, y(s)) - g(s, \bar{y}(s)))\|ds \\
&\leqslant KL_g\left(\sum_{k=0}^{i(t,0)-1} \int_{s_k}^{t_{k+1}} e^{\gamma(t-s)+\omega s}ds + \int_{s_{i(t,0)}}^{t} e^{\gamma(t-s)+\omega s}ds\right)\|y - \bar{y}\|_{PCB} \\
&\leqslant KL_g e^{\omega t}\left(\int_0^t e^{(\omega-\gamma)(s-t)}ds\right)\|y - \bar{y}\|_{PCB} \leqslant e^{\omega t}\frac{KL_g}{\omega - \gamma}\|y - \bar{y}\|_{PCB},
\end{aligned}$$

which implies that

$$e^{-\omega t}\|(Gy)(t) - (G\bar{y})(t)\| \leqslant L_G\|y - \bar{y}\|_{PCB}, \tag{1.28}$$

where $L_G := \frac{KL_g}{\omega - \gamma}$. Obviously, (1.28) yields that

$$\|Gy - G\bar{y}\|_{PCB} \leqslant L_G\|y - \bar{y}\|_{PCB}.$$

Moreover, if $\omega \geqslant \gamma + 2KL_g$ then $L_G \leqslant \frac{1}{2}$, and so G is a contraction mapping.

Using $[H_3]$, similarly, we can derive that

$$\begin{aligned}
\|(G0)(t)\| &\leqslant \|W(t, 0)\|\|y_0\| + \sum_{k=0}^{i(t,0)-1} \int_{s_k}^{t_{k+1}} \|W(t, s)g(s, 0)\|ds \\
&\quad + \int_{s_{i(t,0)}}^{t} \|W(t, s)g(s, 0)\|ds + \sum_{k=1}^{i(t,0)} \|W(t, s_k)\|\|b_k\| \\
&\leqslant K\left(e^{\gamma t}\|y_0\| + \|g\|_\infty \int_0^t e^{\gamma(t-s)}ds + \|b\|_\infty \sum_{k=1}^{i(t,0)} e^{\gamma(t-s_k)^+}\right) \\
&\leqslant Ke^{\gamma t}\left(\|y_0\| + \frac{\|g\|_\infty}{\gamma} + L\|b\|_\infty e^{\omega_0 t}\right),
\end{aligned}$$

which implies

$$\begin{aligned} e^{-\omega t}\|(G0)(t)\| &\leqslant Ke^{(\gamma-\omega)t}\left(\|y_0\| + \frac{\|g\|_\infty}{\gamma}\right) + KL\|b\|_\infty e^{(\gamma-\omega+\omega_0)t} \\ &\leqslant K\left(\|y_0\| + \frac{\|g\|_\infty}{\gamma} + L\|b\|_\infty\right), \end{aligned}$$

when $\omega \geqslant \gamma + \omega_0$. Thus

$$\|G0\|_{PCB} \leqslant K\left(\|y_0\| + \frac{\|g\|_\infty}{\gamma} + L\|b\|_\infty\right).$$

So if (1.27) holds then G: $PC(J, \mathbb{R}^n) \to PC(J, \mathbb{R}^n)$ is a contraction. Its fixed point gives a unique solution $y \in PC(J, \mathbb{R}^n)$ of (1.3) with $y(0) = y_0 \in \mathbb{R}^n$. The proof is complete. □

Now we refine theorem 1.1.16 as follows.

Theorem 1.1.17. *Assume that (1.1) is exponentially stable with constants $K \geqslant 1$ and $\gamma > 0$. Moreover suppose that $[H_2]$ is satisfied along with*

$$[H_3'] \quad \bar{L} := \sup_{i\in\mathbb{N}} \sum \sum_{k=1}^{i} e^{-\gamma(t_{i+1}-s_k)} < \infty.$$

If

$$KL_g < \gamma, \tag{1.29}$$

then (1.3) with $y(0) = y_0 \in \mathbb{R}^n$ has a unique solution $y \in PC(J, \mathbb{R}^n)$ which is also bounded on J with

$$\|y\|_{PC} \leqslant \frac{\gamma K}{\gamma - KL_g}\left(\|y_0\| + \frac{\|g\|_\infty}{\gamma} + (\bar{L} + 1)\|b\|_\infty\right).$$

Proof. By following the first part of the proof of theorem 1.1.16 with $\omega = 0$ and $-\gamma$ instead of γ, we derive

$$\|Gy - G\bar{y}\|_{PC} \leqslant \frac{KL_g}{\gamma}\|y - \bar{y}\|_{PC}$$

for any $y, \bar{y} \in PC(J, \mathbb{R}^n)$. So from (1.29), G is a contraction mapping. Using $[H_3']$, similarly we derive

$$\begin{aligned} \|(G0)(t)\| &\leqslant K\left(e^{-\gamma t}\|y_0\| + \|g\|_\infty \int_0^t e^{-\gamma(t-s)}ds + \|b\|_\infty \sum_{k=1}^{i(t,0)} e^{-\gamma(t-s_k)^+}\right) \\ &\leqslant K\left(\|y_0\| + \frac{\|g\|_\infty}{\gamma} + (\bar{L} + 1)\|b\|_\infty\right), \end{aligned}$$

since $(t - s_k)^+ = t - s_k \geqslant t_{i(t,0)} - s_k > 0$ for $t \in (t_{i(t,0)}, t_{i(t,0)+1}]$, $k = 1, 2, \dots, i(t, 0) - 1$ and $(t - s_{i(t,0)})^+ \geqslant 0$. So we obtain

$$\|G0\|_{PC} \leqslant K\left(\|y_0\| + \frac{\|g\|_\infty}{\gamma} + (\bar{L} + 1)\|b\|_\infty\right).$$

Hence G: $PC(J, \mathbb{R}^n) \to PC(J, \mathbb{R}^n)$ is a contraction. Its fixed point gives a unique solution $y \in PC(J, \mathbb{R}^n)$ of (1.3) with $y(0) = y_0 \in \mathbb{R}^n$. Moreover, we have

$$\|y\|_{PC} \leqslant \frac{\gamma\|G0\|_{PC}}{\gamma - KL_g} \leqslant \frac{\gamma K}{\gamma - KL_g}\left(\|y_0\| + \frac{\|g\|_\infty}{\gamma} + (\bar{L} + 1)\|b\|_\infty\right).$$

The proof is complete. □

Remark 1.1.18. *From bootstrap arguments, equation (1.3) with $y(0) = y_0 \in \mathbb{R}^n$ has a unique solution $y \in PC(J, \mathbb{R}^n)$ under $[H_2]$. Theorems 1.1.16 and 1.1.17 provide its growing properties on J under additional reasonable assumptions.*

Next, motivated by [17], we discuss the Ulam–Hyers–Rassias stability for (1.3).

Let $\varepsilon > 0$, $\psi > 0$ and $\varphi \in PC(J, \mathbb{R}_+)$ be a nondecreasing function, and consider the following inequalities:

$$\begin{cases} \|\hat{y}'(t) - A\hat{y}(t) - g(t, \hat{y}(t))\| \leqslant \varepsilon\varphi(t), \ t \in [s_i, t_{i+1}], \\ \qquad\qquad\qquad\qquad\qquad\qquad i = 0, 1, 2, \dots, \\ \|\hat{y}(t_i^+) - B\hat{y}(t_i^-) - b_i\| \leqslant \varepsilon\psi, \ i = 1, 2, \dots, \\ \|\hat{y}(t) - B\hat{y}(t_i^-) - b_i\| \leqslant \varepsilon\psi, \ t \in (t_i, s_i], \ i = 1, 2, \dots. \end{cases} \tag{1.30}$$

Denote

$$X := PC^1(J, \mathbb{R}^n)\textstyle\bigcap_{i\in\mathbb{N}} C^1([s_i, t_{i+1}], \mathbb{R}^n).$$

Definition 1.1.19. *Equation (1.3) is Ulam–Hyers–Rassias stable with respect to (φ, ψ), if there exists $c_{g,\mathbb{N},\varphi} > 0$ such that for each $\varepsilon > 0$ and for each function $\hat{y} \in X$ satisfying (1.30), there exists a solution $y \in X$ of (1.3) with*

$$\|\hat{y}(t) - y(t)\| \leqslant c_{g,\mathbb{N},\varphi}\varepsilon(\varphi(t) + \psi), \ t \in J. \tag{1.31}$$

Remark 1.1.20. A function $\hat{y} \in X$ is a solution of (1.30) if and only if there is $Z \in \bigcap_{i\in\mathbb{N}} C^1([s_i, t_{i+1}], \mathbb{R}^n)$ and $z \in \bigcap_{i\in\mathbb{N}} C([t_i, s_i], \mathbb{R}^n)$ such that

(i) $\|Z(t)\| \leqslant \varepsilon\varphi(t)$, $t \in \bigcup_{i\in\mathbb{N}}[s_i, t_{i+1}]$ and $\|z(t)\| \leqslant \varepsilon\psi$, $t \in \bigcup_{i\in\mathbb{N}}[t_i, s_i]$;
(ii) $\hat{y}'(t) = A\hat{y}(t) + g(t, \hat{y}(t)) + Z(t)$, $t \in [s_i, t_{i+1}]$, $i \in \mathbb{N}$;
(iii) $\hat{y}(t_i^+) = B\hat{y}(t_i^-) + b_i + z(t_i)$, $i \in \mathbb{N}$;
(iv) $\hat{y}(t) = B\hat{y}(t_i^-) + b_i + z(t)$, $t \in (t_i, s_i]$, $i \in \mathbb{N}$.

We introduce the following assumption:

$[H_4]$ There exists a constant $\tau > 0$ such that

$$\int_0^t e^{-\gamma(t-s)}\varphi(s)ds \leqslant \tau\varphi(t)$$

for each $t \in J$.

Theorem 1.1.21. *Assume that (1.3) is exponentially stable with constants $K \geqslant 1$ and $\gamma > 0$, and (1.26), $[H_3']$ and $[H_4]$ are satisfied. Then (1.3) is Ulam–Hyers–Rassias stable with respect to (φ, ψ) provided that the following holds:*

$$KL_g \max\{\tau, \gamma^{-1}\} < 1. \tag{1.32}$$

Proof. Let $\hat{y} \in X$ satisfy (1.30). Based on remark 1.1.20, we need to consider the solution of the following impulsive problem

$$\begin{cases} \hat{y}'(t) = Ax(t) + g(t, \hat{y}(t)) + Z(s), \ \ t \in [s_i, t_{i+1}], \ \ i \in \mathbb{N}, \\ \hat{y}(t_i^+) = B\hat{y}(t_i^-) + b_i + z(t_i), \ \ i \in \mathbb{N}, \\ \hat{y}(t) = B\hat{y}(t_i^-) + b_i + z(t), \ \ t \in (t_i, s_i], \ \ i \in \mathbb{N}. \end{cases} \tag{1.33}$$

Clearly, the solution $\hat{y}(t, y_0)$ of (1.33) with $\hat{y}(0) = y_0$ can be expressed as follows

$$\begin{aligned} &\hat{y}(t, y_0) \\ &= W(t, 0)y_0 + \sum_{k=0}^{i(t,0)-1} \int_{s_k}^{t_{k+1}} W(t, s)(g(s, \hat{y}(s, y_0)) + Z(s))ds \\ &\quad + \int_{s_{i(t,0)}}^{t} W(t, s)(g(s, \hat{y}(s, y_0)) + Z(s))ds + \sum_{k=1}^{i(t,0)-1} W(t, s_k) \\ &\quad (b_k + z(s_k)) \\ &\quad + W(t, s_{i(t,0)})b_{i(t,0)} + W(t, t_{i(t,0)}^+)z(\max\{t, s_{i(t,0)}\}) \end{aligned} \tag{1.34}$$

$$= (G\hat{y})(t) + \hat{z}(t) \tag{1.35}$$

for

$$\begin{aligned} \hat{z}(t) := &\sum_{k=0}^{i(t,0)-1} \int_{s_k}^{t_{k+1}} W(t, s)Z(s)ds + \int_{s_{i(t,0)}}^{t} W(t, s)Z(s)ds \\ &+ \sum_{k=1}^{i(t,0)-1} W(t, s_k)z(s_k) + W(t, t_{i(t,0)}^+)z(\max\{t, s_{i(t,0)}\}). \end{aligned}$$

It follows from our assumptions that

$$
\begin{aligned}
\|\hat{z}(t)\| \leqslant & \sum_{k=0}^{i(t,0)-1} \int_{s_k}^{t_{k+1}} \|W(t, s)\|\|Z(s)\|ds + \int_{s_{i(t,\,0)}}^{t} \|W(t, s)\|\|Z(s)\|ds \\
& + \sum_{k=1}^{i(t,0)-1} \|W(t, s_k)\|\|z(s_k)\| + \|W(t, t_{i(t,\,0)}^{+})\|\|z(\max\{t, s_{i(t,\,0)}\})\| \\
\leqslant & \varepsilon K\left(\int_0^t e^{-\gamma(t-s)}\varphi(s)ds + \sum_{k=1}^{i(t,0)-1} e^{-\gamma(t-s_k)}\psi + e^{-\gamma(t-t_{i(t,\,0)})}\psi\right) \\
\leqslant & \varepsilon K\left(\tau\varphi(t) + \sum_{k=1}^{i(t,0)-1} e^{-\gamma(t_{i(t,\,0)}-s_k)}\psi + \psi\right) \\
\leqslant & \varepsilon K \max\{\tau, \bar{L} + 1\}(\varphi(t) + \psi),
\end{aligned}
\tag{1.36}
$$

since $t - s_k \geqslant t_{i(t,\,0)} - s_k > 0$ for $t \in (t_{i(t,\,0)}, t_{i(t,\,0)+1}]$, $k = 1, 2, \dots, i(t, 0) - 1$ and $t - t_{i(t,\,0)} \geqslant 0$. Let $y(t, y_0)$ be the unique solution of (1.3) with $y(0) = y_0$. Using a similar computation as above via (1.25) and (1.36), one can arrive at

$$
\begin{aligned}
\|\hat{y}(t, y_0) - y(t, y_0)\| \leqslant & \|\hat{z}(t)\| + \|(G\hat{y})(t) - (Gy)(t)\| \\
\leqslant & \varepsilon K \max\{\tau, \bar{L} + 1\}(\varphi(t) + \psi) \\
& + KL_g \int_0^t e^{-\gamma(t-s)}\|\hat{y}(s, y_0) - y(s, y_0)\|ds.
\end{aligned}
\tag{1.37}
$$

Let $T > 0$ and setting

$$
m(T) := \sup_{t\in[0,\,T]} \frac{\|\hat{y}(t, y_0) - y(t, y_0)\|}{\varphi(t) + \psi},
$$

from (1.37) and $[H_4]$, we have

$$
\begin{aligned}
& \|\hat{y}(t, y_0) - y(t, y_0)\| \\
\leqslant & \varepsilon K \max\{\tau, \bar{L} + 1\}(\varphi(t) + \psi) + KL_g m(T) \\
& \int_0^t e^{-\gamma(t-s)}(\varphi(s) + \psi)ds \\
\leqslant & \varepsilon K \max\{\tau, \bar{L} + 1\}(\varphi(t) + \psi) + KL_g m(T)\left(\tau\varphi(t) + \frac{\psi}{\gamma}\right) \\
\leqslant & \varepsilon K \max\{\tau, \bar{L} + 1\}(\varphi(t) + \psi) + KL_g m(T)\max\{\tau, \gamma^{-1}\} \\
& (\varphi(t) + \psi)
\end{aligned}
\tag{1.38}
$$

for any $t \in [0, T]$. Then (1.38) implies

$$
m(T) \leqslant \varepsilon K \max\{\tau, \bar{L} + 1\} + KL_g m(T)\max\{\tau, \gamma^{-1}\},
$$

which from (1.32) gives

$$m(T) \leqslant \varepsilon \frac{K \max\{\tau, \bar{L} + 1\}}{1 - KL_g \max\{\tau, \gamma^{-1}\}},$$

that is

$$\begin{aligned} \|\hat{y}(t, y_0) - y(t, y_0)\| &\leqslant \varepsilon \frac{K \max\{\tau, \bar{L} + 1\}}{1 - KL_g \max\{\tau, \gamma^{-1}\}}(\varphi(t) + \psi) \\ &\leqslant c_{g, \mathbb{N}, \varphi} \varepsilon(\varphi(t) + \psi) \end{aligned} \tag{1.39}$$

for any $t \in [0, T]$, where $c_{g, \mathbb{N}, \varphi} := \frac{K \max\{\tau, \bar{L} + 1\}}{1 - KL_g \max\{\tau, \gamma^{-1}\}} > 0$. Since (1.39) holds for any $T > 0$, (1.3) is Ulam–Hyers–Rassias stable with respect to (φ, ψ). The proof is complete.

Finally, the following example is used to illustrate the above results.

Example 1.1.22. *Consider (1.3) with* $s_i = i + \frac{1}{2}$, $t_i = i$, $i = 1, 2, \ldots$, *and*

$$A = \begin{pmatrix} -4 & 0 \\ 0 & -4 \end{pmatrix} \in \mathbb{R}^{2\times 2}, \quad B = \begin{pmatrix} 1 & -1 \\ 0 & 1 \end{pmatrix} \in \mathbb{R}^{2\times 2}, \quad x_0 = \begin{pmatrix} 1 \\ 0 \end{pmatrix} \in \mathbb{R}^2,$$

$$g(t, x(t)) = \begin{pmatrix} \sin t \\ \cos t \end{pmatrix} \in \mathbb{R}^2, \quad b_i = \begin{pmatrix} 0 \\ 1 \end{pmatrix} \in \mathbb{R}^2.$$

By taking the norm $\|(x_1, x_2)\| = \max\{|x_1|, |x_2|\}$, *we obtain* $\|B\| = 2$. *Now we have* $i(t, 0) = \lceil t \rceil - 1$. *So (1.4) gives*

$$\begin{aligned} \| W(t, s)\| &\leqslant 2^{\lceil t \rceil - \lceil s \rceil} e^{-4\left[(t - \left\lceil t \right\rceil + 1)^+ - (s - \left\lceil s \right\rceil + 1)^+ + \frac{\lceil t \rceil - \lceil s \rceil}{2}\right]} \leqslant e^4 e^{-(2 - \ln 2)[\lceil t \rceil - \lceil s \rceil]} \\ &\leqslant e^4 e^{-[\lceil t \rceil - \lceil s \rceil]} \leqslant e^5 e^{-(t-s)} \end{aligned}$$

since $(t - \lceil t \rceil + 1)^+ - (s - \lceil s \rceil + 1)^+ \geqslant -1$, $2 - \ln 2 > 1$ *and* $\lceil t \rceil - \lceil s \rceil \geqslant t - s - 1$. *Hence the above linear system is exponentially stable for* $\gamma = 1$ *and* $K = e^5$. *Next, (1.26) holds for* $L_g = 0$ *and (1.32) holds trivially. For the condition [*H_3'*], we have* $\bar{L} = \frac{\sqrt{e}}{e - 1}$. *Set* $\varphi(t) = e^{\frac{t}{2}}$ *and* $\psi = 1$, *then [*H_4*] is satisfied with* $\tau = \frac{2}{3}$. *Thus, applying theorems 1.1.16 and 1.1.21, equation (1.3) has a unique solution and it is Ulam–Hyers–Rassias stable with respect to* $\varphi(t) = (e^{\frac{t}{2}}, 1)^T$ *with* $c_{g, M, \tau} = e^5(\frac{\sqrt{e}}{e - 1} + 1)$.

1.1.5 Notes and remarks

For (1.1), we introduce the notation of non-instantaneous impulsive Cauchy matrix $W(\cdot, \cdot)$ (see (1.4) for the explicit formula) and analyze its exponential structure in terms of eigenvalues of the matrix via the distance between impulsive points and junction points. This is an natural extension of work in [7, chapter 2.1] and [7, chapter 2.2, equation (2.24) or (2.25)]. We also give sufficient conditions for the stability of solutions of (1.1) by investigating the structure and behavior of the non-instantaneous impulsive Cauchy matrix and in terms of the eigenvalues of A, B via the distance between impulsive points $\{t_i\}$ and junction points $\{s_i\}$.

For (1.2), we establish a uniform framework to deal with asymptotic stability and present many sufficient conditions to guarantee the asymptotic stability of solutions under mild sufficient conditions by virtue of non-instantaneous impulsive Cauchy matrix $W(\cdot, \cdot)$.

We prove the existence and uniqueness results and provide an approach to find approximate solution to (1.3) by placing certain conditions, such as exponential growth or stable conditions for the non-instantaneous impulsive Cauchy matrix, and utilizing a set of inequalities in the sense of Ulam–Hyers–Rassias stability.

The results in section 1.1 are motivated by [26].

1.2 Lyapunov regularity

1.2.1 Introduction

It is well known that the notation of the fundamental matrix and its exponential stability will reduce the further study required on the Lyapunov exponent, exponential dichotomy and periodic solutions, which play a central role, for example, in the Hadamard–Perron theory of invariant manifolds and Poincaré mapping of impulsive dynamical systems involving continuous and discrete time. In addition, some prior contributions have shown the relationship between exponential dichotomy and Hyers–Ulam stability for linear continuous/discrete differential systems [31–33], which seems to open the way to obtaining many other results in this field. Goebel *et al* considered modeling, stability and robustness analysis of instantaneous impulsive dynamical systems (hybrid dynamical systems) in [37], where known theoretical results for classical nonlinear dynamical systems were unified and generalized to hybrid dynamical systems. Recently, Liu and Teel applied the Lyapunov–Krasovskii method in [38, 39] to establish sufficient conditions for the stability of impulsive dynamical systems with memory (which is an extension of the hybrid inclusion model). Davó *et al* [40] used the Lyapunov–Krasovskii method and looped functionals to establish sufficient conditions for asymptotic stability in the form of linear matrix inequalities for impulsive delay dynamical systems with a linear and time-invariant part.

The regularity of the Lyapunov exponent is the key issue of dynamical systems [41–44]. Goebel *et al* [37] considered Lyapunov conditions, robustness and asymptotic stability for a class of hybrid dynamical systems described by instantaneous impulsive differential equations. The existence of smooth Lyapunov functions and structural properties of solution sets were obtained in [37]. However, the Lyapunov regularity, the relation between the Lyapunov characteristic exponent and stability, and the nonuniform exponential behavior for the new class of hybrid dynamical systems described by non-instantaneous linear impulsive differential equations have not yet been investigated in the literature from the point of view of dynamical systems and ergodic theory.

We consider a non-instantaneous linear impulsive differential system as follows:

$$\begin{cases} x'(t) = A(t)x(t), \ t \in (s_i, t_{i+1}], \ i = 0, 1, 2, \ldots, \\ x(t_i^+) = B_i(t_i)x(t_i^-), \ i = 1, 2, \ldots, \\ x(t) = B_i(t)x(t_i^-), \ t \in (t_i, s_i], \ i = 1, 2, \ldots, \\ x(s_i^+) = x(s_i^-) = x(s_i), \ i = 1, 2, \ldots, \end{cases} \tag{1.40}$$

in an n-dimensional Euclidean space, where the $n \times n$ matrices $A(t)$, $B_i(t)$ vary continuously for $t \geqslant 0$, and here $t_0 = s_0 = 0 < t_1 < s_1 < t_2 < \cdots < s_i < t_{i+1} \cdots$. Here we call t_i the impulsive point and s_i the junction point. The symbols $x(t_i^+)$ and $x(t_i^-)$ represent the right and left limits of $x(t)$ at $t = t_i$, respectively. In addition, we set $x(t_i^-) = x(t_i)$ and assume that $B_i^{-1}(t)$ exist.

Next, we introduce the Cauchy matrix $W(\cdot, \cdot)$ for (1.40), where $W(t, s)$: $\mathbb{R}^+ \times \mathbb{R}^+ \to \mathbb{R}^{n\times n}$ is given by

$$W(t, s)$$

$$= \begin{cases} B_{i+k}(t)U(t_{i+k}, s_{i+k-1})\prod_{v=1}^{k-1} [B_{i+v}(s_{i+v})U(t_{i+v}, s_{i+v-1})]B_i(s_i)U(t_i, s), \\ \text{for} s_{i-1} \leqslant s < t_i < s_i < t_{i+1} < s_{i+1} < t_{i+k} < t \leqslant s_{i+k}, \ k = 0, 1, \ldots, \\ U(t, s_{i+k})\prod_{v=1}^{k} [B_{i+v}(s_{i+v})U(t_{i+v}, s_{i+v-1})]B_i(s_i)U(t_i, s), \\ \text{for} s_{i-1} \leqslant s < t_i < s_i < t_{i+k} < s_{i+k} < t \leqslant t_{i+k+1}, \ k = 0, 1, \ldots, \\ B_{i+k}(t)U(t_{i+k}, s_{i+k-1})\prod_{v=1}^{k-1} [B_{i+v}(s_{i+v})U(t_{i+v}, s_{i+v-1})]B_i(s_i)B_i^{-1}(s), \\ \text{for } t_i \leqslant s < s_i < t_{i+1} < s_{i+1} < t_{i+k} < t \leqslant s_{i+k}, \ k = 0, 1, \ldots, \\ U(t, s_{i+k})\prod_{v=1}^{k} [B_{i+v}(s_{i+v})U(t_{i+v}, s_{i+v-1})]B_i(s_i)B_i^{-1}(s), \\ \text{for } t_i \leqslant s < s_i < t_{i+1} < s_{i+1} < t_{i+k} < s_{i+k} < t \leqslant t_{i+k+1}, \ k = 0, 1, \ldots; \end{cases} \tag{1.41}$$

here we let $U(t, s)$ be the Cauchy matrix for the linear equation $x'(t) = A(t)x(t)$, $t \geqslant 0$.

Set $PC(\mathbb{R}^+, \mathbb{R}^n) := \{x: \mathbb{R}^+ \to \mathbb{R}^n: x \in C((t_i, t_{i+1}], \mathbb{R}^n), i = 0, 1, 2, \ldots$ and there exist $x(t_i^-)$ and $x(t_i^+)$ with $x(t_i^-) = x(t_i)\}$ with the norm $\|x\|_{PC} := \sup_{t\in\mathbb{R}^+}\|x(t)\|$, and $C(\mathbb{R}^+, \mathbb{R}^n)$ denotes the Banach space of vector-valued continuous functions from $R^+ \to \mathbb{R}^n$ endowed with the norm $\|x\|_{C(\mathbb{R}^+)} = \sup\limits_{t\in\mathbb{R}^+} \|x(t)\|$ for a norm $\|\cdot\|$ on $\mathbb{R}^n$.

A solution $x(t, x_0) \in PC(\mathbb{R}^+, \mathbb{R}^n)$ of (1.40) with initial value condition $x(t_0) = x_0 \in \mathbb{R}^n$ has the form

$$x(t, x_0) = W(t, t_0)x_0, \quad \forall\ 0 \leqslant t_0 \leqslant t.$$

The rest of this section is organized as follows. In section 1.2.2 we study the Lyapunov regularity of (1.40). We present an upper and lower bound for the solution of (1.40) in theorem 1.2.3 and we give sufficient conditions to guarantee any nontrivial solution of (1.40) has a finite Lyapunov exponent in theorem 1.2.5. We show that (1.40) is regular in theorem 1.2.8 under mild conditions. In section 1.2.3 we apply the Lyapunov exponent to characterize the stability of (1.40). Analyzing the spectrum for the set of Lyapunov exponents and showing that this set consists of finite numbers of distinct elements in theorem 1.2.10, we give sufficient conditions to guarantee that all the solutions of (1.40) are asymptotically stable in theorem 1.2.11. In section 1.2.4 we give a Perron theorem for (1.39) in theorem 1.2.15 by introducing the associated adjoint system and we present criteria for the existence of nonuniform exponential behavior in theorem 1.2.17. In section 1.2.5 we consider the stability of a perturbed impulsive system in theorem 1.2.19. In section 1.2.6 we adopt ideas from [43] to establish a bound for the regularity coefficient in theorem 1.2.23.

1.2.2 Lyapunov regularity

For any vector-valued function $x = (x_1, x_2, \ldots, x_n)^T \in \mathbb{R}^n$, let $\|\cdot\|^2 = \langle\cdot, \cdot\rangle$ where $\langle\cdot, \cdot\rangle$ is the usual inner product in $\mathbb{R}^n$.

Now, we recall the notion of a classical Lyapunov exponent of a vector function in [45].

Definition 1.2.1. *For a function f: $\mathbb{R}^+ \to \mathbb{R}^n$, the classical Lyapunov exponent of f is defined by*

$$\chi(f) := \limsup_{t\to\infty} \frac{1}{t} \log \|f(t)\|,$$

where χ: $\mathbb{R}^n \to \mathbb{R} \cup \{-\infty\}$.

Lemma 1.2.2. *(See [7, equations (2.45)–(2.48), p 61].) Let f_1, f_2: $\mathbb{R}^+ \to \mathbb{R}^n$ be arbitrary functions. Then, the following statements hold:*

(i) $\chi(cf_1) = \chi(f_1)$ *for* $c \in \mathbb{R}/\{0\}$.
(ii) $\chi(f_1 + f_2) \leqslant \max\{\chi(f_1), \chi(f_2)\}$.
(iii) $\chi(f_1 + f_2) = \max\{\chi(f_1), \chi(f_2)\}$, *for each* $f_1, f_2 \in \mathbb{R}^n$ *with* $\chi(f_1) \neq \chi(f_2)$.
(iv) $\chi(0) = -\infty$ *(normalization property).*

In this section, we give a bound for the solution of (1.40) and define the Lyapunov exponent χ: $\mathbb{R}^n \to \mathbb{R} \cup \{-\infty\}$ associated with (1.40) by

$$\chi(x_0) = \limsup_{t\to+\infty} \frac{1}{t} \log \|x(t)\|, \tag{1.42}$$

where $x(t)$ is the solution of (1.40) with $x(t_0) = x_0$; we write $\chi(x(.,x_0))$ or $\chi(x(.))$ (when x_0 is clear) for $\chi(x_0)$ as well.

Theorem 1.2.3. *For any solution $x(t, x_0)$ of (1.40) with $x(t_0, x_0) = x_0$, and $t \geqslant t_0 = 0$, the following inequalities hold:*

(I) *Let* $t \in (s_i, t_{i+1}]$, $i = 0, 1, 2, \ldots$ *and we have*

$$\begin{aligned}&\prod_{j=1}^{i} \left(\lambda_j e^{2\int_{s_{j-1}}^{t_j} \lambda(\tau)d\tau}\right) e^{2\int_{s_i}^{t} \lambda(\tau)d\tau} \|x(0)\|^2 \leqslant \|x(t)\|^2 \\ &\leqslant \prod_{j=1}^{i} \left(\Lambda_j e^{2\int_{s_{j-1}}^{t_j} \Lambda(\tau)d\tau}\right) e^{2\int_{s_i}^{t} \Lambda(\tau)d\tau} \|x(0)\|^2;\end{aligned} \tag{1.43}$$

(II) *Let* $t \in (t_{i+1}, s_{i+1}]$, $i = 1, 2, \ldots$ *and we have*

$$\prod_{j=1}^{i+1} \left(\lambda_j e^{2\int_{s_{j-1}}^{t_j} \lambda(\tau)d\tau}\right) \|x(0)\|^2 \leqslant \|x(t)\|^2 \leqslant \prod_{j=1}^{i+1} \left(\Lambda_j e^{2\int_{s_{j-1}}^{t_j} \Lambda(\tau)d\tau}\right) \|x(0)\|^2, \tag{1.44}$$

where $\|x(t)\|^2 = \langle x(t), x(t)\rangle$, $\lambda(t)$ *and* $\Lambda(t)$ *are the largest and the smallest eigenvalues of the matrix* $\widetilde{A(t)} = \frac{1}{2}(A(t) + A^{\top}(t))$, *respectively, and* $A^{\top}(t)$ *is the transpose of the matrix* $A(t)$ *and* $B_i^{\top}(t)$ *is the transpose of the matrix* $B_i(t)$, *and* $\lambda_i = \inf_{t\in(t_i, s_i]}(\lambda_B(t))$, $\Lambda_i = \sup_{t\in(t_i, s_i]} (\Lambda_B(t))$; *here* $\lambda_B(t)$ *and* $\Lambda_B(t)$ *are the largest and the smallest eigenvalues of the matrix* $B_i^{\top}(t)B_i(t)$, *respectively.*

Proof. If $x(t, x_0) = 0$, then the result holds immediately. Let $x(t, t_0) = x(t)$ be a nontrivial solution of (1.40). We consider two cases:

(I) Let $t \in (s_i, t_{i+1}]$ and $i = 0, 1, 2, \ldots$, and we have

$$\begin{aligned}\frac{d}{dt}\|x(t)\|^2 &= \frac{d}{dt}\langle x(t), x(t)\rangle = \left\langle \frac{dx(t)}{dt}, x(t)\right\rangle + \left\langle x(t), \frac{dx(t)}{dt}\right\rangle \\ &= \langle A(t)x(t), x(t)\rangle + \langle x(t), A(t)x(t)\rangle \\ &= \langle A(t)x(t), x(t)\rangle + \langle A^{\top}(t)x(t), x(t)\rangle \\ &= \langle (A(t) + A^{\top}(t))x(t), x(t)\rangle.\end{aligned}$$

Let $\widetilde{A(t)} = \frac{1}{2}(A(t) + A^{\top}(t))$, and we have

$$\frac{d}{dt}\|x(t)\|^2 = 2\langle \widetilde{A(t)}x(t), x(t)\rangle. \tag{1.45}$$

Since the matrix $\widetilde{A(t)}$ is symmetric, we have

$$\lambda(t)\langle x(t), x(t)\rangle \leqslant \langle \widetilde{A(t)}x(t), x(t)\rangle \leqslant \Lambda(t)\langle x(t), x(t)\rangle,$$

where $\lambda(t)$ and $\Lambda(t)$ are the largest and the smallest eigenvalues of the matrix $\widetilde{A(t)}$. Hence it follows from (1.45) that we have

$$2\lambda(t)\|x(t)\|^2 \leqslant \frac{d}{dt}\|x(t)\|^2 \leqslant 2\Lambda(t)\|x(t)\|^2. \tag{1.46}$$

For $s_i < t \leqslant t_{i+1}$, we have from (1.46) that

$$e^{2\int_{s_i}^{t}\lambda(\tau)d\tau}\|x(s_i^+)\|^2 \leqslant \|x(t)\|^2 \leqslant e^{2\int_{s_i}^{t}\Lambda(\tau)d\tau}\|x(s_i^+)\|^2.$$

According to $x(s_i^+) = x(s_i^-) = x(s_i)$, we have

$$e^{2\int_{s_i}^{t}\lambda(\tau)d\tau}\|x(s_i)\|^2 \leqslant \|x(t)\|^2 \leqslant e^{2\int_{s_i}^{t}\Lambda(\tau)d\tau}\|x(s_i)\|^2. \tag{1.47}$$

(i) Let $t \in (s_0, t_1]$, and we have

$$e^{2\int_{s_0}^{t}\lambda(\tau)d\tau}\|x(0)\|^2 \leqslant \|x(t)\|^2 \leqslant e^{2\int_{s_0}^{t}\Lambda(\tau)d\tau}\|x(0)\|^2. \tag{1.48}$$

(ii) Let $t \in (s_1, t_2]$, and we have

$$e^{2\int_{s_1}^{t}\lambda(\tau)d\tau}\|x(s_1)\|^2 \leqslant \|x(t)\|^2 \leqslant e^{2\int_{s_1}^{t}\Lambda(\tau)d\tau}\|x(s_1)\|^2. \tag{1.49}$$

From $x(s_1) = B_1(s_1)x(t_1^-) = B_1(s_1)x(t_1)$, we see that

$$\begin{aligned}\|x(s_1)\|^2 &= \langle B_1(s_1)x(t_1),\ B_1(s_1)x(t_1)\rangle \\ &= \langle B_1^{\mathsf{T}}(s_1)B_1(s_1)x(t_1),\ x(t_1)\rangle,\end{aligned}$$

and therefore,

$$\begin{aligned}&\lambda_1\|x(t_1)\|^2 \leqslant \langle B_1^{\mathsf{T}}(s_1)B_1(s_1)x(t_1),\ x(t_1)\rangle \leqslant \Lambda_1\|x(t_1)\|^2 \\ &\lambda_1\|x(t_1)\|^2 \leqslant \|x(s_1)\|^2 \leqslant \Lambda_1\|x(t_1)\|^2.\end{aligned}$$

According to inequality (1.48) we have

$$\lambda_1 e^{2\int_{s_0}^{t_1}\lambda(\tau)d\tau}\|x(0)\|^2 \leqslant \|x(s_1)\|^2 \leqslant \Lambda_1 e^{2\int_{s_0}^{t_1}\Lambda(\tau)d\tau}\|x(0)\|^2. \tag{1.50}$$

From inequalities (1.49) and (1.50) we have

$$\lambda_1 e^{2\int_{s_1}^{t}\lambda(\tau)d\tau+2\int_{s_0}^{t_1}\lambda(\tau)d\tau}\|x(0)\|^2 \leqslant \|x(t)\|^2 \leqslant \Lambda_1 e^{2\int_{s_1}^{t}\Lambda(\tau)d\tau+2\int_{s_0}^{t_1}\Lambda(\tau)d\tau}\|x(0)\|^2. \tag{1.51}$$

(iii) Let $t \in (s_2, t_3]$, and we have

$$e^{2\int_{s_2}^{t}\lambda(\tau)d\tau}\|x(s_2)\|^2 \leqslant \|x(t)\|^2 \leqslant e^{2\int_{s_2}^{t}\Lambda(\tau)d\tau}\|x(s_2)\|^2.$$

From $x(s_2) = B_2(s_2)x(t_2^-) = B_2(s_2)x(t_2)$, we see that

$$\|x(s_2)\|^2 = \langle B_2(s_2)x(t_2),\ B_2(s_2)x(t_2)\rangle$$
$$= \langle B_2^{\mathrm{T}}(s_2)B_2(s_2)x(t_2),\ x(t_2)\rangle,$$

and therefore,

$$\lambda_2\|x(t_2)\|^2 \leqslant \|x(s_2)\|^2 \leqslant \Lambda_2\|x(t_2)\|^2.$$

According to inequality (1.51), we have

$$\lambda_1 e^{2\int_{s_1}^{t_2}\lambda(\tau)d\tau+2\int_{s_0}^{t_1}\lambda(\tau)d\tau}\|x(0)\|^2 \leqslant \|x(t_2)\|^2 \leqslant \Lambda_1 e^{2\int_{s_1}^{t_2}\Lambda(\tau)d\tau+2\int_{s_0}^{t_1}\Lambda(\tau)d\tau}\|x(0)\|^2.$$

Hence, we have

$$\lambda_1\lambda_2 e^{2\int_{s_1}^{t_2}\lambda(\tau)d\tau+2\int_{s_0}^{t_1}\lambda(\tau)d\tau}\|x(0)\|^2 \leqslant \Lambda_1\Lambda_2 e^{2\int_{s_1}^{t_2}\Lambda(\tau)d\tau+2\int_{s_0}^{t_1}\Lambda(\tau)d\tau}\|x(0)\|^2.$$

Therefore, we obtain the following inequality,

$$\prod_{j=1}^{2}\left(\lambda_j e^{2\int_{s_{j-1}}^{t_j}\lambda(\tau)d\tau}\right)e^{2\int_{s_2}^{t}\lambda(\tau)d\tau}\|x(0)\|^2 \leqslant \|x(t)\|^2 \leqslant \prod_{j=1}^{2}\left(\Lambda_j e^{2\int_{s_{j-1}}^{t_j}\Lambda(\tau)d\tau}\right)e^{2\int_{s_2}^{t}\Lambda(\tau)d\tau}\|x(0)\|^2.$$

Now, by using mathematical induction, it is easy to see from (1.47) that for all $i = 0, 1, 2, \ldots$, we obtain (1.43).

(II) Let $t \in (t_{i+1}, s_{i+1}]$ and $i = 1, 2, 3, \ldots$, and one can apply the same ideas in (I) to obtain

$$\|x(t)\|^2 = \|B_{i+1}(t)x(t_{i+1}^-)\|^2 = \|B_{i+1}(t)x(t_{i+1})\|^2$$
$$= \langle B_{i+1}^{\mathrm{T}}(t)B_{i+1}(t)x(t_{i+1}),\ x(t_{i+1})\rangle.$$

Therefore, we can obtain the following inequality,

$$\lambda_{i+1}\|x(t_{i+1})\|^2 \leqslant \|x(t)\|^2 \leqslant \Lambda_{i+1}\|x(t_{i+1})\|^2. \tag{1.52}$$

From inequality (1.43) we derive (1.44). The proof is complete. □

Remark 1.2.4. *Note that [7, theorem 13] is a corollary of theorem* 1.2.3 *by setting* $s_i = t_i$.

Next, we suppose additionally that $A(t)$ and $B_i(t)$ are bounded for $t \geqslant 0$, i.e.

$$\bar{A} = \sup_{t\geqslant 0}\|A(t)\| < \infty,\quad \overline{B_i} = \sup_{t\geqslant 0}\|B_i(t)\| < \infty. \tag{1.53}$$

Theorem 1.2.5. *Suppose that condition (1.53) holds. The relation* $\lim_{t\to\infty}\frac{r(t)}{t} = \rho$ *exists and is finite, where* $r(t)$ *is the number of impulsive points* $t_i \in [0, t)$, $t > 0$. *Then any nontrivial solution* $x(t, x_0) = x(t)$ *of system (1.40) has a finite Lyapunov exponent.*

Proof. First, we need a bounded estimate for the nontrivial solution $x(t, x_0) = x(t)$. For a nontrivial solution $x(t, x_0) = x(t)$, $x_0 \neq 0$, we consider two cases:

(I) Let $t \in (s_i, t_{i+1}]$, $i = 0, 1, \ldots$, and consider

$$\begin{cases} x'(t) = A(t)x(t), \\ x(s_i^+) = x(s_i^-) = x(s_i), \end{cases}$$

so we obtain

$$x(t) = x(s_i) + \int_{s_i^+}^{t} A(\tau)x(\tau)d\tau.$$

Thus,

$$\|x(t)\| \leqslant \|x(s_i)\| + \left\| \int_{s_i^+}^{t} A(\tau)x(\tau)d\tau \right\| \leqslant \|x(s_i)\| + \bar{A}\int_{s_i^+}^{t} \|x(\tau)\|d\tau,$$

which together with Grönwall's inequality for the solution, yields

$$\|x(t)\| \leqslant \|x(s_i)\|e^{\bar{A}(t-s_i)}. \tag{1.54}$$

(i) Let $t \in (s_1, t_2]$, and we obtain

$$\|x(s_1)\| = \|B_1(s_1)x(t_1^-)\| \leqslant \|B_1(s_1)x(t_1)\| \leqslant \overline{B_1}\|x(t_1)\|.$$

It follows from (1.54) that

$$\begin{aligned} \|x(t)\| &\leqslant \overline{B_1}\|x(t_1)\|e^{\bar{A}(t-s_1)} \\ &\leqslant \overline{B_1}\|x(0)\|e^{\bar{A}t_1}e^{\bar{A}(t-s_1)} \\ &\leqslant \overline{B_1}\|x(0)\|e^{\bar{A}((t-s_1)+(t_1-s_0))}. \end{aligned}$$

(ii) Let $t \in (s_2, t_3]$, and we obtain

$$\|x(s_2)\| \leqslant \|B_2(s_2)x(t_2^-)\| \leqslant \|B_2(s_2)x(t_2)\| \leqslant \overline{B_2}\|x(t_2)\|. \tag{1.55}$$

Thus, using (1.55), we have

$$\begin{aligned} \|x(t)\| &\leqslant \|x(s_2)\|e^{\bar{A}(t-s_2)} \\ &\leqslant \overline{B_2}\|x(t_2)\|e^{\bar{A}(t-s_2)} \\ &\leqslant \overline{B_2}\ \overline{B_1}\|x(0)\|e^{\bar{A}((t_2-s_1)+(t_1-s_0))}e^{\bar{A}(t-s_2)} \\ &\leqslant \overline{B_2}\ \overline{B_1}\|x(0)\|e^{\bar{A}((t-s_2)+(t_2-s_1)+(t_1-s_0))}, \end{aligned}$$

where,

$$\|x(t_2)\| \leqslant \overline{B_1}\|x(0)\|e^{\bar{A}(t_1-s_1)}e^{\bar{A}t_2}.$$

Let $B = \max\{\overline{B_1}, \overline{B_2}\}$, and we have

$$\|x(t)\| \leqslant B^2\|x(0)\|e^{\bar{A}(t_1-s_1)}e^{\bar{A}(t_2-s_2)}e^{\bar{A}t}$$
$$\leqslant B^2\|x(0)\|e^{\bar{A}((t-s_2)+\sum_{j=1}^{2}(t_j-s_{j-1}))}.$$

By using mathematic induction, it is easy to see from (1.54) that for all $t \in (s_i, t_{i+1}]$, $i = 0, 1, 2, \ldots$, we have

$$\|x(t)\| \leqslant B^i\|x(0)\|e^{\bar{A}((t-s_i)+\sum_{j=1}^{i}(t_j-s_{j-1}))},$$

where, $B = \max\{\overline{B_1}, \overline{B_2}\ldots, \overline{B_i}\}$.

(II) Let $t \in (t_{i+1}, s_{i+1}]$, $i = 1, 2, \ldots$, and we have

$$\|x(t)\| = \|B_{i+1}(t)x(t_{i+1}^-)\|$$
$$\leqslant \|B_{i+1}(t)\|\|x(t_{i+1})\|$$
$$\leqslant B^{i+1}\|x(0)\|e^{\bar{A}((t_{i+1}-s_i)+\sum_{j=1}^{i}(t_j-s_{j-1}))}.$$

Next, we prove that any nontrivial solution $x(t, x_0) = x(t)$ of (1.40) has a finite Lyapunov exponent. Let $t \in (s_k, t_{k+1}]$, and we have $r(t) = k$. Therefore, according to the Lyapunov exponent definition (1.42) and lemma 1.2.2, we have

$$\chi\left(B^k\|x(0)\|e^{\bar{A}((.-s_k)+\sum_{j=1}^{k}(t_j-s_{j-1}))}\right)$$
$$\leqslant \limsup_{t\to\infty}\frac{1}{t}\left(\log B^k + \bar{A}\left((t - s_k) + \sum_{j=1}^{k}(t_j - s_{j-1})\right)\right) \tag{1.56}$$
$$\leqslant \limsup_{t\to\infty}\frac{k\log B}{t} + \limsup_{t\to\infty}\frac{\bar{A}t}{t}$$
$$\leqslant \rho\log B + \bar{A};$$

here we use

$$\limsup_{t\to\infty}\frac{k}{t} = \limsup_{t\to\infty}\frac{r(t)}{t} = \rho \text{ and } t - s_k + \sum_{j=1}^{k}(t_j - s_{j-1}) \leqslant t.$$

When $t \in (t_{k+1}, s_{k+1}]$ and $r(t) = k + 1$, one can use the same ideas to derive

$$\chi(x(.)) \leqslant \rho\log B + \bar{A} < \infty.$$

The proof is complete. □

Lemma 1.2.6. *Let $X(t)$ be a fundamental solution of (1.40). Then:*

(I) *For $t_0 = s_0 = 0 < t_1 < s_1 < t_2 < s_2\cdots<t_i < t \leqslant s_i$, $i = 1, 2, \ldots$, we have*

$$\frac{\det X(t)}{\det X(0)} = e^{\sum_{j=1}^{i}\int_{s_{j-1}}^{t_j} trA(s)ds} \det B_i(t)\prod_{v=1}^{i-1} \det B_v(s_v);$$

(II) *For* $t_0 = s_0 = 0 < t_1 < s_1 < t_2 < s_2 \cdots s_i < t \leqslant t_{i+1},\ i = 1, 2, \ldots,$ *we have*

$$\frac{\det X(t)}{\det X(0)} = e^{\int_{s_i}^{t} trA(s)ds + \sum_{j=1}^{i}\int_{s_{j-1}}^{t_j} trA(s)ds} \prod_{v=1}^{i} \det B_v(s_v).$$

Proof. Any solution $X(t)$ of (1.40) can be represented as follows:

(I) For $t_0 = s_0 = 0 < t_1 < s_1 < t_2 < s_2 \cdots < t_i < t \leqslant s_i,\ i = 1, 2, \ldots,$ we obtain

$$X(t) = B_i(t)U(t_i, s_{i-1})\prod_{v=1}^{i-1} [B_v(s_v)U(t_v, s_{v-1})]X(0). \tag{1.57}$$

By using the Liouville formula, we obtain from (1.57) that

$$\frac{\det X(t)}{\det X(0)} = e^{\sum_{j=1}^{i}\int_{s_{j-1}}^{t_j} \operatorname{tr}A(s)ds} \det B_i(t)\prod_{v=1}^{i-1} \det B_v(s_v).$$

(II) For $t_0 = s_0 = 0 < t_1 < s_1 < t_2 < s_2 \cdots s_i < t \leqslant t_{i+1},\ i = 1, 2, \ldots,$ we use the same ideas and obtain

$$\frac{\det X(t)}{\det X(0)} = e^{\int_{s_i}^{t} \operatorname{tr}A(s)ds + \sum_{j=1}^{i}\int_{s_{j-1}}^{t_j} \operatorname{tr}A(s)ds} \prod_{v=1}^{i} \det B_v(s_v).$$

The proof is complete. □

Next, we study the regularity of (1.40). Let $X(t)$ be a fundamental matrix of (1.40) and

$$\sigma_X = \sum_{i=1}^{n} \chi(x^i(\cdot)) = \sum_{k=1}^{m} n_k \eta_k$$

be the sum of the Lyapunov exponents of all the solutions from $X(t)$, where $n_k (n_k \geqslant 1)$ is the number of solutions with Lyapunov exponent η_k (see [7, theorem 17]). The Lyapunov inequality (see [7]) is given by the following formula

$$\sigma_X \geqslant \limsup_{t\to+\infty} \frac{1}{t} \log \|X(t)\|.$$

Definition 1.2.7. *(See* [46]*.) If*

$$\sigma_X\ (\equiv\sigma) = \liminf_{t\to+\infty} \frac{1}{t}\log\|X(t)\|,$$

then we say (1.40) is Lyapunov regular.

For $t_0 = s_0 = 0 < t_1 < s_1 < t_2 < s_2 \cdots < t_i < t \leqslant s_i,\ i = 1, 2, \ldots$, we obtain

$$\sigma = \liminf_{t\to+\infty} \frac{1}{t}\left(\sum_{j=1}^{i}\int_{s_{j-1}}^{t_j} \mathrm{tr}A(s)ds + \log|\det B_i(t)| + \sum_{v=1}^{i-1}\log|\det B_v(s_v)|\right).$$

For $t_0 = s_0 = 0 < t_1 < s_1 < t_2 < s_2 \cdots < s_i < t \leqslant t_{i+1},\ i = 1, 2, \ldots$, we obtain

$$\sigma = \liminf_{t\to+\infty} \frac{1}{t}\left(\int_{s_i}^{t} \mathrm{tr}A(s)ds + \sum_{j=1}^{i}\int_{s_{j-1}}^{t_j} \mathrm{tr}A(s)ds + \sum_{v=1}^{i}\log|\det B_v(s_v)|\right).$$

Necessary and sufficient conditions for (1.40) to be regular are given in the following theorem.

Theorem 1.2.8. *The system (1.40) is regular if and only if (the limits exist)*

(I) *For* $t_0 = s_0 = 0 < t_1 < s_1 < t_2 < s_2 \cdots < t_i < t \leqslant s_i,\ i = 1, 2, \ldots$, *we obtain*

$$\lim_{t\to+\infty} \frac{1}{t}\left(\sum_{j=1}^{i}\int_{s_{j-1}}^{t_j} trA(s)ds + \log|\det B_i(t)| + \sum_{v=1}^{i-1}\log|\det B_v(s_v)|\right) = s;$$

(II) *For* $t_0 = s_0 = 0 < t_1 < s_1 < t_2 < s_2 \cdots < s_i < t \leqslant t_{i+1},\ i = 1, 2, \ldots$, *we obtain*

$$\lim_{t\to+\infty} \frac{1}{t}\left(\int_{s_i}^{t} trA(s)ds + \sum_{j=1}^{i}\int_{s_{j-1}}^{t_j} trA(s)ds + \sum_{v=1}^{i}\log|\det B_v(s_v)|\right) = s;$$

and the Lyapunov equality $\sigma = s$ *holds.*

Proof. It is evident that these condition are sufficient for the regularity of (1.40). Let us prove that they are also necessary. Suppose that (1.40) is regular and

(I) For $t_0 = s_0 = 0 < t_1 < s_1 < t_2 < s_2 \cdots < t_i < t \leqslant s_i,\ i = 1, 2, \ldots$, we have

$$\underline{s} = \liminf_{t\to+\infty} \frac{1}{t}\left(\sum_{j=1}^{i}\int_{s_{j-1}}^{t_j} \mathrm{tr}A(s)ds + \log|\det B_i(t)| + \sum_{v=1}^{i-1}\log|\det B_v(s_v)|\right),$$

$$\bar{s} = \limsup_{t\to+\infty} \frac{1}{t}\left(\sum_{j=1}^{i}\int_{s_{j-1}}^{t_j} \mathrm{tr}A(s)ds + \log|\det B_i(t)| + \sum_{v=1}^{i-1}\log|\det B_v(s_v)|\right).$$

From the definition of a regular system and by using the Lyapunov inequality, we have that $\bar{s} \leqslant \sigma = \underline{s}$. However $\underline{s} \leqslant \bar{s}$, so we obtain $\underline{s} = \bar{s} = s = \sigma$.

(II) For $t_0 = s_0 = 0 < t_1 < s_1 < t_2 < s_2 \cdots < s_i < t \leqslant t_{i+1}$, $i = 1, 2, \ldots$, we have

$$\underline{s} = \liminf_{t \to +\infty} \frac{1}{t}\left(\int_{s_i}^{t} \mathrm{tr}A(s)ds + \sum_{j=1}^{i} \int_{s_{j-1}}^{t_j} \mathrm{tr}A(s)ds + \sum_{\nu=1}^{i} \log|\det B_\nu(s_\nu)| \right),$$

$$\bar{s} = \limsup_{t \to +\infty} \frac{1}{t}\left(\int_{s_i}^{t} \mathrm{tr}A(s)ds + \sum_{j=1}^{i} \int_{s_{j-1}}^{t_j} \mathrm{tr}A(s)ds + \sum_{\nu=1}^{i} \log|\det B_\nu(s_\nu)| \right).$$

From the definition of a regular system and by using the Lyapunov inequality, we have that $\bar{s} \leqslant \sigma = \underline{s}$. However $\underline{s} \leqslant \bar{s}$, so we obtain $\underline{s} = \bar{s} = s = \sigma$. The proof is complete. □

1.2.3 Lyapunov exponents and stability

In this section, we study the stability of (1.40) using Lyapunov exponents.

Definition 1.2.9. *We say that (1.40) is asymptotically stable if its zero solution is asymptotically stable.*

Theorem 1.2.10. *Consider the set of Lyapunov exponents of all nontrivial solutions of (1.40), and we have*

$$\Sigma := \{\chi(x(\cdot, x_i)) : x_i \in \mathbb{R}^n/\{0\}, i = 1, 2, \ldots n\}. \tag{1.58}$$

Then its spectrum consists of finite numbers of distinct elements

$$\eta_1 \leqslant \eta_2 \leqslant \cdots \leqslant \eta_{i-1} \leqslant \eta_i \leqslant \rho \log B + \bar{A}, \quad 1 \leqslant i \leqslant n.$$

Proof. We first show that $\eta = \chi(x(\cdot, x_i)) \leqslant \rho \log B + \bar{A}$ for any $x_i \in \mathbb{R}^n/\{0\}$. From theorem 1.2.5, we obtain

$$\eta = \chi(x(\cdot, x_i)) = \limsup_{t \to \infty} \frac{1}{t} \log \|x(t, x_i)\| \leqslant \rho \log B + \bar{A}.$$

Next we show that Σ has at most n distinct elements. Suppose the opposite, i.e. suppose that Σ consist of $n + 1$ distinct values $\eta_1 \leqslant \eta_2 \leqslant \cdots \leqslant \eta_n \leqslant \eta_{n+1}$. From the definition of Σ in (1.58), there exist $x_1, x_2, \ldots, x_{n+1} \in \mathbb{R}^n/\{0\}$, such that

$$\chi(x(\cdot, x_i)) = \eta_i, \quad \text{for } i = 1, 2, \ldots, n + 1. \tag{1.59}$$

Since $\dim(\mathbb{R}^n) = n$ it follows that $\sum_{j=1}^{n+1} \alpha_j x_j = 0$ for some $\alpha_1, \alpha_2, \ldots, \alpha_{n+1} \in \mathbb{R}$ such that $\alpha_i \neq 0$ for some $i \in \{1, 2, 3, \ldots, n + 1\}$, and let $l := \min\{i \in 1, 2, \ldots,$

$n+1: \alpha_i \neq 0\}$. Thus, we have $x_l = \sum_{i=l+1}^{n+1} -\frac{\alpha_i}{\alpha_l}x_i$, and due to the linearity of the map $x(t, \cdot)$ we obtain

$$x(t, x_l) = x\left(t, \sum_{i=l+1}^{n+1} -\frac{\alpha_i}{\alpha_l}x_i\right) = \sum_{i=l+1, \alpha_l\neq 0}^{n+1} -\frac{\alpha_i}{\alpha_l}x(t, x_i),$$

where, $x(t, ax_i + bx_j) = ax(t, x_i) + bx(t, x_j)$, for $a, b \in \mathbb{R}$, $x_i, x_j \in \mathbb{R}^n/\{0\}$. Therefore,

$$\begin{aligned}\chi(x(\cdot, x_l)) &\leqslant \max\left\{\chi\left(-\frac{\alpha_j}{\alpha_l}x(\cdot, x_j)\right): j \in \{l+1, l+2, \ldots, n+1\} \text{ with } \alpha_j \neq 0\right\} \\ &= \max\left\{\chi\left(x(\cdot, x_j)\right): j \in \{l+1, l+2, \ldots, n+1\} \text{ with } \alpha_j \neq 0\right\}.\end{aligned}$$

Therefore, from (1.59) we obtain $\eta_l \leqslant \max\{\eta_j: j \in \{l+1, l+2, \ldots, n+1\}\}$ and this leads to a contradiction. The proof is complete. □

Theorem 1.2.11. *Consider all nontrivial solutions of (1.40). Then, the following statements hold:*

(I) *If (1.40) is stable, then $\Sigma \subset (-\infty, 0]$, where Σ is defined in (1.58).*
(II) *If $\Sigma \subset (-\infty, 0)$, then (1.40) is asymptotically stable.*

Proof. (I) Suppose that system (1.40) is stable. Let $x_0 \in \mathbb{R}^n/\{0\}$ be arbitrary and we need to show that $\chi(x(\cdot, x_0)) \leqslant 0$. Suppose the opposite, i.e. suppose $\chi(x(\cdot, x_0)) \geqslant \varepsilon_0$ for some $\varepsilon_0 > 0$, where

$$\chi(x(\cdot, x_0)) = \limsup_{t\to\infty} \frac{1}{t} \log \|x(t, x_0)\| \geqslant \varepsilon_0.$$

There exists a sequence $\{t_n\}_{n=1}^{\infty}$ with $\lim_{n\to\infty} t_n = \infty$, such that

$$\frac{1}{t_n} \log \|x(t_n, x_0)\| \geqslant 2\varepsilon_0 \Rightarrow \log \|x(t_n, x_0)\| \geqslant 2t_n\varepsilon_0 \Rightarrow \|x(t_n, x_0)\| \geqslant e^{2t_n\varepsilon_0}.$$

This implies that

$$\lim_{n\to\infty} \|x(t_n, x_0)\| \geqslant \lim_{n\to\infty} e^{2t_n\varepsilon_0} \to \infty.$$

Thus, system (1.40) is not stable, which is a contradiction.

(II) Suppose $\Sigma \subset (-\infty, 0)$. Let $x(t, x_0)$ be a nontrivial solution of system (1.40) and $\eta = \max \eta_j < 0$. Choose $\varepsilon = -\frac{\eta}{2} > 0$. Note

$$\eta = \chi(x(\cdot, x_0)) = \limsup_{t\to\infty} \frac{1}{t} \log \|x(t, x_0)\|.$$

From the definition of the solution for the Lyapunov exponent, we obtain

$$\frac{1}{t}\log\|x(t, x_0)\| \leqslant \eta + \varepsilon \Rightarrow \log\|x(t, x_0)\| \leqslant e^{(\eta+\varepsilon)t} = e^{\frac{\eta}{2}t} \text{ for } t \text{ large.}$$

Therefore,

$$\lim_{t\to\infty}\|x(t, x_0)\| \leqslant \lim_{t\to\infty} e^{\frac{\eta}{2}t} = 0.$$

This implies that system (1.40) is asymptotically stable. The proof is complete. □

Theorem 1.2.12. *Let system (1.40) satisfy the conditions: the greatest eigenvalue of the matrix $\widetilde{A(t)}$ satisfies*

$$\Lambda(t) \leqslant \gamma,$$

for all $t \geqslant t_0$, for all $i = 1, 2, \ldots$, and the greatest eigenvalues of the matrix $B_i^{\top}(t)B_i(t)$ is such that

$$\Lambda_i \leqslant \Lambda,$$

and the limit

$$\lim_{t\to\infty}\frac{r(t)}{t} = \rho$$

exists and is finite.

If

$$\frac{1}{2}\rho\log\Lambda + \gamma < 0,$$

then all solutions of (1.40) are asymptotically stable.

Proof. From theorem 1.2.3, we have the following inequalities:

(i) Let $t \in (s_i, t_{i+1}]$, and we have

$$\|x(t)\|^2 \leqslant \prod_{j=1}^{i}\left(\Lambda_j e^{2\int_{s_{j-1}}^{t_j}\Lambda(\tau)d\tau}\right)e^{2\int_{s_i}^{t}\Lambda(\tau)d\tau}\|x(0)\|^2. \tag{1.60}$$

From the definition of the Lyapunov exponent, we obtain

$$2\chi(x(\cdot)) = \limsup_{t\to\infty}\frac{1}{t}\log\|x(t)\|^2 \Longrightarrow \chi(x(\cdot)) = \frac{1}{2}\limsup_{t\to\infty}\frac{1}{t}\chi\|x(t)\|^2.$$

Therefore, use (1.60) and the definition of the Lyapunov exponent to derive

$$
\begin{aligned}
&\chi(x(\cdot)) \\
&\leqslant \frac{1}{2}\limsup_{t\to\infty}\frac{1}{t}\log\left(\prod_{j=1}^{i}(\Lambda_j e^{2\int_{s_{j-1}}^{t_j}\Lambda(\tau)d\tau})e^{2\int_{s_i}^{t}\Lambda(\tau)d\tau}\|x(0)\|^2\right) \\
&= \frac{1}{2}\limsup_{t\to\infty}\frac{1}{t}\left(\log\prod_{j=1}^{i}(\Lambda_j) + \log\prod_{j=1}^{i}(e^{2\int_{s_{j-1}}^{t_j}\Lambda(\tau)d\tau})\right. \\
&\quad \left. + \log e^{2\int_{s_i}^{t}\Lambda(\tau)d\tau} + \log\|x(0)\|^2\right) \\
&\leqslant \frac{1}{2}\limsup_{t\to\infty}\frac{\log\prod_{j=1}^{i}(\Lambda_j)}{t} + \frac{1}{2}\limsup_{t\to\infty}\frac{2\int_{s_i}^{t}\Lambda(\tau)d\tau + 2\sum_{j=1}^{i}(\int_{s_{j-1}}^{t_j}\Lambda(\tau)d\tau)}{t} \\
&\leqslant \frac{1}{2}\limsup_{t\to\infty}\frac{\sum_{j=1}^{i}\log(\Lambda_j)}{t} + \frac{1}{2}\limsup_{t\to\infty}\frac{2\gamma t}{t} \\
&\leqslant \frac{1}{2}\rho\log\Lambda + \gamma,
\end{aligned} \tag{1.61}
$$

where we use $r(t) = i$ and

$$
\int_{s_i}^{t}\Lambda(\tau)d\tau + \sum_{j=1}^{i}\left(\int_{s_{j-1}}^{t_j}\Lambda(\tau)d\tau\right) \leqslant \gamma\left(\int_{s_i}^{t}d\tau + \sum_{j=1}^{i}\left(\int_{s_{j-1}}^{t_j}d\tau\right)\right) < \gamma t.
$$

Therefore, from (1.61), we have

$$
\chi(x(\cdot)) \leqslant \frac{1}{2}\rho\log\Lambda + \gamma.
$$

(ii) Let $t \in (t_{i+1}, s_{i+1}]$, and we have

$$
\|x(t)\|^2 \leqslant \prod_{j=1}^{i+1}(\Lambda_j e^{2\int_{s_{j-1}}^{t_j}\Lambda(\tau)d\tau})\|x(0)\|^2.
$$

One can use the same ideas to obtain

$$
\begin{aligned}
\chi(x(\cdot)) &\leqslant \frac{1}{2}\limsup_{t\to\infty}\frac{1}{t}\log\left(\prod_{j=1}^{i+1}(\Lambda_j e^{2\int_{s_{j-1}}^{t_j}\Lambda(\tau)d\tau})\|x(0)\|^2\right) \\
&= \frac{1}{2}\limsup_{t\to\infty}\frac{1}{t}\left(\log\prod_{j=1}^{i+1}(\Lambda_j e^{2\int_{s_{j-1}}^{t_j}\Lambda(\tau)d\tau}) + \log\|x(0)\|^2\right) \\
&\leqslant \frac{1}{2}\limsup_{t\to\infty}\frac{1}{t}\left(\log\prod_{j=1}^{i+1}(\Lambda_j)e^{2\sum_{j=1}^{i+1}(\int_{s_{j-1}}^{t_j}\Lambda(\tau)d\tau)}\right) \\
&\leqslant \frac{1}{2}\limsup_{t\to\infty}\frac{\log\prod_{j=1}^{i+1}(\Lambda_j)}{t} + \frac{1}{2}\limsup_{t\to\infty}\frac{2\sum_{j=1}^{i+1}(\int_{s_{j-1}}^{t_j}\Lambda(\tau)d\tau)}{t} \\
&\leqslant \frac{1}{2}\rho\log\Lambda + \gamma.
\end{aligned}
\tag{1.62}
$$

Therefore, from (1.62), we have

$$
\chi(x(\cdot)) \leqslant \frac{1}{2}\rho\log\Lambda + \gamma < 0.
$$

This implies that all solutions of system (1.40) are asymptotically stable. The proof is complete. □

1.2.4 Criteria for nonuniform exponential behavior

In this section, we describe criteria for the existence of nonuniform exponential behavior. Using the properties of the adjoint system, we consider the regularity of (1.40). Let $\langle\cdot,\cdot\rangle$ be the standard inner product in $\mathbb{R}^n$. We recall that two bases $x_1, \dots, x_n$ and $y_1, \dots, y_n$ of $\mathbb{R}^n$ are said to be dual if $\langle x_i, y_j\rangle = \delta_{ij}$ for all i and j, where δ_{ij} is the Kronecker symbol.

Consider the adjoint system to (1.40):

$$
\begin{cases}
y'(t) = -A(t)^{\top}y(t), & t \in (s_i, t_{i+1}],\ i = 0, 1, 2, \dots, \\
y(t_i^+) = (B_i(t_i)^{\top})^{-1}y(t_i^-), & i = 1, 2, \dots, \\
y(t) = (B_i(t)^{\top})^{-1}y(t_i^-), & t \in (t_i, s_i],\ i = 1, 2, \dots, \\
y(s_i^+) = y(s_i^-) = y(s_i), & i = 1, 2, \dots.
\end{cases}
\tag{1.63}
$$

We define the Lyapunov exponent $\mu\colon \mathbb{R}^n \to \mathbb{R} \cup \{-\infty\}$ associated with (1.63) by

$$
\mu(y_0) = \limsup_{t\to+\infty}\frac{1}{t}\log\|y(t)\|,
\tag{1.64}
$$

where $y(t)$ is the solution of (1.63) with $y(t_0) = y_0$; we write $\mu(y(\cdot, y_0))$ for $\mu(y_0)$ as well.

We define the regularity coefficient of Lyapunov exponents χ and μ by [43]

$$\omega(\chi, \mu) = \min\max\{\chi(x_i) + \mu(y_i): i = 1, 2, \dots, n\}, \tag{1.65}$$

where the minimum is taken over all dual bases $x_1, \dots, x_n$ and $y_1, \dots, y_n$ of $\mathbb{R}^n$; here $\chi(x_i) = \chi(x(\cdot, x_i))$ and $\mu(y_i) = \mu(y(\cdot, y_i))$.

Lemma 1.2.13. *For any two solutions $x(t)$ and $y(t)$ of the mutually adjoint systems (1.40) and (1.63), we have*

$$\langle x(t), y(t)\rangle = c, \tag{1.66}$$

where c is some constant. For any fundamental matrices of the solutions of these systems $X(t)$ and $Y(t)$, the following holds:

$$Y(t)^{\mathsf{T}} X(t) = C, \tag{1.67}$$

where C is a constant matrix.

Proof. That the scalar product of the solution of adjoint systems is constant on every interval $s_i < t \leqslant t_{i+1}$ follows because, on this interval, we have

$$\frac{dx}{dt} = A(t)x(t), \qquad \frac{dy}{dt} = -A(t)^{\mathsf{T}} y(t),$$

so we obtain

$$\left\langle \frac{dx}{dt}, y(t) \right\rangle = \langle A(t)x(t), y(t)\rangle, \qquad \left\langle x(t), \frac{dy}{dt} \right\rangle = \langle x(t), -A(t)^{\mathsf{T}} y(t)\rangle,$$

and adding them we obtain

$$\begin{aligned} \frac{d}{dt}\langle x(t), y(t)\rangle &= \left\langle \frac{dx}{dt}, y(t) \right\rangle + \left\langle x(t), \frac{dy}{dt} \right\rangle \\ &= \langle A(t)x(t), y(t)\rangle + \langle x(t), -A(t)^{\mathsf{T}} y(t)\rangle = 0. \end{aligned}$$

Therefore, we obtain $\langle x(t), y(t)\rangle = c_i$.

It remains to show that the constants c_i are all equal, i.e. $c_i = c$ for all $i = 1, 2, \dots$. For $t = t_i$, $i = 1, 2, \dots$, we obtain

$$\begin{aligned} \Delta\langle x(t), y(t)\rangle|_{t=t_i} &= \langle x(t_i^+), y(t_i^+)\rangle - \langle x(t_i), y(t_i)\rangle \\ &= \langle B_i(t_i)x(t_i), (B_i(t_i)^{\mathsf{T}})^{-1} y(t_i)\rangle - \langle x(t_i), y(t_i)\rangle \\ &= x(t_i)^{\mathsf{T}} B_i(t_i)^{\mathsf{T}} (B_i(t_i)^{\mathsf{T}})^{-1} y(t_i) - x(t_i)^{\mathsf{T}} y(t_i) = 0. \end{aligned}$$

For $t \in (t_i, s_i]$, $i = 1, 2, \dots$, we have

$$\begin{aligned} \langle x(t), y(t)\rangle &= \langle B_i(t)x(t_i), (B_i(t)^{\mathsf{T}})^{-1} y(t_i)\rangle \\ &= x(t_i)^{\mathsf{T}} B_i(t)^{\mathsf{T}} (B_i(t)^{\mathsf{T}})^{-1} y(t_i) \\ &= \langle x(t_i), y(t_i)\rangle = c_i. \end{aligned}$$

For $t = s_i$, $i = 1, 2, \ldots$, we obtain

$$\Delta\langle x(t), y(t)\rangle|_{t=s_i} = \langle x(s_i^+), y(s_i^+)\rangle - \langle x(s_i), y(s_i)\rangle = 0.$$

It follows from here that $c_{i+1} = c_i$ for all $i = 1, 2, \ldots$, and so $\langle x(t), y(t)\rangle = c$ for all $t \geqslant t_0$.

The fundamental matrices $X(t)$ and $Y(t)$ satisfy the corresponding systems (1.40) and (1.63). Therefore, one can apply a similar method to complete the proof of equality (1.67). The proof is complete. □

Lemma 1.2.14. *If $Y(t)^\mathsf{T} X(t) = C$ holds, the matrix C is nonsingular, and $X(t)$ is a fundamental matrix of (1.40), then $Y(t)$ is a fundamental matrix of the adjoint system (1.63).*

Proof. Suppose that the identity $Y(t)^\mathsf{T} X(t) = C$ holds, where C is a nonsingular matrix and $X(t)$ is a fundamental matrix of (1.40). We now show that $Y(t)$ is a fundamental matrix of system (1.63). Note from $Y(t)^\mathsf{T} X(t) = C$ that

$$Y(t) = (X(t)^\mathsf{T})^{-1} C^\mathsf{T}. \tag{1.68}$$

The matrix $X(t)^\mathsf{T}$ satisfies the system of equations

$$\frac{dX^\mathsf{T}(t)}{dt} = X(t)^\mathsf{T} A(t)^\mathsf{T},\ t \in (s_i, t_{i+1}], \quad X(t)^\mathsf{T} = X(t_i^-)^\mathsf{T} B(t)^\mathsf{T},\ t \in (t_i, s_i].$$

Hence, from (1.68) and the expression for the derivative of the inverse of a matrix, we obtain the following:

For $t \in (s_i, t_{i+1}]$ and $i = 0, 1, \ldots$, we have

$$\begin{aligned}\frac{dY(t)}{dt} &= -(X(t)^\mathsf{T})^{-1} \frac{dX(t)^\mathsf{T}}{dt} (X(t)^\mathsf{T})^{-1} C^\mathsf{T}\\ &= -(X(t)^\mathsf{T})^{-1} X(t)^\mathsf{T} A(t)^\mathsf{T} (X(t)^\mathsf{T})^{-1} C^\mathsf{T}\\ &= -A(t)^\mathsf{T} Y(t);\end{aligned}$$

For $t = t_i^+$ and $i = 1, 2, \ldots$, we have

$$\begin{aligned}Y(t_i^+) &= (X(t_i^+)^\mathsf{T})^{-1} C^\mathsf{T} = (X(t_i^-)^\mathsf{T} B_i(t_i)^\mathsf{T})^{-1} C^\mathsf{T}\\ &= (B_i(t_i)^\mathsf{T})^{-1} (X(t_i^-)^\mathsf{T})^{-1} C^\mathsf{T}\\ &= (B_i(t_i)^\mathsf{T})^{-1} Y(t_i^-);\end{aligned}$$

For $t \in (t_i, s_i]$ and $i = 1, 2, \ldots$, we have

$$\begin{aligned}Y(t) &= (X(t)^\mathsf{T})^{-1} C^\mathsf{T} = (X(t_i^-)^\mathsf{T} B_i(t)^\mathsf{T})^{-1} C^\mathsf{T}\\ &= (B_i(t)^\mathsf{T})^{-1} (X(t_i^-)^\mathsf{T})^{-1} C^\mathsf{T}\\ &= (B_i(t)^\mathsf{T})^{-1} Y(t_i^-);\end{aligned}$$

For $t = s_i$, $i = 1, 2, \ldots$, we have

$$Y(s_i^+) = (X(s_i^+)^{\mathrm{T}})^{-1}C^{\mathrm{T}} = (X(s_i^-)^{\mathrm{T}})^{-1}C^{\mathrm{T}} = Y(s_i^-).$$

Note that

$$\det Y(t) = \det((X(t)^{\mathrm{T}})^{-1})\det C^{\mathrm{T}} = \det(X(t))^{-1}\det C^{\mathrm{T}} \neq 0.$$

Therefore, the $Y(t)$ is a fundamental matrix of system (1.63). The proof is complete. □

Now we present a Perron theorem for (1.40).

Theorem 1.2.15. *The desired system (1.40) is regular if and only if its complete spectrum*

$$\eta_1 \leqslant \eta_2 \leqslant \cdots \leqslant \eta_n \tag{1.69}$$

and the complete spectrum of the adjoint system (1.62)

$$\mu_1 \geqslant \mu_2 \geqslant \cdots \geqslant \mu_n \tag{1.70}$$

are symmetric with respect to zero, i.e.

$$\eta_k + \mu_k = 0, \quad k = 1, 2, \ldots, n. \tag{1.71}$$

Proof. Let system (1.40) be regular and $X(t) = \{x_{jk}(t)\}$ be its normal fundamental matrix which consists of the solutions $x_k(t)$ such that $\chi(x_k(t)) = \eta_k$, where the numbers η_k satisfy (1.69). The matrix

$$Y(t) = \{y_{jk}(t)\} = [X(t)^{\mathrm{T}}]^{-1} \tag{1.72}$$

is a fundamental matrix of the adjoint system (1.63) and $Y(t)^{\mathrm{T}}X(t) = \mathrm{Id}$. Noting that $\langle x_k(t), y_k(t)\rangle = 1$, we find that $\chi(1) = 0 = \chi(y_k(t)) + \chi(x_k(t))$ i.e.

$$\mu_k + \eta_k \geqslant 0. \tag{1.73}$$

If $X_{jk}(t)$ is the algebraic complement of x_{jk} in $\det X(t)$, then

$$y_{jk}(t) = \frac{1}{\det X(t)}\{X_{jk}(t)\},$$

where $\det X(t)$ is given by the following formulas:

For $t_0 = s_0 = 0 < t_1 < s_1 < t_2 < s_2 \cdots < t_i < t \leqslant s_i$, $i = 1, 2, \ldots$, we have

$$\det X(t) = e^{\sum_{j=1}^{i}\int_{s_{j-1}}^{t_j} \mathrm{tr}A(s)ds} \det B_i(t)\prod_{v=1}^{i-1} \det B_v(s_v)\det X(0) \neq 0;$$

For $t_0 = s_0 = 0 < t_1 < s_1 < t_2 < s_2 \cdots s_i < t \leqslant t_{i+1},\ i = 1, 2, \ldots,$ we have

$$\det X(t) = e^{\int_{s_i}^{t} \mathrm{tr}A(s)ds + \sum_{j=1}^{i}\int_{s_{j-1}}^{t_j} \mathrm{tr}A(s)ds} \prod_{v=1}^{i} \det B_v(s_v) \det X(0) \neq 0.$$

From these it follows that:

For $t_0 = s_0 = 0 < t_1 < s_1 < t_2 < s_2 \cdots < t_i < t \leqslant s_i,\ i = 1, 2, \ldots,$ we obtain

$$\begin{aligned}
&\mu(y_{jk}(t)) \\
&= \limsup_{t\to+\infty} \frac{1}{t} \log \|y_{jk}(t)\| \\
&\leqslant \limsup_{t\to+\infty} \frac{1}{t} \log \|\frac{1}{\det X(t)}\| + \chi(X_{jk}(t)) \\
&\leqslant -\limsup_{t\to+\infty} \frac{1}{t}\left(\sum_{j=1}^{i}\int_{s_{j-1}}^{t_j} \mathrm{tr}A(s)ds + \log|\det B_i(t)| + \sum_{v=1}^{i-1} \log|\det B_v(s_v)|\right) \\
&+\chi(X_{jk}(t)).
\end{aligned}$$

For $t_0 = s_0 = 0 < t_1 < s_1 < t_2 < s_2 \cdots < s_i < t \leqslant t_{i+1},\ i = 1, 2, \ldots,$ we obtain

$$\begin{aligned}
&\mu(y_{jk}(t)) \\
&= \limsup_{t\to+\infty} \frac{1}{t} \log \|y_{jk}(t)\| \\
&\leqslant \limsup_{t\to+\infty} \frac{1}{t} \log \|\frac{1}{\det X(t)}\| + \chi(X_{jk}(t)) \\
&\leqslant -\limsup_{t\to+\infty} \frac{1}{t}\left(\int_{s_i}^{t} \mathrm{tr}A(s)ds + \sum_{j=1}^{i}\int_{s_{j-1}}^{t_j} \mathrm{tr}A(s)ds + \sum_{v=1}^{i} \log|\det B_v(s_v)|\right) \\
&+\chi(X_{jk}(t)).
\end{aligned}$$

Note that (1.40) is regular. Therefore, the Lyapunov equality holds:

$$-\limsup_{t\to+\infty} \frac{1}{t}\left(\sum_{j=1}^{i}\int_{s_{j-1}}^{t_j} \mathrm{tr}A(s)ds + \log|\det B_i(t)| + \sum_{v=1}^{i-1} \log|\det B_v(s_v)|\right) = -\sigma,$$

and

$$-\limsup_{t\to+\infty} \frac{1}{t}\left(\int_{s_i}^{t} \mathrm{tr}A(s)ds + \sum_{j=1}^{i}\int_{s_{j-1}}^{t_j} \mathrm{tr}A(s)ds + \sum_{v=1}^{i} \log|\det B_v(s_v)|\right) = -\sigma.$$

Note $\chi(X_{jk}(t)) \leqslant \sigma - \eta_k$. Thus

$$\mu(y_{jk}(t)) \leqslant -\sigma + \sigma - \eta_k = -\eta_k$$

and $\mu_k = \max_i \mu(y_{ik}(t)) \leqslant -\eta_k$, i.e.

$$\mu_k + \eta_k \leqslant 0. \tag{1.74}$$

Comparing this inequality with (1.73) yields

$$\mu_k + \eta_k = 0, \quad k = 1, 2, \dots, n. \tag{1.75}$$

It remains to show that the fundamental matrix $Y(t)$ is normal and the numbers $\mu_1, \mu_2, \dots, \mu_n$ realize the whole spectrum of the adjoint system. Indeed from (1.75), we have:

For $t_0 = s_0 = 0 < t_1 < s_1 < t_2 < s_2 \cdots < t_i < t \leqslant s_i$, $i = 1, 2, \dots$, we obtain

$$\begin{aligned}
\sigma_Y &= \sum_{k=1}^{n} \mu_k = -\sum_{k=1}^{n} \eta_k \\
&= -\lim_{t\to+\infty} \frac{1}{t}\left(\sum_{j=1}^{i} \int_{s_{j-1}}^{t_j} \mathrm{tr}A(s)ds + \log|\det B_i(t)| + \sum_{v=1}^{i-1} \log|\det B_v(s_v)| \right) \\
&= \lim_{t\to+\infty} \frac{1}{t}\left(\sum_{j=1}^{i} \int_{s_{j-1}}^{t_j} \mathrm{tr}(-A^{\mathsf{T}}(s))ds + \log|(\det B_i^{\mathsf{T}}(t))^{-1}| \right. \\
&\quad \left. + \sum_{v=1}^{i-1} \log|(\det B_v^{\mathsf{T}}(s_v))^{-1}| \right),
\end{aligned}$$

For $t_0 = s_0 = 0 < t_1 < s_1 < t_2 < s_2 \cdots s_i < t \leqslant t_{i+1}$, $i = 1, 2, \dots$, we obtain

$$\begin{aligned}
\sigma_Y &= \sum_{k=1}^{n} \mu_k = -\sum_{k=1}^{n} \eta_k \\
&= -\lim_{t\to+\infty} \frac{1}{t}\left(\int_{s_i}^{t} \mathrm{tr}A(s)ds + \sum_{j=1}^{i} \int_{s_{j-1}}^{t_j} \mathrm{tr}A(s)ds + \sum_{v=1}^{i} \log|\det B_v(s_v)| \right) \\
&= \lim_{t\to+\infty} \frac{1}{t}\left(\int_{s_i}^{t} \mathrm{tr}(-A^{\mathsf{T}}(s))ds + \sum_{j=1}^{i} \int_{s_{j-1}}^{t_j} \mathrm{tr}(-A^{\mathsf{T}}(s))ds \right. \\
&\quad \left. + \sum_{v=1}^{i} \log|(\det B_v^{\mathsf{T}}(s_v))^{-1}| \right).
\end{aligned}$$

Hence, the Lyapunov equality holds for the fundamental matrix $Y(t)$ of the adjoint system (1.63) and this matrix is normal.

Now we show that if (1.71) holds, then system (1.40) is regular. From the Lyapunov inequality, we have the following:

For $t_0 = s_0 = 0 < t_1 < s_1 < t_2 < s_2 \cdots < t_i < t \leqslant s_i,\ i = 1, 2, \ldots,$ we obtain

$$\sigma_X = \sum_{k=1}^{n} \eta_k \geqslant \limsup_{t \to +\infty} \frac{1}{t} \left(\sum_{j=1}^{i} \int_{s_{j-1}}^{t_j} \mathrm{tr} A(s) ds + \log|\det B_i(t)| + \sum_{v=1}^{i-1} \log|\det B_v(s_v)| \right) = \bar{s},$$

and

$$\begin{aligned} \sigma_Y &= \sum_{k=1}^{n} \mu_k \\ &\geqslant \limsup_{t \to +\infty} \frac{1}{t} \left(\sum_{j=1}^{i} \int_{s_{j-1}}^{t_j} \mathrm{tr}(-A^{\top}(s)) ds + \log|(\det B_i^{\top}(t))^{-1}| + \sum_{v=1}^{i-1} \log|(\det B_v^{\top}(s_v))^{-1}| \right) \\ &\geqslant -\liminf_{t \to +\infty} \frac{1}{t} \left(\sum_{j=1}^{i} \int_{s_{j-1}}^{t_j} \mathrm{tr} A(s) ds + \log|\det B_i(t)| + \sum_{v=1}^{i-1} \log|\det B_v(s_v)| \right) = -\underline{s} \end{aligned}$$

For $t_0 = s_0 = 0 < t_1 < s_1 < t_2 < s_2 \cdots < s_i < t \leqslant t_{i+1},\ i = 1, 2, \ldots,$ we obtain

$$\sigma_X = \sum_{k=1}^{n} \eta_k \geqslant \limsup_{t \to +\infty} \frac{1}{t} \left(\int_{s_i}^{t} \mathrm{tr} A(s) ds + \sum_{j=1}^{i} \int_{s_{j-1}}^{t_j} \mathrm{tr} A(s) ds + \sum_{v=1}^{i} \log|\det B_v(s_v)| \right) = \bar{s},$$

and

$$\begin{aligned} \sigma_Y &= \sum_{k=1}^{n} \mu_k \\ &\geqslant \limsup_{t \to +\infty} \frac{1}{t} \left(\int_{s_i}^{t} \mathrm{tr}(-A^{\top}(s)) ds + \sum_{j=1}^{i} \int_{s_{j-1}}^{t_j} \mathrm{tr}(-A^{\top}(s)) ds + \sum_{v=1}^{i} \log|(\det B_v^{\top}(s_v))^{-1}| \right) \\ &\geqslant -\liminf_{t \to +\infty} \frac{1}{t} \left(\int_{s_i}^{t} \mathrm{tr} A(s) ds + \sum_{j=1}^{i} \int_{s_{j-1}}^{t_j} \mathrm{tr} A(s) ds + \sum_{v=1}^{i} \log|\det B_v(s_v)| \right) = -\underline{s}. \end{aligned}$$

Therefore, adding these inequalities and using (1.71), we obtain $\bar{s} = \underline{s}$. Thus

$$s = \lim_{t\to+\infty}\frac{1}{t}\left(\sum_{j=1}^{i}\int_{s_{j-1}}^{t_j}\operatorname{tr}A(s)ds + \log|\det B_i(t)| + \sum_{v=1}^{i-1}\log|\det B_v(s_v)|\right),$$

and

$$s = \lim_{t\to+\infty}\frac{1}{t}\left(\int_{s_i}^{t}\operatorname{tr}A(s)ds + \sum_{j=1}^{i}\int_{s_{j-1}}^{t_j}\operatorname{tr}A(s)ds + \sum_{v=1}^{i}\log|\det B_v(s_v)|\right).$$

Moreover, we have $\sum_{k=1}^{n}\eta_k = s$. From theorem 1.2.8 we see that (1.40) is regular. The proof is complete. □

Next, we describe criteria for the existence of nonuniform exponential contractions. We write the unique left-continuous solution of (1.40) in the form $x(t) = W(t, s)x(s)$ for $t, s \geqslant t_0$.

Definition 1.2.16. *(See [43].) We say that (1.40) admits a (strong) nonuniform exponential contraction if there exist constants*

$$\underline{a} \leqslant \bar{a} < 0, \ \ a \geqslant 0, \ \ \text{and } K > 0,$$

such that

$$\|W(t, s)\| \leqslant Ke^{\bar{a}(t-s)+as}, \qquad \|W(t, s)^{-1}\| \leqslant Ke^{-\underline{a}(t-s)+at}, \tag{1.76}$$

for every $t \geqslant s \geqslant t_0$.

Theorem 1.2.17. *For any $\varepsilon > 0$, (1.40) admits a nonuniform exponential contraction with*

$$\underline{a} = \underline{\eta} + \varepsilon, \ \ \bar{a} = \bar{\eta} + \varepsilon, \ \ a = \omega(\chi, \mu) + 2\varepsilon,$$

where $\bar{\eta} = \max\{\eta_j: j = 1, 2, \ldots n\}$ and $\underline{\eta} = \min\{\eta_j: j = 1, 2, \ldots n\}$.

Proof. We consider a fundamental solution of (1.40), that is, an $n \times n$ matrix $X(t)$ whose columns $x_1(t), \ldots, x_n(t)$ form a basis for the space of solution of (1.40). Given a basis $x_1, \ldots, x_n \in \mathbb{R}^n$, let $x_1(t), \ldots, x_n(t)$ be the solutions of (1.40) with $x_i(t_0) = x_i$ for $i = 1, 2, \ldots, n$. Now we observe that

$$W(t, s) = X(t)X(s)^{-1} = X(t)Y(s)^{\top},$$

where $X(t)$ is a fundamental solution of (1.40) and $Y(t) = [X(t)^{\top}]^{-1}$. For any $t \in (s_i, t_{i+1}]$ and $i = 0, 1, 2, \ldots$, taking derivatives in

$$X(t)X(t)^{-1} = X(t)Y(t)^{\top} = \mathrm{Id},$$

we obtain

$$X'(t)X(t)^{-1} + X(t)X'(t)^{-1} = 0 \Rightarrow X'(t)Y(t)^{\mathsf{T}} + X(t)Y'(t)^{\mathsf{T}} = 0,$$

thus yielding

$$X(t)Y'(t)^{\mathsf{T}} = -X'(t)Y(t)^{\mathsf{T}} = -A(t)X(t))X(t)^{-1} = -A(t).$$

Therefore,

$$Y'(t)^{\mathsf{T}} = -X(t)^{-1}A(t) \Rightarrow Y'(t) = -A(t)^{\mathsf{T}}Y(t).$$

For $t = t_i,\ i = 1, 2, \ldots,$ we have

$$\begin{aligned} Y(t_i^+) &= [X(t_i^+)^{\mathsf{T}}]^{-1} = [(B_i(t_i)x(t_i^-))^{\mathsf{T}}]^{-1} = [x(t_i^-)^{\mathsf{T}} B_i(t_i)^{\mathsf{T}}]^{-1} \\ &= [B_i(t_i)^{\mathsf{T}}]^{-1}[x(t_i^-)^{\mathsf{T}}]^{-1} = [B_i(t_i)^{\mathsf{T}}]^{-1} Y(t_i^-). \end{aligned}$$

For any $t \in (t_i, s_i]$ and $i = 1, 2, \ldots,$ we have

$$Y(t) = [X(t)^{\mathsf{T}}]^{-1} = [(B_i(t)x(t_i))^{\mathsf{T}}]^{-1} = [B_i(t)^{\mathsf{T}}]^{-1}[x(t_i)^{\mathsf{T}}]^{-1} = [B_i(t)^{\mathsf{T}}]^{-1} Y(t_i).$$

For $t = s_i,\ i = 1, 2, \ldots,$ we have

$$Y(s_i^+) = [X(s_i^+)^{\mathsf{T}}]^{-1} = [X(s_i^-)^{\mathsf{T}}]^{-1} = Y(s_i^-).$$

This shows that $Y(t)$ is a fundamental solution of (1.63). Now let $x_1(t), \ldots, x_n(t)$ be the columns of $X(t)$, and let $y_1(t), \ldots, y_n(t)$ be the columns of $Y(t)$. For $j = 1, 2, \ldots, n$, we have

$$\eta_j = \chi(x_j(t_0)) \quad \text{and} \quad \mu_j = \mu(y_j(t_0)),$$

where χ and μ are the Lyapunov exponents given in (1.42) and (1.64). Given $\varepsilon > 0$ there is a constant $K > 0$ such that

$$\|x_j(t)\| \leqslant Ke^{(\eta_j+\varepsilon)t} \quad \text{and} \quad \|y_j(t)\| \leqslant Ke^{(\mu_j+\varepsilon)t},$$

for any $j = 1, 2, \ldots, n$ and $t \geqslant t_0$. It follows from the identity $X(t)Y(t)^{\mathsf{T}} = \mathrm{Id}$ that the basis $x_1(t_0), \ldots, x_n(t_0)$ and $y_1(t_0), \ldots, y_n(t_0)$ are dual. Without loss of generality assume

$$\omega(\chi, \mu) = \max\{\eta_j + \mu_j : j = 1, 2, \ldots, n\}.$$

The entries $w_{ik}(t, s)$ of the matrix $W(t, s) = X(t)Y(s)^{\mathsf{T}}$ are given by

$$w_{ik}(t, s) = \sum_{j=1}^{n} x_{ij}(t) y_{kj}(s),$$

where $x_{ij}(t)$ is the ith coordinate of $x_j(t)$, and $y_{kj}(s)$ is the kth coordinate of $y_j(s)$. Therefore, we have

$$\begin{aligned}|w_{ik}(t, s)| &\leqslant \sum_{j=1}^{n}|x_{ij}(t)||y_{kj}(s)| \leqslant \sum_{j=1}^{n}\|x_j(t)\|\|y_j(s)\| \\ &\leqslant \sum_{j=1}^{n} Ke^{(\eta_j+\varepsilon)t}Ke^{(\mu_j+\varepsilon)s} \leqslant \sum_{j=1}^{n} K^2e^{(\eta_j+\varepsilon)(t-s)+(\eta_j+\mu_j+2\varepsilon)s} \\ &\leqslant nK^2e^{(\bar{\eta}+\varepsilon)(t-s)+(\omega(\chi,\,\mu)+2\varepsilon)s}.\end{aligned}$$

This yields the first inequality in (1.76).

Now we consider the matrix $W(t, s)^{-1} = X(s)Y(t)^{\mathrm{T}}$. Its entries $\widetilde{w_{ik}}(t, s)$ are given by

$$\widetilde{w_{ik}}(s, t) = \sum_{j=1}^{n} x_{ij}(s)y_{kj}(t).$$

Therefore, we have

$$\begin{aligned}|\widetilde{w_{ik}}(s, t)| &\leqslant \sum_{j=1}^{n}|x_{ij}(s)||y_{kj}(t)| \leqslant \sum_{j=1}^{n}\|x_j(s)\|\|y_j(t)\| \\ &\leqslant \sum_{j=1}^{n} Ke^{(\eta_j+\varepsilon)s}Ke^{(\mu_j+\varepsilon)t} \leqslant \sum_{j=1}^{n} K^2e^{-(\eta_j+\varepsilon)(t-s)+(\eta_j+\mu_j+2\varepsilon)t} \\ &\leqslant nK^2e^{-(\underline{\eta}+\varepsilon)(t-s)+(\omega(\chi,\,\mu)+2\varepsilon)t}.\end{aligned}$$

This yields the second inequality in (1.76). The proof is complete. □

Remark 1.2.18. *According to theorem 1.2.17, if (1.40) admits a nonuniform exponential contraction, then*

$$\chi(x_0) \leqslant \bar{a} < 0, \quad \textit{for every } x(t_0) = x_0 \in \mathbb{R}^n.$$

1.2.5 Stability for an impulsive perturbed system

We consider the following non-instantaneous nonlinear impulsive perturbed differential system

$$\begin{cases} x'(t) = A(t)x(t) + f(t, x(t)), \; t \in (s_i, t_{i+1}], \; i = 0, 1, 2, \dots, \\ x(t_i^+) = B_i(t_i)x(t_i^-) + g_i(x(t_i^-)), \; i = 1, 2, \dots, \\ x(t) = B_i(t)x(t_i^-) + g_i(x(t_i^-)), \; t \in (t_i, s_i], \; i = 1, 2, \dots, \\ x(s_i^+) = x(s_i^-) = x(s_i), \; i = 1, 2, \dots, \end{cases} \tag{1.77}$$

where f: $\mathbb{R}^+ \times \mathbb{R}^n \to \mathbb{R}^n$ and g_i: $\mathbb{R}^n \to \mathbb{R}^n$ for $i = 1, 2, \dots$ are continuous functions satisfying $f(t, 0) = g_i(0) = 0$. Moreover, we assume:

For every $t \geqslant t_0$, there exist constants L, $p > 0$ such that

$$\|f(t, x) - f(t, y)\| \leqslant L\|x - y\|(\|x\|^p + \|y\|^p),$$

and

$$\|g_i(x) - g_i(y)\| \leqslant L\|x - y\|(\|x\|^p + \|y\|^p),$$

for any $x, y \in \mathbb{R}^n$.

Using the method of variation of constants, the solution of (1.77) with $x(s) = x_s \in \mathbb{R}^n$, $t_0 < s < t_1$ has the following form

$$x(t, x_s) = \begin{cases} W(t, s)x_s + \int_s^t W(t, \tau)f(\tau, x(\tau))d\tau, & s < t \leqslant t_1, \\ W(t, s)x_s + \int_s^{t_1} W(t, \tau)f(\tau, x(\tau))d\tau \\ + \sum_{k=1}^{r(t,s)-1} \int_{s_k}^{t_{k+1}} W(t, \tau)f(\tau, x(\tau))d\tau \\ + \int_{s_{r(t,s)}}^{t} W(t, \tau)f(\tau, x(\tau))d\tau \\ + \sum_{k=1}^{r(t,s)} W(t, s_k)g_k(x(t_k^-)), & s_{r(t,s)} < t \leqslant t_{r(t)+1}, \\ B_i(t)x(t_i^-) + g_i(x(t_i^-)), & t_{r(t,s)+1} < t \leqslant s_{r(t,s)+1}, \end{cases}$$

where $r(t, s)$ is the number of impulsive points $t_i \in [s, t)$, $t > s$.

Let $B(\delta) \subset \mathbb{R}^n$ be the open ball of radius $\delta > 0$ centered at zero and $PC_\delta(\mathbb{R}^+) := \{x: x \in PC(\mathbb{R}^+), \|x\|_{PC_\delta} \leqslant \delta e^{-a(1+\frac{1}{p})s}\}$, where (the norm) $\|x\|_{PC_\delta} = \frac{1}{D}\sup\{\|x(t)\|e^{-\bar{a}(t-s)-as}: t \geqslant s > t_0, D > 0\}$. Now, $PC_\delta(\mathbb{R}^+)$ is a complete metric space with the norm $\|\cdot\|_{PC_\delta}$.

Consider the operator $\Lambda: PC_\delta(\mathbb{R}^+) \to PC_\delta(\mathbb{R}^+)$ defined by

$$\Lambda(x)(t) = \begin{cases} W(t, s)x_s + \int_s^t W(t, \tau)f(\tau, x(\tau))d\tau, & s < t \leqslant t_1, \\ W(t, s)x_s + \int_s^{t_1} W(t, \tau)f(\tau, x(\tau))d\tau \\ + \sum_{k=1}^{r(t,s)-1} \int_{s_k}^{t_{k+1}} W(t, \tau)f(\tau, x(\tau))d\tau \\ + \int_{s_{r(t,s)}}^{t} W(t, \tau)f(\tau, x(\tau))d\tau \\ + \sum_{k=1}^{r(t,s)} W(t, s_k)g_k(x(t_k^-)), & s_{r(t,s)} < t \leqslant t_{r(t,s)+1}, \\ B_i(t)x(t_i^-) + g_i(x(t_i^-)), & t_{r(t,s)+1} < t \leqslant s_{r(t,s)+1}. \end{cases}$$

Given $x_1, x_2 \in PC_\delta(\mathbb{R}^+)$, and setting $\xi = 2D^{p+1}L\delta^p$, we obtain

$$\|f(\tau, x_1(\tau)) - f(\tau, x_2(\tau))\| \leqslant \xi e^{\bar{a}(p+1)(\tau-s)}\|x_1 - x_2\|_{PC_\delta},$$

and

$$\|g_k(x_1(t_k^-)) - g_k(x_2(t_k^-))\| \leqslant \xi e^{\bar{a}(p+1)(t_k-s)}\|x_1 - x_2\|_{PC_\delta}.$$

Now we present a result analogous to [43, theorem 6] (we omit some of the details since they are similar to those in [43, theorem 6]).

Theorem 1.2.19. *Suppose that (1.40) admits a nonuniform exponential contraction and* $p\bar{a} + a < 0$. *Then there exist* $a\delta > 0$ *and* $aD > 0$ *such that for any* $x_s \in B(\delta e^{-a(1+\frac{1}{p})s})$, *equation (1.77) with* $x(s) = x_s$ *has a unique solution* $x(\cdot, x_s) \in PC_\delta(\mathbb{R}^+)$ *satisfying*

$$\|x(t, x_s)\| \leqslant De^{\bar{a}(t-s)+as}\|x_s\|, \quad t \geqslant s. \tag{1.78}$$

Proof. Without loss of generality, for $s_{r(t,\,s)} < t \leqslant t_{r(t,\,s)+1}$,

$$\begin{aligned}
\|\Lambda(x_1)(t) - \Lambda(x_2)(t)\| \leqslant & \int_s^{t_1} \|W(t, \tau)\|\|f(\tau, x_1(\tau)) - f(\tau, x_2(\tau))\|d\tau \\
& + \sum_{k=1}^{r(t)-1} \int_{s_k}^{t_{k+1}} \|W(t, \tau)\|\|f(\tau, x_1(\tau)) - f(\tau, x_2(\tau))\|d\tau \\
& + \int_{s_{r(t,\,s)}}^{t} \|W(t, \tau)\|\|f(\tau, x_1(\tau)) - f(\tau, x_2(\tau))\|d\tau \\
& + \sum_{k=1}^{r(t,s)} \|W(t, s_k)\|\|g_k(x_1(t_k^-)) - g_k(x_2(t_k^-))\| \\
\leqslant & \int_s^{t} \|W(t, \tau)\|\|f(\tau, x_1(\tau)) - f(\tau, x_2(\tau))\|d\tau \\
& + \sum_{k=1}^{r(t,s)} \|W(t, s_k)\|\|g_k(x_1(t_k^-)) - g_k(x_2(t_k^-))\|,
\end{aligned} \tag{1.79}$$

and repeating the same computation as in [43, p 1608], then (1.79) reduces to

$$\|\Lambda(x_1)(t) - \Lambda(x_2)(t)\| \leqslant 2\frac{\xi K}{|p\bar{a} + a|}e^{\bar{a}(t-s)+as}\|x_1 - x_2\|_{PC_\delta},$$

which implies that

$$\|\Lambda(x_1) - \Lambda(x_2)\|_{PC_\delta} \leqslant 2\frac{\xi K}{|p\bar{a} + a|}\|x_1 - x_2\|_{PC_\delta}.$$

We can choose $2\frac{\xi K}{|p\bar{a}+a|}$ to be less than $\frac{1}{2}$ by choosing a sufficiently small δ. Also (see [43, equation (30)]) one can obtain $\|\Lambda(x)\|_{PC_\delta} \leqslant \delta e^{-a(1+\frac{1}{p})s}$, provided $D > 2K$.

Finally, repeat the procedure in the proof of [43, theorem 6] to obtain (1.78).

1.2.6 Bounds for the regularity coefficient

In this section, we adopt ideas from [43] to establish lower and upper bounds for the regularity coefficient. Note that the regularity coefficient measures the nonuniformity of the exponential behavior.

Consider (1.40), for some continuous function A: $(0, \infty) \to \mathbb{R}^{n\times n}$ such that

$$\limsup_{t\to+\infty} \frac{1}{t} \log^+\|A(t)\| = 0, \tag{1.80}$$

where $\log^+ x = \max\{0, \log x\}$, and some matrices $B_i(t)$ satisfying

$$\limsup_{i\to+\infty} \frac{1}{s_i} \log^+\|B_i(t)\| = 0, \quad \text{and} \quad \inf_i|\det B_i(t)| > 0. \tag{1.81}$$

We first obtain a lower bound for the regularity coefficient.

Theorem 1.2.20. *The following results hold:*

(I) *For* $t_0 = s_0 = 0 < t_1 < s_1 < t_2 < s_2 \cdots < t_i < t \leqslant s_i$, $i = 1, 2, \ldots,$ *we have*

$$n\omega(\chi, \mu) \geqslant$$

$$\limsup_{t\to+\infty} \frac{1}{t}\left(\sum_{j=1}^{i}\int_{s_{j-1}}^{t_j} trA(s)ds + \log|\det B_i(t)| + \sum_{v=1}^{i-1}\log|\det B_v(s_v)|\right)$$

$$-\liminf_{t\to+\infty} \frac{1}{t}\left(\sum_{j=1}^{i}\int_{s_{j-1}}^{t_j} trA(s)ds + \log|\det B_i(t)| + \sum_{v=1}^{i-1}\log|\det B_v(s_v)|\right);$$

(II) *For* $t_0 = s_0 = 0 < t_1 < s_1 < t_2 < s_2 \cdots s_i < t \leqslant t_{i+1}$, $i = 1, 2, \ldots,$ *we have*

$$n\omega(\chi, \mu) \geqslant$$

$$\limsup_{t\to+\infty} \frac{1}{t}\left(\int_{s_i}^{t} trA(s)ds + \sum_{j=1}^{i}\int_{s_{j-1}}^{t_j} trA(s)ds + \sum_{v=1}^{i}\log|\det B_v(s_v)|\right)$$

$$-\liminf_{t\to+\infty} \frac{1}{t}\left(\int_{s_i}^{t} trA(s)ds + \sum_{j=1}^{i}\int_{s_{j-1}}^{t_j} trA(s)ds + \sum_{v=1}^{i}\log|\det B_v(s_v)|\right).$$

Proof. Calculating the determinant, we obtain

$$\det X(t) = \sum_{p_1,\ldots,p_n} (-1)^* x_{p_1^1} x_{p_2^2} \cdots x_{p_n^n}, \tag{1.82}$$

where the summation is taken over all permutations $(p_1, \ldots, p_n)$ of n elements and $(-1)^*$ is the signature of the permutation. From (1.82) we have

$$\chi(\det X(t)) \leqslant \max_{(p_1,\ldots,p_n)} (\chi(x_{p_1^1}(t)) + \cdots \chi(x_{p_n^n}(t))) \leqslant \sum_{i=1}^{n} \chi(x_i(0)), \tag{1.83}$$

and it follows from lemma 1.2.6 that we have:

(I) For $t_0 = s_0 = 0 < t_1 < s_1 < t_2 < s_2 \cdots < t_i < t \leqslant s_i,\ i = 1, 2, \ldots,$

$$\limsup_{t\to+\infty} \frac{1}{t} \log \left| \frac{\det X(t)}{\det X(0)} \right|$$

$$= \limsup_{t\to+\infty} \frac{1}{t} \left(\sum_{j=1}^{i} \int_{s_{j-1}}^{t_j} \mathrm{tr} A(s) ds + \log|\det B_i(t)| + \sum_{v=1}^{i-1} \log|\det B_v(s_v)| \right) \leqslant \sum_{i=1}^{n} \chi(x_i(0)),$$

and similarly, we have

$$\liminf_{t\to+\infty} \frac{1}{t} \left(\sum_{j=1}^{i} \int_{s_{j-1}}^{t_j} \mathrm{tr} A(s) ds + \log|\det B_i(t)| + \sum_{v=1}^{i-1} \log|\det B_v(s_v)| \right)$$

$$= \liminf_{t\to+\infty} \frac{1}{t} \left(\sum_{j=1}^{i} \int_{s_{j-1}}^{t_j} \mathrm{tr} A(s) ds - \log|(\det B_i^{\top}(t))^{-1}| - \sum_{v=1}^{i-1} \log|(\det B_v^{\top}(s_v))^{-1}| \right)$$

$$= -\limsup_{t\to+\infty} \frac{1}{t} \left(\sum_{j=1}^{i} \int_{s_{j-1}}^{t_j} \mathrm{tr}(-A^{\top}(s)) ds + \log|(\det B_i^{\top}(t))^{-1}| + \sum_{v=1}^{i-1} \log|(\det B_v^{\top}(s_v))^{-1}| \right)$$

$$\geqslant -\sum_{i=1}^{n} \mu(y_i(0)),$$

where $y_1(t), \ldots, y_n(t)$ are the columns of $Y(t) = [X(t)^{\top}]^{-1}$.

Therefore,

$$\limsup_{t\to+\infty} \frac{1}{t} \left(\sum_{j=1}^{i} \int_{s_{j-1}}^{t_j} \mathrm{tr} A(s) ds + \log|\det B_i(t)| + \sum_{v=1}^{i-1} \log|\det B_v(s_v)| \right)$$

$$- \liminf_{t\to+\infty} \frac{1}{t} \left(\sum_{j=1}^{i} \int_{s_{j-1}}^{t_j} \mathrm{tr} A(s) ds + \log|\det B_i(t)| + \sum_{v=1}^{i-1} \log|\det B_v(s_v)| \right)$$

$$\leqslant \sum_{i=1}^{n} [\chi(x_i(0)) + \mu(y_i(0))].$$

Without loss of generality, we can always assume that the minimum in (1.65) is attained, that is

$$\omega(\chi, \mu) = \max\{\chi(x_i(0)) + \mu(y_i(0): 1 \leqslant i \leqslant n\}.$$

Therefore,

$$\sum_{i=1}^{n}[\chi(x_i(0)) + \mu(y_i(0))] \leqslant n \max\{\chi(x_i(0)) + \mu(y_i(0): 1 \leqslant i \leqslant n\} = n\omega(\chi, \mu).$$

(II) For $t_0 = s_0 = 0 < t_1 < s_1 < t_2 < s_2 \cdots s_i < t \leqslant t_{i+1},\ i = 1, 2, \ldots,$

$$\limsup_{t\to+\infty} \frac{1}{t} \log \left| \frac{\det X(t)}{\det X(0)} \right|$$

$$= \limsup_{t\to+\infty} \frac{1}{t}\left(\int_{s_i}^{t} \operatorname{tr}A(s)ds + \sum_{j=1}^{i} \int_{s_{j-1}}^{t_j} \operatorname{tr}A(s)ds + \sum_{v=1}^{i} \log|\det B_v(s_v)| \right) \leqslant \sum_{i=1}^{n} \chi[x_i(0)].$$

The same ideas yield

$$\limsup_{t\to+\infty} \frac{1}{t}\left(\int_{s_i}^{t} \operatorname{tr}A(s)ds + \sum_{j=1}^{i} \int_{s_{j-1}}^{t_j} \operatorname{tr}A(s)ds + \sum_{v=1}^{i} \log|\det B_v(s_v)| \right)$$

$$- \liminf_{t\to+\infty} \frac{1}{t}\left(\int_{s_i}^{t} \operatorname{tr}A(s)ds + \sum_{j=1}^{i} \int_{s_{j-1}}^{t_j} \operatorname{tr}A(s)ds + \sum_{v=1}^{i} \log|\det B_v(s_v)| \right)$$

$$\leqslant \sum_{i=1}^{n}[\chi(x_i(0)) + \mu(y_i(0))] \leqslant n\omega(\chi, \mu).$$

The proof is complete. □

Next, we obtain an upper bound for the regularity coefficient. We consider only triangular matrices. Let $x_1(t), \ldots, x_n(t)$ be a basis of the space of solutions of (1.40). Applying the Gram–Schmidt orthogonalization procedure we obtain an orthonormal basis $\phi_1(t), \ldots, \phi_n(t)$ such that any $\phi_k(t)$ depends only on the solutions $x_1(t), \ldots, x_n(t)$. The fundamental matrices $X(t)$ and $\Phi(t)$ whose columns are, respectively, $x_1(t), \ldots, x_n(t)$ and $\phi_1(t), \ldots, \phi_n(t)$ satisfy $\Phi(t) = X(t)S(t)$, where $S(t)$ is a triangular matrix for any $t \geqslant t_0$. Then by using the change of variables $x = \Phi(t)z$, one can transform (1.40) into

$$\begin{cases} z'(t) = -S(t)^{-1}S'(t)z(t), & t \in (s_i, t_{i+1}],\ i = 0, 1, 2, \ldots, \\ z(t_i^+) = S(t_i^+)^{-1}S(t_i)z(t_i^-), & i = 1, 2, \ldots, \\ z(t) = S(t)^{-1}S(t)z(t_i^-), & t \in (t_i, s_i],\ i = 1, 2, \ldots, \\ z(s_i^+) = z(s_i^-), & i = 1, 2, \ldots. \end{cases}$$

Since $S(t)$ is triangular, we can verify that $S(t)^{-1}S(t)$ is also triangular. Moreover, $\Phi(t)$ is orthogonal for any t, the values of the Lyapunov exponents and thus also the regularity coefficient remain unchanged under the change of variables $x = \Phi(t)z$. Therefore, we consider the triangular impulsive system.

Suppose that in (1.40) the matrices $A(t) = \{a_{\alpha\beta}(t)\}$ and $B_i(t) = \{b_i^{\alpha\beta}(t)\}$ are triangular. We will consider only lower triangular matrices: $a_{\alpha\beta}(t) = 0$, $b_i^{\alpha\beta}(t) = 0$ for $\alpha < \beta$ and for any $t \geqslant 0$, $i = 1, 2, \ldots$. With these assumptions (1.40) can be written as

$$\begin{cases} \dfrac{dx_\alpha}{dt} = \sum\limits_{\beta\leqslant\alpha} a_{\alpha\beta}(t)x_\beta(t), \ t \in (s_i, t_{i+1}], \ i = 0, 1, 2, \ldots, \\ x_\alpha(t_i^+) = \sum\limits_{\beta\leqslant\alpha} b_i^{\alpha\beta}(t_i)x_\beta(t_i^-), \ i = 1, 2, \ldots, \\ x_\alpha(t) = \sum\limits_{\beta\leqslant\alpha} b_i^{\alpha\beta}(t)x_\beta(t_i^-), \ t \in (t_i, s_i], \ i = 1, 2, \ldots, \\ x_\alpha(s_i^+) = x_\alpha(s_i^-), \ i = 1, 2, \ldots. \end{cases} \tag{1.84}$$

For $k = 1, 2, \ldots, n$, we consider the following cases:

(i) For $t_i < t \leqslant s_i$, we have

$$A_k(t) = e^{\sum_{j=1}^{i}\int_{s_{j-1}}^{t_j} a_{kk}(s)ds} b_i^{kk}(t)\prod_{v=1}^{i-1} b_v^{kk}(s_v),$$

$$\bar{\gamma}_k = \limsup_{t\to+\infty}\left(\sum_{j=1}^{i}\int_{s_{j-1}}^{t_j} a_{kk}(s)ds + \log|b_i^{kk}(t)| + \sum_{v=1}^{i-1}\log|b_v^{kk}(s_v)|\right),$$

and

$$\underline{\gamma}_k = \liminf_{t\to+\infty}\left(\sum_{j=1}^{i}\int_{s_{j-1}}^{t_j} a_{kk}(s)ds + \log|b_i^{kk}(t)| + \sum_{v=1}^{i-1}\log|b_v^{kk}(s_v)|\right).$$

(ii) For $s_i < t \leqslant t_{i+1}$, we have

$$A_k(t) = e^{\int_{s_i}^{t} a_{kk}(s)ds + \sum_{j=1}^{i}\int_{s_{j-1}}^{t_j} a_{kk}(s)ds} \prod_{v=1}^{i} b_v^{kk}(s_v),$$

$$\bar{\gamma}_k = \limsup_{t\to+\infty}\left(\int_{s_i}^{t} a_{kk}(s)ds + \sum_{j=1}^{i}\int_{s_{j-1}}^{t_j} a_{kk}(s)ds + \sum_{v=1}^{i}\log|b_v^{kk}(s_v)|\right),$$

and

$$\underline{\gamma}_k = \liminf_{t\to+\infty}\left(\int_{s_i}^{t} a_{kk}(s)ds + \sum_{j=1}^{i}\int_{s_{j-1}}^{t_j} a_{kk}(s)ds + \sum_{v=1}^{i}\log|b_v^{kk}(s_v)|\right).$$

Note that (1.84) can be successively integrated, and we see that its fundamental matrix is $X(t, t_0) = \{x_{\alpha\beta}(t)\}$, $X(t_0, t_0) = \mathrm{Id}$, where

$$x_{\alpha\beta}(t) = 0, \ \text{for } \alpha < \beta, \quad x_{\alpha\alpha}(t) = A_\alpha(t),$$

$$x_{\alpha\beta}(t) = A_\alpha(t)\left(\int_{h_{\alpha\beta}}^{t} A_\alpha^{-1}(s)\sum_{j=\beta}^{\alpha-1} a_{\alpha j}(s)x_{j\beta}(s)ds + \mathrm{sign}(t - h_{\alpha\beta})\sum_{j=\beta}^{\alpha-1} A_\alpha^{-1}(s_i)b_i^{\alpha j}(s_i)x_{j\beta}(t_i)\right),$$

$$\text{for } s_i < t \leqslant t_{i+1}, \ \alpha > \beta \text{ and } \alpha, \beta = 1, 2, \dots, n,$$

$$x_{\alpha\beta}(t) = \sum_{j=\beta}^{\alpha-1} b_i^{\alpha j}(t)x_{j\beta}(t_i), \ \text{for } t_{i+1} < t \leqslant s_{i+1}, \ \alpha > \beta \text{ and } \alpha, \beta = 1, 2, \dots, n,$$

for some $h_{\alpha,\beta}$. Taking

$$c_{\alpha\beta} = \bar{\gamma}_\beta - \underline{\gamma}_\alpha + \sum_{m=\beta+1}^{\alpha-1}(\bar{\gamma}_m - \underline{\gamma}_m),$$

for any $\alpha > \beta$, we set

$$h_{\alpha\beta} = \begin{cases} s_i, & \text{if } c_{\alpha\beta} \geqslant 0, \\ +\infty, & \text{if } c_{\alpha\beta} \leqslant 0. \end{cases}$$

One can verify that $X(t) = \{x_{\alpha\beta}(t)\}$ is a fundamental solution of system (1.84), and the columns of $X(t)$ are

$$x_\beta(t) = (x_{1\beta}(t), x_{2\beta}(t), \dots, x_{n\beta}(t))^{\mathrm{T}},$$

for $\beta = 1, 2, \dots, n$. Given $\alpha, \beta = 1, 2, \dots, n$, we write

$$\chi(x_{\alpha\beta}) = \limsup_{t\to+\infty}\frac{1}{t}\log|x_{\alpha\beta}(t)|.$$

Lemma 1.2.21. *For any $\alpha, \beta = 1, 2, \dots, n$ we have*

$$\chi(x_{\alpha\alpha}) = \bar{\gamma}_\alpha \ \text{and} \ \chi(x_{\alpha\beta}) \leqslant \bar{\gamma}_\beta + \sum_{k=\beta+1}^{\alpha}(\bar{\gamma}_k - \underline{\gamma}_k).$$

Proof. Note, $\chi(x_{\alpha\alpha}) = \bar{\gamma}_\alpha$, for $\alpha = 1, 2, \ldots, n$. Now we proceed by induction on α. For a given $\alpha > 1$, we assume that

$$\chi(x_{k\beta}) \leqslant \bar{\gamma}_\beta + \sum_{m=\beta+1}^{k} (\bar{\gamma}_m - \underline{\gamma}_m), \quad \text{for } \beta \leqslant k \leqslant \alpha - 1. \tag{1.85}$$

It follows from (1.80), (1.81) and (1.85) that given $\varepsilon > 0$ there exists $D > 0$ such at

$$|a_{\alpha\beta}(t)| \leqslant De^{t\varepsilon}, \quad |A_\alpha^{-1}(t)| \leqslant De^{(-\underline{\gamma}_\alpha+\varepsilon)t}, \quad |b_i^{\alpha\beta}(t)| \leqslant De^{\varepsilon s_i},$$

and

$$|x_{k\beta}(t)| \leqslant De^{(\bar{\gamma}_\beta + \sum_{m=\beta+1}^{k} (\bar{\gamma}_m - \underline{\gamma}_m) + \varepsilon)t}.$$

Consider $t \geqslant 0$ and $\beta \leqslant k \leqslant \alpha - 1$. We consider the following cases:

Case 1. $c_{\alpha\beta} \geqslant 0$ and $s_i < t$. We have

$$\begin{aligned}
\chi(x_{\alpha\beta}) &\leqslant \limsup_{t\to+\infty} \frac{1}{t} \log|A_\alpha(t)| + \limsup_{t\to+\infty} \frac{1}{t} \log \Bigg| \int_{s_i}^{t} |A_\alpha^{-1}(s)| \sum_{j=\beta}^{\alpha-1} |a_{\alpha j}(s)||x_{j\beta}(s)| ds \\
&\quad + \sum_{j=\beta}^{\alpha-1} |A_\alpha^{-1}(s_i)||b_i^{\alpha j}(s_i)||x_{j\beta}(t_i)| \Bigg| \\
&\leqslant \bar{\gamma}_\alpha + \limsup_{t\to+\infty} \frac{1}{t} \log \Bigg| \int_{s_i}^{t} D^3 \sum_{j=\beta}^{\alpha-1} e^{(\bar{\gamma}_\beta + \sum_{m=\beta+1}^{j} (\bar{\gamma}_m - \underline{\gamma}_m) - \underline{\gamma}_\alpha + 3\varepsilon)s} ds \\
&\quad + \sum_{j=\beta}^{\alpha-1} D^3 e^{(-\underline{\gamma}_\alpha + 3\varepsilon)s_i + (\bar{\gamma}_\beta + \sum_{m=\beta+1}^{j} (\bar{\gamma}_m - \underline{\gamma}_m) + \varepsilon)t_i} \Bigg| \\
&\leqslant \bar{\gamma}_\alpha + \limsup_{t\to+\infty} \frac{1}{t} \log \Bigg| \int_{0}^{t} D^3 \sum_{j=\beta}^{\alpha-1} e^{(\bar{\gamma}_\beta + \sum_{m=\beta+1}^{j} (\bar{\gamma}_m - \underline{\gamma}_m) - \underline{\gamma}_\alpha + 3\varepsilon)s} ds \\
&\quad + \sum_{j=\beta}^{\alpha-1} D^3 e^{(\bar{\gamma}_\beta + \sum_{m=\beta+1}^{j} (\bar{\gamma}_m - \underline{\gamma}_m) - \underline{\gamma}_\alpha + 3\varepsilon)t} \Bigg| \\
&\leqslant \bar{\gamma}_\alpha + \limsup_{t\to+\infty} \frac{1}{t} \log \left| \int_{0}^{t} D^3 n e^{(c_{\alpha\beta}+3\varepsilon)s} ds + D^3 n e^{(c_{\alpha\beta}+3\varepsilon)t} \right| \\
&\leqslant \bar{\gamma}_\alpha + \limsup_{t\to+\infty} \frac{1}{t} \log \left| \frac{D^3 n e^{(c_{\alpha\beta}+3\varepsilon)t}}{c_{\alpha\beta} + 3\varepsilon} + D^3 n e^{(c_{\alpha\beta}+3\varepsilon)t} \right| \\
&\leqslant \bar{\gamma}_\alpha + \bar{\gamma}_\beta - \underline{\gamma}_\alpha + \sum_{m=\beta+1}^{\alpha-1} (\bar{\gamma}_m - \underline{\gamma}_m) + 3\varepsilon \\
&\leqslant \bar{\gamma}_\beta + \sum_{m=\beta+1}^{\alpha} (\bar{\gamma}_m - \underline{\gamma}_m) + 3\varepsilon.
\end{aligned}$$

Case 2. $c_{\alpha\beta} \leqslant 0$ and $s_i \geqslant t$. Proceeding as above, and taking $\varepsilon > 0$ such that $c_{\alpha\beta} + 3\varepsilon < 0$, we obtain

$$\chi(x_{\alpha\beta}) \leqslant \limsup_{t\to+\infty} \frac{1}{t}\log|A_\alpha(t)| + \limsup_{t\to+\infty}\frac{1}{t}\log\left|\int_{+\infty}^{t} |A_\alpha^{-1}(s)|\sum_{j=\beta}^{\alpha-1}|a_{\alpha j}(s)||x_{j\beta}(s)|ds\right.$$
$$\left. + \sum_{j=\beta}^{\alpha-1}|A_\alpha^{-1}(s_i)||b_i^{\alpha j}(s_i)||x_{j\beta}(t_i)|\right|$$
$$\leqslant \bar{\gamma}_\alpha + \limsup_{t\to+\infty}\frac{1}{t}\log\left|\int_t^{+\infty} D^3 n e^{(c_{\alpha\beta}+3\varepsilon)s}ds + D^3 n e^{(c_{\alpha\beta}+3\varepsilon)s_i}\right|$$
$$\leqslant \bar{\gamma}_\alpha + \limsup_{t\to+\infty}\frac{1}{t}\log\left|\frac{D^3 n e^{(c_{\alpha\beta}+3\varepsilon)t}}{c_{\alpha\beta}+3\varepsilon}\right|$$
$$\leqslant \bar{\gamma}_\alpha + \bar{\gamma}_\beta - \underline{\gamma}_\alpha + \sum_{m=\beta+1}^{\alpha-1}(\bar{\gamma}_m - \underline{\gamma}_m) + 3\varepsilon$$
$$\leqslant \bar{\gamma}_\beta + \sum_{m=\beta+1}^{\alpha}(\bar{\gamma}_m - \underline{\gamma}_m) + 3\varepsilon.$$

Since ε is arbitrary. The proof is complete. □

Now we consider the adjoint equation (1.84). Let $\alpha, \beta = 1, 2, \dots, n$ and $t \geqslant 0$. From lemma 1.2.14, we obtain

$$Y(t) = (X(t)^\top)^{-1}C^\top.$$

Therefore, $Y(t) = \{y_{\alpha\beta}(t)\}$ is a fundamental solution of the adjoint system (1.84). We obtain

$$y_{\alpha\beta}(t) = 0, \text{ for } \alpha > \beta, \quad y_{\alpha\alpha}(t) = A_\alpha^{-1}(t)c_{\alpha\alpha},$$

and

$$y_{\alpha\beta}(t) = (x_{\alpha\beta}(t)^\top)^{-1}c_{\alpha\beta}^\top$$
$$= A_\alpha^{-1}(t)\left(\int_{h_{\alpha\beta}}^{t} A_\alpha(s)\sum_{j=\alpha+1}^{\beta} a_{j\alpha}(s)y_{j\beta}(s)ds\right.$$
$$\left. + \operatorname{sign}(t-h_{\alpha\beta})\sum_{j=\alpha+1}^{\beta} A_\alpha(s_i)b_i^{j\alpha}(s_i)y_{j\beta}(t_i)\right)^{-1} c_{\alpha\beta},$$

for $s_i < t \leqslant t_{i+1}$, $\alpha < \beta$ and $\alpha, \beta = 1, 2, \dots, n$,

and

$$y_{\alpha\beta}(t) = \sum_{j=\alpha+1}^{\beta} b_i^{j\alpha}(t) y_{j\beta}(t_i), \quad \text{for} t_{i+1} < t \leqslant s_{i+1}, \quad \alpha < \beta \text{ and } \alpha, \beta = 1, 2, \ldots, n.$$

Lemma 1.2.22. *For any $\alpha, \beta = 1, 2, \ldots, n$, we have*

$$\mu(y_{\alpha\alpha}) = -\underline{\gamma}_\alpha \text{ and } \mu(y_{\alpha\beta}) \leqslant -\underline{\gamma}_\beta + \sum_{k=\alpha}^{\beta-1} (\bar{\gamma}_k - \underline{\gamma}_k).$$

Proof. One can use the same ideas as in lemma 1.2.21 to obtain the conclusion so we omit the details.

Theorem 1.2.23. *If $A(t)$ and $B_i(t)$ are lower triangular matrices for every $t \geqslant 0$, then*

$$\omega(\chi, \mu) \leqslant \sum_{k=1}^{n} (\bar{\gamma}_k - \underline{\gamma}_k).$$

Proof. From lemmas 1.2.21 and 1.2.22 we obtain

$$\chi(x_j) = \max\{\chi(x_{lj}): l = 1, 2, \ldots, n\} \leqslant \bar{\gamma}_j + \sum_{m=j+1}^{n} (\bar{\gamma}_m - \underline{\gamma}_m),$$

and

$$\mu(y_j) = \max\{\mu(y_{lj}): l = 1, 2, \ldots, n\} \leqslant -\underline{\gamma}_j + \sum_{m=1}^{j-1} (\bar{\gamma}_m - \underline{\gamma}_m).$$

Therefore, we have

$$\chi(x_j) + \mu(y_j) \leqslant \sum_{k=1}^{n} (\bar{\gamma}_k - \underline{\gamma}_k), \quad \text{for all } j = 1, 2, \ldots, n.$$

From the definition of the regularity coefficient $\omega(\chi, \mu)$ it is sufficient to show that the bases $x_1, \ldots, x_n$ and $y_1, \ldots, y_n$ are dual. From lemma 1.2.13, we obtain

$$\frac{d}{dt}\langle x_l(t), y_j(t)\rangle = \langle A(t)x_l(t), y_j(t)\rangle + \langle x_l(t), -A(t)^\top y_j(t)\rangle = 0,$$

for $t \neq t_i$. Moreover,

$$\begin{aligned}
\Delta\langle x_l(t), y_j(t)\rangle|_{t=t_i} &= \langle x_l(t_i^+), y_j(t_i^+)\rangle - \langle x_l(t_i), y_j(t_i)\rangle \\
&= \langle B_i(t_i)x_l(t_i), (B_i(t_i)^\top)^{-1} y_j(t_i)\rangle - \langle x_l(t_i), y_j(t_i)\rangle \\
&= x_l(t_i)^\top B_i(t_i)^\top (B_i(t_i)^\top)^{-1} y_j(t_i) - x_l(t_i)^\top y_j(t_i) = 0.
\end{aligned}$$

For $t \in (t_i, s_i]$, $i = 1, 2, \ldots$, we have

$$\begin{aligned}\langle x_l(t), y_j(t)\rangle &= \langle B_i(t)x_l(t_i), (B_i(t)^{\mathrm{T}})^{-1}y_j(t_i)\rangle \\ &= x_l(t_i)^{\mathrm{T}} B_i(t)^{\mathrm{T}} (B_i(t)^{\mathrm{T}})^{-1} y_j(t_i) \\ &= \langle x_l(t_i), y_j(t_i)\rangle = c_i.\end{aligned}$$

For $t = s_i$, $i = 1, 2, \ldots$, we obtain

$$\Delta\langle x_l(t), y_j(t)\rangle|_{t=s_i} = \langle x_l(s_i^+), y_j(s_i^+)\rangle - \langle x_l(s_i), y_j(s_i)\rangle = 0.$$

This shows that

$$\langle x_l(t), y_j(t)\rangle = \langle x_l(t_0), y_j(t_0)\rangle, \quad \text{for every } t \geqslant t_0.$$

Clearly, $\langle x_l(t_0), y_j(t_0)\rangle = 0$, for $l > j$. Furthermore, for $l = j$ and $l = 1, 2, \ldots, n$ we have

$$\langle x_l(t_0), y_l(t_0)\rangle = \sum_{j=1}^{n} x_{jl}(t_0) y_{jl}(t_0) = x_{ll}(t_0) y_{ll}(t_0) = 1.$$

For $l < j$ and $t \geqslant t_0$, we have

$$\langle x_l(t), y_j(t)\rangle = \sum_{k=1}^{n} x_{kl}(t) y_{kj}(t) = \sum_{k=l}^{j} x_{kl}(t) y_{kj}(t) = 0.$$

We conclude that $\langle x_l(t), y_j(t)\rangle = \delta_{lj}$ for every l and j. The proof is complete. □

1.2.7 Notes and remarks

We discuss [37, 43] in section 1.2. The hybrid dynamical system we consider is more complex than the classical hybrid dynamical system in [37, 43]. Note that there is a new algebraic system associated with a continuous-time dynamical system, and a discrete-time dynamical system is involved. In [37, chapters 3 and 4], the topic of asymptotic stability and state perturbations overlap in some sense with this section. However, here we adopt different tools and a different approach to our problem. For the asymptotic stability of the zero solution, we do not use the Lyapunov–Krasovskii method. We consider the method developed in [43] and consider the set of Lyapunov exponents and sufficient conditions for the asymptotic stability of the zero solution. We use the eigenvalues of a matrix and a certain limit formula on the proportion of the number of impulsive points and time variables to obtain some sufficient conditions. Concerning the stability of the state for the perturbed system, we consider the idea in [43] to show that the state for the nonlinear perturbed system can have a nonuniform exponential contraction behaviour if we assume the original hybrid dynamical system admits a nonuniform exponential contraction and we restrict the initial value in a certain sphere. This is different to [37, chapter 4]. In this section we present a series of results on Lyapunov regularity, a new version of Perron's theorem, criteria for nonuniform exponential behavior, and lower and upper bounds for regularity coefficients, and our theory generalizes the theory in [43].

The results in section 1.2 are motivated by [27].

References

[1] Myshkis A D and Samoilenko A M 1967 Systems with impulsive at fixed moments of time *Mat. Sb.* **74** 202–8

[2] Hernández E and O'Regan D 2013 On a new class of abstract impulsive differential equations *Proc. Amer. Math. Soc.* **141** 1641–9

[3] Bainov D D and Simeonov P S 1993 *Impulsive Differential Equations: Periodic Solutions and Applications* (Boca Raton, FL: CRC Press)

[4] Bainov D D and Simeonov P S 1995 *Theory of Impulsive Differential EquationsSeries in Modern Applied Mathematics* vol 6 (Singapore: World Scientific)

[5] Bainov D D and Simeonov P S 1998 *Oscillation Theory of Impulsive Differential Equations* (Orlando, FL: International Publications)

[6] Lakshmikantham V, Bainov D D and Simeonov P S 1989 *Theory of Impulsive Differential Equations* Series in Modern Applied Mathematics vol 6 (Singapore: World Scientific)

[7] Samoilenko A M, Perestyuk N A and Chapovsky Y 1995 *Impulsive Differential Equations* (Singapore: World Scientific)

[8] Benchohra M, Henderson J and Ntouyas S 2006 *Impulsive Differential Equations and InclusionsContemporary Mathematics and its Applications* vol 2 (New York, NY, USA: Hindawi)

[9] Akhmet M U, Alzabut J and Zafer A 2006 Perron's theorem for linear impulsive differential equations with distributed delay *J. Comput. Appl. Math.* **193** 204–18

[10] Agarwal R P, Benchohra M and Hamani S 2010 A survey on existence results for boundary value problems of nonlinear fractional differential equations and inclusions *Acta. Appl. Math.* **109** 973–1033

[11] Shao Y, Li Y and Xu C 2011 Periodic solutions for a class of nonautonomous differential system with impulses and time-varying delays *Acta. Appl. Math.* **115** 105–21

[12] Yuan X, Xia Y H and O'Regan D 2014 Nonautonomous impulsive systems with unbounded nonlinear terms *Appl. Math. Comput.* **245** 391–403

[13] Sun J, Chu J and Chen H 2013 Periodic solution generated by impulses for singular differential equations *J. Math. Anal. Appl.* **404** 562–9

[14] Fan Z and Li G 2010 Existence results for semilinear differential equations with nonlocal and impulsive conditions *J. Funct. Anal.* **258** 1709–27

[15] Pierri M, O'Regan D and Rolnik V 2013 Existence of solutions for semi-linear abstract differential equations with not instantaneous impulses *Appl. Math. Comput.* **219** 6743–9

[16] Fečkan M, Wang J and Zhou Y 2014 Existence of periodic solutions for nonlinear evolution equations with non-instantaneous impulses *Nonauton. Dyn. Syst.* **1** 93–101

[17] Wang J and Fečkan M 2015 A general class of impulsive evolution equations *Topol. Meth. Nonlinear Anal.* **46** 915–34

[18] Wang J, Fečkan M and Zhou Y 2016 A survey on impulsive fractional differential equations *Fract. Calc. Appl. Anal.* **19** 806–31

[19] Pierri M, Henríquez H R and Prokczyk A 2016 Global solutions for abstract differential equations with non-instantaneous impulses *Mediterr. J. Math.* **13** 1685–708

[20] Hernández E, Pierri M and O'Regan D 2015 On abstract differential equations with non-instantaneous impulses *Topol. Methods Nonlinear Anal.* **46** 1067–85

[21] Agarwal R, O'Regan D and Hristova S 2017 Stability by Lyapunov like functions of nonlinear differential equations with non-instantaneous impulses *J. Appl. Math. Comput.* **53** 147–68

[22] Agarwal R, O'Regan D and Hristova S 2016 Stability by Lyapunov functions of Caputo fractional differential equations with non-instantaneous impulses *Electron. J. Differ. Equ.* **2016** 1–22

[23] Wang J, Zhou Y and Lin Z 2014 On a new class of impulsive fractional differential equations *Appl. Math. Comput.* **242** 649–57

[24] Wang J, Lin Z and Zhou Y 2015 On the stability of new impulsive differential equations *Topol. Meth. Nonlinear Anal.* **45** 303–14

[25] Wang J, Fečkan M and Zhou Y 2016 Random noninstantaneous impulsive models for studying periodic evolution processes in pharmacotherapy *Mathematical Modeling and Applications in Nonlinear Dynamics, Nonlinear Systems and Complexity*vol 14 ed A Luo and H Merdan (Cham: Springer)

[26] Wang J, Fečkan M and Tian Y 2017 Stability analysis for a general class of non-instantaneous impulsive differential equations *Mediter. J. Math.* **46** 1–21

[27] Wang J, Li M and O'Regan D 2017 Lyapunov regularity and stability of linear non-instantaneous impulsive differential systems submitted

[28] Abbas S and Benchohra M 2015 Uniqueness and Ulam stabilities results for partial fractional differential equations with not instantaneous impulses *Appl. Math. Comput.* **257** 190–8

[29] Gautam G R and Dabas J 2015 Mild solutions for class of neutral fractional functional differential equations with not instantaneous impulses *Appl. Math. Comput.* **259** 480–9

[30] Yan Z and Lu F 2016 The optimal control of a new class of impulsive stochastic neutral evolution integro-differential equations with infinite delay *Int. J. Control* **89** 1592–612

[31] Zada A, Shah O and Shah R 2015 Hyers–Ulam stability of non-autonomous systems in terms of boundedness of Cauchy problems *Appl. Math. Comput.* **271** 512–8

[32] Barbu D, Buşe C and Tabassum A 2015 Hyers–Ulam stability and discrete dichotomy *J. Math. Anal. Appl.* **42** 1738–52

[33] Buşe C, O'Regan D, Saierli O and Tabassum A 2016 Hyers–Ulam stability and discrete dichotomy for difference periodic systems *Bull. Sci. Math.* **140** 908–34

[34] Ortega J M and Rheinboldt W C 1970 *Iterative Solution of Nonlinear Equations in Several Variables* (New York: Academic)

[35] Liu Z, Wu J and Cheke R A 2010 Coexistence and partial extinction in a delay competitive system subject to impulsive harvesting and stocking *IMA J. Appl. Math.* **75** 777–95

[36] Stamov G and Stamova I 2015 Second method of Lyapunov and almost periodic solutions for impulsive differential systems of fractional order *IMA J. Appl. Math.* **80** 1619–33

[37] Goebel R, Sanfelice R G and Teel A R 2012 *Hybrid Dynamical Systems: Modeling, Stability and Robustness* (Princeton, NJ: Princeton University Press)

[38] Liu J and Teel A R 2016 Invariance principles for hybrid systems with memory *Nonlinear Anal.:Hybrid Systems* **21** 130–8

[39] Liu J and Teel A R 2016 Lyapunov-based sufficient conditions for stability of hybrid systems with memory *IEEE Trans. Aut. Cont.* **61** 1057–62

[40] Davó M A, Baños A, Gouaisbaut F, Tarbouriech S and Seuret A 2017 Stability analysis of linear impulsive delay dynamical systems via looped-functionals *Automatica* **81** 107–14

[41] Barreira L and Pesin Y 2002 *Lyapunov Exponents and Smooth Ergodic Theory University Lecture Series* vol 23 (Providence, RI: American Mathematical Society)

[42] Barreira L and Valls C 2007 Stability theory and Lyapunov regularity *J. Differ. Equ.* **232** 675–701

[43] Barreira L and Valls C 2010 Lyapunov regularity of impulsive differential equations *J. Differ. Equ.* **249** 1596–619

[44] Chu J and Zhu H 2013 Lyapunov regularity for random dynamical systems *Bull. Sci. Math.* **137** 671–87

[45] Adrianova L Y 1995 *Introduction to Linear Systems of Differential Equations Translations of Mathematical Monographs* vol 46 (Providence, RI: American Mathematical Society)

[46] Lie E B and Markus L 1972 *Fundamentals of Optimum Control Theory* (Moscow: Nauka)

IOP Publishing

Non-Instantaneous Impulsive Differential Equations
Basic theory and computation
JinRong Wang and Michal Fečkan

Chapter 2

Nonlinear differential equations

In this chapter we establish a unified framework to investigate the continuous dependence, stability and differentiability of solutions for nonlinear differential equations. In section 2.1 we discuss the continuous dependence and stability of the solutions with respect to initial conditions and junction parameters in the sense of the Euclid metric, and provide sufficient conditions to guarantee that the solutions are continuously depending and stable. Section 2.2 is devoted to studying the orbital Hausdorff continuous dependence of solutions, which extends some results in section 2.1. In section 2.3 we study the differentiability of solutions for parameter equations.

2.1 Continuous dependence and stability of solutions

2.1.1 Introduction

This section discusses the asymptotic properties of the solutions of integer order nonlinear non-instantaneous impulsive differential equations

$$\begin{cases} x'(t) = f(t, x(t)), & t \in (s_i, t_{i+1}],\ i \in \{0\} \cup \mathbb{N}, \\ x(t_i^+) = g_i(t_i, x(t_i^-)), & i \in \mathbb{N}, \\ x(t) = g_i(t, x(t_i^-)), & t \in (t_i, s_i],\ i \in \mathbb{N}, \\ x(0) = x_0, \end{cases} \tag{2.1}$$

and fractional order nonlinear non-instantaneous impulsive differential equations

$$\begin{cases} {}^c\mathbf{D}^{\alpha}_{s_i, t} x(t) = f(t, x(t)), & t \in (s_i, t_{i+1}],\ i \in \{0\} \cup \mathbb{N},\ \alpha \in (0, 1), \\ x(t_i^+) = g_i(t_i, x(t_i^-)), & i \in \mathbb{N}, \\ x(t) = g_i(t, x(t_i^-)), & t \in (t_i, s_i],\ i \in \mathbb{N}, \\ x(0) = x_0, \end{cases} \tag{2.2}$$

here ${}^c\mathbf{D}^{\alpha}_{s_i, t}$ denotes the classical Caputo fractional derivative of order α by changing the lower limit s_i [1], t_i acts as an impulsive point and s_i acts as a junction point

doi:10.1088/2053-2563/aada21ch2

satisfying $s_i < t_{i+1} \to \infty$ with $t_0 = s_0 = 0$. Now $x(t_i^+) = \lim_{\varepsilon\to 0^+} x(t_i + \varepsilon)$ and $x(t_i^-) = \lim_{\varepsilon\to 0^+} x(t_i - \varepsilon) := x(t_i)$. The function $f \in C([0, \infty) \times \mathbb{R}, \mathbb{R})$ and $g_i \in C^1([t_i, s_i] \times \mathbb{R}, \mathbb{R})$, $i \in \mathbb{N}$.

Fractional calculus extends the classical integer order calculus to an arbitrary order case. It is a powerful mathematical tool for describing complex movements, irregularities and memory features. Fractional differential equations have historical importance and global relevance, and can also reflect the basic properties of a material more accurately, which plays an important role in different research areas, such as engineering, physics, finance and signal analysis. There are many important monographs on fractional differential equations and their applications, we recommend [2–13] and reference therein to the reader.

Together with (2.1) and (2.2), we consider the corresponding impulsive perturbed problems of the following form:

$$\begin{cases} \hat{x}'(t) = f(t, \hat{x}(t)), & t \in (\hat{s}_i, \hat{t}_{i+1}],\ i \in \{0\} \cup \mathbb{N}, \\ \hat{x}((\hat{t}_i)^+) = g_i(\hat{t}_i, \hat{x}((\hat{t}_i)^-)), & i \in \mathbb{N}, \\ \hat{x}(t) = g_i(t, \hat{x}((\hat{t}_i)^-)), & t \in (\hat{t}_i, \hat{s}_i],\ i \in \mathbb{N}, \\ \hat{x}(0) = \hat{x}_0, \end{cases} \tag{2.3}$$

and

$$\begin{cases} {}^{c}\mathbf{D}^{\alpha}_{\hat{s}_i, t}\hat{x}(t) = f(t, \hat{x}(t)), & t \in (\hat{s}_i, \hat{t}_{i+1}],\ i \in \{0\} \cup \mathbb{N},\ \alpha \in (0, 1), \\ \hat{x}((\hat{t}_i)^+) = g_i(\hat{t}_i, \hat{x}((\hat{t}_i)^-)), & i \in \mathbb{N}, \\ \hat{x}(t) = g_i(t, \hat{x}((\hat{t}_i)^-)), & t \in (\hat{t}_i, \hat{s}_i],\ i \in \mathbb{N}, \\ \hat{x}(0) = \hat{x}_0, \end{cases} \tag{2.4}$$

where $\hat{t}_0 = \hat{s}_0 = 0,\ \hat{s}_i < \hat{t}_{i+1} \to \infty$.

The representations of piecewise continuous solutions to the problems (2.1) and (2.3), which we denote, respectively, by $x(\cdot;0, x_0) \in PC([0, \infty), \mathbb{R})$ and $\hat{x}(\cdot;0, \hat{x}_0) \in PC([0, \infty), \mathbb{R})$, are as follows:

$$x(t; 0, x_0) = \begin{cases} x_0 + (\mathbf{I}^1_{0,t} f)(t, x), & t \in [0, t_1], \\ g_i(t, x(t_i^-)), & t \in (t_i, s_i],\ i \in \mathbb{N}, \\ g_i(s_i, x(t_i^-)) + (\mathbf{I}^1_{s_i, t} f)(t, x), & t \in (s_i, t_{i+1}],\ i \in \mathbb{N}, \end{cases} \tag{2.5}$$

and

$$\hat{x}(t; 0, \hat{x}_0) = \begin{cases} \hat{x}_0 + (\mathbf{I}^1_{0,t} f)(t, \hat{x}), & t \in [0, \hat{t}_1], \\ g_i(t, \hat{x}((\hat{t}_i)^-)), & t \in (\hat{t}_i, \hat{s}_i],\ i \in \mathbb{N}, \\ g_i(\hat{s}_i, \hat{x}((\hat{t}_i)^-)) + (\mathbf{I}^1_{\hat{s}_i, t} f)(t, \hat{x}), & t \in (\hat{s}_i, \hat{t}_{i+1}],\ i \in \mathbb{N}, \end{cases} \tag{2.6}$$

where $(\mathbf{I}^p_{a,t} f)(t, x) := \frac{1}{\Gamma(p)} \int_a^t \frac{f(\sigma, x(\sigma))}{(t-\sigma)^{1-p}} d\sigma,\ p > 0$.

Similarly, we can obtain the representation of piecewise continuous solutions of the problems (2.2) and (2.4), respectively:

$$x(t;0,x_0)=\begin{cases} x_0+(\mathbf{I}^{\alpha}_{0,t}f)(t,x), & t\in[0,t_1],\\ g_i(t,x(t_i^-)), & t\in(t_i,s_i],\ i\in\mathbb{N},\\ g_i(s_i,x(t_i^-))+(\mathbf{I}^{\alpha}_{s_i,t}f)(t,x) & t\in(s_i,t_{i+1}],\ i\in\mathbb{N}, \end{cases} \tag{2.7}$$

and

$$\hat{x}(t;0,\hat{x}_0)=\begin{cases} \hat{x}_0+(\mathbf{I}^{\alpha}_{0,t}f)(t,\hat{x}), & t\in[0,\hat{t}_1],\\ g_i(t,\hat{x}((\hat{t}_i)^-)), & t\in(\hat{t}_i,\hat{s}_i],\ i\in\mathbb{N},\\ g_i(\hat{s}_i,\hat{x}((\hat{t}_i)^-))+(\mathbf{I}^{\alpha}_{\hat{s}_i,t}f)(t,\hat{x}), & t\in(\hat{s}_i,\hat{t}_{i+1}],\ i\in\mathbb{N}. \end{cases} \tag{2.8}$$

The main objective is to extend the ideas in [14] and introduce continuous dependence and stability about the initial condition, impulsive point and junction point for (2.1) and (2.2).

In section 2.1.2 we introduce the definition of continuous dependence and stability for our problems. In addition, we introduce the notion of a gravitational constant such as that for fractional order differential equations, which will be used to consider the stability of solutions. In sections 2.1.3 and 2.1.4 we establish the continuous dependence and stability of solutions for (2.1) and (2.2) by discussing several possible cases depending on where the time variable falls in the interval between impulsive points and junction points. Examples are given in section 2.1.5 to illustrate our results.

2.1.2 Preliminaries

Let $J=[0,\infty)$ and $C^{(p)}(J,\mathbb{R})=\{f: f: J\to\mathbb{R}\text{is continuous and} f^{(p)}\text{is continuous}\}$, $p=0$ or 1. If $p=0$, we write $C^{(0)}(J,\mathbb{R})=C(J,\mathbb{R})$. We also recall the piecewise continuous function space $PC(J,\mathbb{R}):=\{\forall\,\chi: J\to\mathbb{R}: \chi\in C((\tau_\varsigma,\tau_{\varsigma+1}],\mathbb{R}),\ \varsigma=0,1,\ldots$ and $\exists\ \chi(\tau_\varsigma^+)$ and $\chi(\tau_\varsigma^-)$, $\varsigma=1,2,\ldots$with $\chi(\tau_\varsigma^-)=\chi(\tau_\varsigma)\}$ with the norm $\|\chi\|_{PC}:=\sup_{\tau\in J}|\chi(\tau)|$.

Let $x(t;0,x_0)$ and $\hat{x}(t;0,\hat{x}_0)$ be solutions of the original problem (2.1), (2.2) and perturbed problem (2.3), (2.4), respectively.

Now we introduce the definitions of the continuous dependence and stability of the solution to problems (2.1) and (2.2) linking with figure 2.1.

Definition 2.1.1. *The solution of (2.1) and (2.2) depends continuously on the initial point $(0,x_0)$, the impulsive points $t_1,t_2,\ldots$ and the junction points $s_1,s_2,\ldots$, if the following is satisfied:* $\forall\,\varepsilon>0,\forall\,T>0,\exists\,\delta=\delta(\varepsilon,T)>0$, *for* $\forall\,(0,\hat{x}_0)\in[0,T]\times\mathbb{R}$, $|\hat{x}_0-x_0|<\delta$ *and* $\forall\,\hat{t}_i,\hat{s}_i\in\mathbb{R}^+$, $i=1,2,\ldots$, *satisfying* $\hat{t}_0=\hat{s}_0=0<\hat{t}_1<\hat{s}_1<\hat{t}_2<\hat{s}_2<\cdots$, $|\hat{t}_i-t_i|<\delta$, $|\hat{s}_i-s_i|<\delta$, $i=1,2,\ldots$, *then*

$$|\hat{x}(t;0,\hat{x}_0)-x(t;0,x_0)|<\varepsilon,\quad t\in[0,T]/\left[\left(\bigcup_{i=1,2,\ldots}\langle\hat{t}_i,t_i\rangle\right)\bigcup\left(\bigcup_{i=1,2,\ldots}\langle\hat{s}_i,s_i\rangle\right)\right],$$

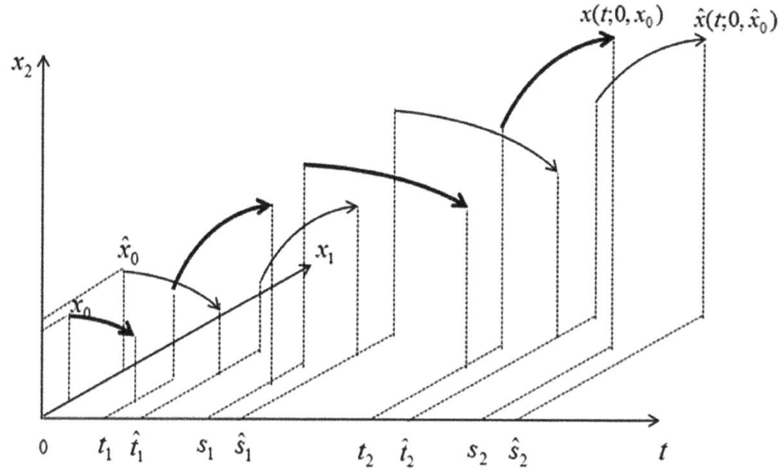

Figure 2.1. The thick line $x(t;\ 0,\ x_0)$ denotes the solution of (2.1) or (2.2) and the thin line $\hat{x}(t;\ 0,\ \hat{x}_0)$ denotes the solution of (2.3) or (2.4). Reprinted from [22]. Copyright (2017), with permission from Elsevier.

where $x(t;\ 0,\ x_0)$ and $\hat{x}(t;\ 0,\ \hat{x}_0)$ are reflected in (2.5) and (2.7), and (2.6) and (2.8), respectively. Here by $\langle a,\ b\rangle$ we mean $(a,\ b]$ if $a < b$, $(b,\ a]$ if $a > b$, and the empty set if $a = b$.

Definition 2.1.2. *The solution of (2.1) and (2.2) is stable with respect to the initial point $(0,\ x_0)$, the impulsive points $t_1,\ t_2,\ \ldots$ and the junction points $s_1,\ s_2,\ \ldots$, if $\forall\ \varepsilon > 0,\ \exists\ \delta = \delta(\varepsilon) > 0,$ for $\forall\ (0,\ \hat{x}_0) \in [0,\ \infty) \times \mathbb{R},$ $|\hat{x}_0 - x_0| < \delta$ and $\forall\ \hat{t}_i,\ \hat{s}_i \in \mathbb{R}^+, i = 1,2,\ \ldots,$ satisfying $\hat{t}_0 = \hat{s}_0 = 0 < \hat{t}_1 < \hat{s}_1 < \hat{t}_2 < \hat{s}_2 < \ldots,$ $|\hat{t}_i - t_i| < \delta_i,\ |\hat{s}_i - s_i| < \delta_i,\ i = 1,\ 2,\ \ldots$ and $\delta,\ \delta_i$ are sufficiently small, $i = 1,\ 2,\ \ldots,$ then*

$$|\hat{x}(t;\ 0,\ \hat{x}_0) - x(t;\ 0,\ x_0)| < \varepsilon, \quad t \in [0,\ \infty)/\left[\left(\bigcup_{i=1,2,\ldots} \langle \hat{t}_i,\ t_i\rangle\right) \bigcup \left(\bigcup_{i=1,2,\ldots} \langle \hat{s}_i,\ s_i\rangle\right)\right].$$

Consider the possible location of the distribution of the impulsive points t_i, $\hat{t}_i$ and the junction points s_i, $\hat{s}_i$ for the problems (2.1) and (2.2). We have the following cases (see figures 2.2, 2.3 and 2.4):

Definition 2.1.3. *(See [14, definition 1.3].) Let $X(t;\ 0,\ X_0)$ be the solution of*

$$\begin{cases} X'(t) = f(t,\ X(t)), & t \geqslant 0, \\ X(0) = X_0. \end{cases} \tag{2.9}$$

The system (2.9) is gravitating with a constant κ, if for arbitrary $t > 0$ and for each $\hat{X}_0,\ X_0 \in \mathbb{R}$, then

$$|X(t;\ 0,\ \hat{X}_0) - X(t;\ 0,\ X_0)| \leqslant \kappa|\hat{X}_0 - X_0|.$$

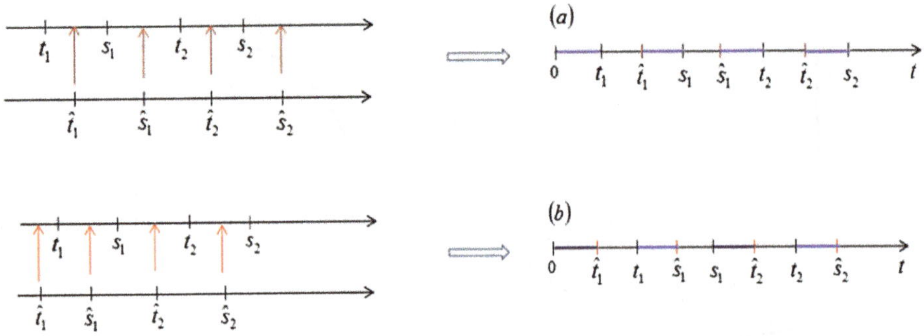

Figure 2.2. The argument presented below for cases (a) and (b) will be similar. Reprinted from [22]. Copyright (2017), with permission from Elsevier.

Definition 2.1.4. *Let* $Y(t;\, 0,\, Y_0)$ *be the solution of*

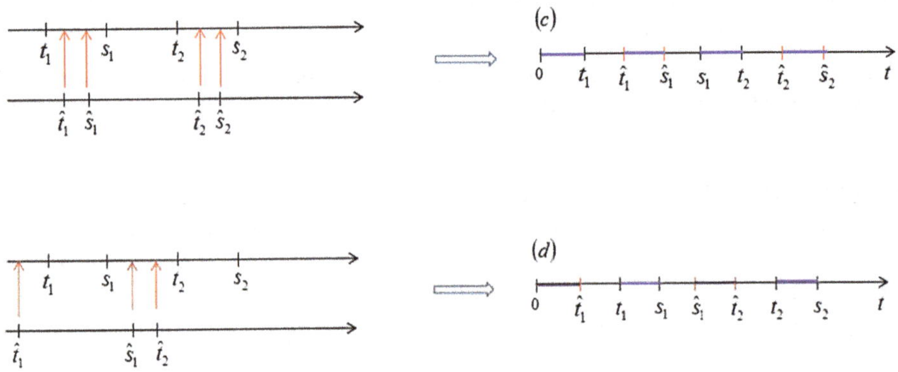

Figure 2.3. The argument presented below for cases (c) and (d) will be similar. Reprinted from [22]. Copyright (2017), with permission from Elsevier.

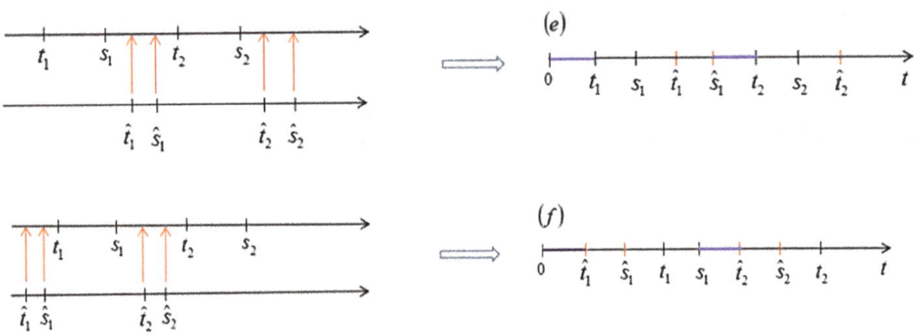

Figure 2.4. The argument presented below for cases (e) and (f) will be similar. Reprinted from [22]. Copyright (2017), with permission from Elsevier.

$$\begin{cases} {}^cD_{t_i,t}^{\alpha}Y(t) = f(t, Y(t)), & t \in (t_i, t_{i+1}],\ i \in \{0\} \bigcup \mathbb{N},\ \alpha \in (0, 1), \\ Y(0) = Y_0. \end{cases} \tag{2.10}$$

The system (2.10) is gravitating with a constant κ_α, if for arbitrary $t > t_i$, $i \in \{0\} \cup \mathbb{N}$ and for each $\hat{Y}_0$, $Y_0 \in \mathbb{R}$, then

$$|Y(t; 0, \hat{Y}_0) - Y(t; 0, Y_0)| \leqslant \kappa_\alpha |\hat{Y}_0 - Y_0|.$$

We list our assumptions:

[H_1] The function $f: J \times \mathbb{R} \to \mathbb{R}$ is continuous and $g_i \in C^1([t_i, s_i] \times \mathbb{R}, \mathbb{R})$, $i \in \mathbb{N}$.

[H_2] There exists a positive constant M such that $|f(t, x)| \leqslant M$, for any $(t, x) \in J \times \mathbb{R}$.

[H_3] There exists a positive constant L_f such that $|f(t, x) - f(t, y)| \leqslant L_f|x - y|$, for each $t \in [s_i, t_{i+1}]$, $i \in \{0\} \cup \mathbb{N}$, for all $x, y \in \mathbb{R}$.

[H_4] There exists a positive constant L_{gi}, $i \in N$ such that $|g_i(t, x) - g_i(t, y)| \leqslant L_{g_i}|x - y|$, for each $t \in [t_i, s_i]$, $i \in \mathbb{N}$, for all $x, y \in \mathbb{R}$.

For brevity, we set $t_i^{\min} = \min\{\hat{t}_i, t_i\}$, $t_i^{\max} = \max\{\hat{t}_i, t_i\}$, $s_i^{\min} = \min\{\hat{s}_i, s_i\}$ and $s_i^{\max} = \max\{\hat{s}_i, s_i\}$. Thus $\langle \hat{t}_i, t_i \rangle = (t_i^{\min}, t_i^{\max}]$, $\langle \hat{s}_i, s_i \rangle = (s_i^{\min}, s_i^{\max}]$, $i = 1, 2, \ldots$.

2.1.3 Continuous dependence of solutions

We begin by studying the continuous dependence on the initial point $(0, x_0)$, the impulsive points $t_1, t_2, \ldots$ and the junction points $s_1, s_2, \ldots$ of the solutions to (2.1).

From the continuity of $x(\cdot)$, if we set η to be an arbitrary positive number, we will suppose $|x(s_1^{\max}) - x(s_1^{\min})| < \eta$, if η is a sufficiently small constant, then δ is small enough.

Theorem 2.1.5. *Assume [H_1]–[H_4] hold. Then, the solution of (2.1) depends continuously on the initial point, the impulsive points t_i, $i = 1, 2, \ldots$ and the junction points s_i, $i = 1, 2, \ldots$.*

Proof. We divide our proofs into several cases.

Case 1. Let $t_i^{\min} = t_i$, $t_i^{\max} = \hat{t}_i$, $s_i^{\min} = s_i$, $s_i^{\max} = \hat{s}_i$, $i = 1, 2, \ldots$; see case (a) in figure 2.2 (the case $t_i^{\min} = \hat{t}_i$, $t_i^{\max} = t_i$, $s_i^{\min} = \hat{s}_i$, $s_i^{\max} = s_i$, $i = 1, 2, \ldots$ can be considered similarly; see case (b) in figure 2.2).

Since $t_i \to \infty$ $(i \to \infty)$, then $\exists\ k \in \mathbb{N}$ such that $s_k < T \leqslant t_{k+1}$. Assume that $|\hat{x}_0 - x_0| < \delta$, $|\hat{t}_i - t_i| < \delta$, $|\hat{s}_i - s_i| < \delta$, $i = 1, 2, \ldots$. If $\delta > 0$ is sufficiently small, then $T < s_{k+1}^{\min}$ and $s_{i-1}^{\max} < t_i^{\min}$, $t_i^{\max} < s_i^{\min}$, $i = 1, 2, \ldots, k + 1$.

Moreover, we have

$$
\begin{aligned}
&[0, T]/\left[\left(\bigcup_{i=1,2,\ldots} \langle \hat{t}_i, t_i \rangle\right) \bigcup \left(\bigcup_{i=1,2,\ldots} \langle \hat{s}_i, s_i \rangle\right)\right] \\
&= [0, T]/\left[\left(\bigcup_{i=1,2,\ldots} (t_i^{\min}, t_i^{\max}]\right) \bigcup \left(\bigcup_{i=1,2,\ldots} (s_i^{\min}, s_i^{\max}]\right)\right] \\
&= [0, t_1^{\min}] \bigcup \left(\bigcup_{i=1,2,\ldots,k} (t_i^{\max}, s_i^{\min}]\right) \bigcup \left(\bigcup_{i=1,2,\ldots,k-1} (s_i^{\max}, t_{i+1}^{\min}]\right) \\
&\quad \bigcup (s_k^{\max}, T^{\min}] \\
&= [0, t_1^{\min}] \bigcup \left(\bigcup_{i=1,2,\ldots,k} (t_i^{\max}, s_i^{\min}]\right) \bigcup \left(\bigcup_{i=1,2,\ldots,k} (s_i^{\max}, t_{i+1}^{\min}]\right),
\end{aligned}
\tag{2.11}
$$

where $T^{\min} = \min\{T, t_{k+1}^{\min}\}$. For convenience, we assume that $T^{\min} = t_{k+1}^{\min}$. If the inequality $T < t_{k+1}^{\min}$ is satisfied, i.e. $T^{\min} = T$, then replace the point $t_{k+1}^{\min}$ with T in the rest of the section on case 1 below.

Next we need to estimate $|\hat{x}(t; 0, \hat{x}_0) - x(t; 0, x_0)|$ in each subinterval, here $x(t; 0, x_0)$ and $\hat{x}(t; 0, \hat{x}_0)$ are given in (2.5) and (2.6), respectively.

For the interval $[0, t_1^{\min}]$, one can follow the procedure in [14, theorem 1.2, p 6] and use [17, theorem 1.1] to derive

$$|\hat{x}(t; 0, \hat{x}_0) - x(t; 0, x_0)| \leqslant \delta e^{L_f t_1} \leqslant \delta e^{L_f T}. \tag{2.12}$$

From (2.12), if $\delta > 0$ is sufficiently small, then

$$|\hat{x}(t; 0, \hat{x}_0) - x(t; 0, x_0)| < \delta_{01}, \quad t \in [0, t_1^{\min}], \tag{2.13}$$

where δ_{01} denotes an arbitrary positive number.

For $(t_1^{\max}, s_1^{\min}]$, in the interval, the representation of the solutions of problems (2.1) and (2.3) are given by

$$x(t; 0, x_0) = g_1(t, x(t_1^-)) \quad \text{and} \quad \hat{x}(t; 0, \hat{x}_0) = g_1(t, \hat{x}((\hat{t}_1)^-)).$$

Thus

$$|\hat{x}(t; 0, \hat{x}_0) - x(t; 0, x_0)| \leqslant L_{g_1} |\hat{x}((\hat{t}_1)^-) - x(t_1^-)|. \tag{2.14}$$

Note that

$$\hat{x}((\hat{t}_1)^-) = \hat{x}_0 + (\mathbf{I}^1_{0,\, t_1^{\max}} f)(t_1^{\max}, \hat{x}),$$

$$x(t_1^-) = x_0 + (\mathbf{I}^1_{0,\, t_1^{\min}} f)(t_1^{\min}, x).$$

Therefore

$$\begin{aligned}
&|\hat{x}((\hat{t}_1)^-) - x(t_1^-)| \\
&\leqslant |\hat{x}_0 - x_0| + |(\mathbf{I}^1_{0,\, t_1^{\min}} f)(t_1^{\min}, \hat{x}) - (\mathbf{I}^1_{0,\, t_1^{\min}} f)(t_1^{\min}, x)| \\
&\quad + |(\mathbf{I}^1_{t_1^{\min},\, t_1^{\max}} f)(t_1^{\max}, \hat{x})| \\
&\leqslant \delta + M\delta + L_f(\mathbf{I}^1_{0,\, t_1^{\min}} |\hat{x} - x|)(t_1^{\min}) \\
&\leqslant (1 + M + L_f t_1 e^{L_f t_1})\delta,
\end{aligned} \tag{2.15}$$

and let δ_{t_1} be an arbitrary positive number, if δ is small enough, then (2.15) becomes

$$|\hat{x}((\hat{t}_1)^-) - x(t_1^-)| < \delta_{t_1}.$$

Thus we obtain

$$|\hat{x}(t;\, 0,\, \hat{x}_0) - x(t;\, 0,\, x_0)| \leqslant L_{g_1}\delta_{t_1},$$

and let δ_{11} be an arbitrary positive number, so if δ_{t_1} is sufficiently small, then derive that

$$|\hat{x}(t;\, 0,\, \hat{x}_0) - x(t;\, 0,\, x_0)| < \delta_{11}, \quad t \in (t_1^{\max},\, s_1^{\min}]. \tag{2.16}$$

For $(s_1^{\max},\, t_2^{\min}]$, we have the expression of the solutions for problems (2.1) and (2.3), respectively,

$$x(t;\, 0,\, x_0) = x(s_1^{\max}) + (\mathbf{I}^1_{s_1^{\max},\, t} f)(t,\, x),$$

and

$$\hat{x}(t;\, 0,\, \hat{x}_0) = \hat{x}(s_1^{\max}) + (\mathbf{I}^1_{s_1^{\max},\, t} f)(t,\, \hat{x}),$$

then

$$|\hat{x}(t;\, 0,\, \hat{x}_0) - x(t;\, 0,\, x_0)| \leqslant |\hat{x}(s_1^{\max}) - x(s_1^{\max})| + L_f(\mathbf{I}^1_{s_1^{\max},\, t}|\hat{x} - x|)(t). \tag{2.17}$$

In the interval $(s_1^{\min},\, t_2^{\min}]$, we have

$$x(t;\, 0,\, x_0) = x(s_1^{\min}) + (\mathbf{I}^1_{s_1^{\min},\, t} f)(t,\, x),$$

and using the impulsive condition, we note that

$$x(s_1^{\min}) = g_1(s_1^{\min},\, x(t_1^-)),$$

therefore

$$x(t;\, 0,\, x_0) = g_1(s_1^{\min},\, x(t_1^-)) + (\mathbf{I}^1_{s_1^{\min},\, t} f)(t,\, x).$$

Let $t = s_1^{\max}$, then

$$x(s_1^{\max}; 0, x_0) = g_1(s_1^{\min}, x(t_1^-)) + (\mathbf{I}^1_{s_1^{\min}, s_1^{\max}} f)(s_1^{\max}, x). \tag{2.18}$$

Note

$$\hat{x}(s_1^{\max}; 0, \hat{x}_0) = g_1(s_1^{\max}, \hat{x}((\hat{t}_1)^-)). \tag{2.19}$$

From (2.18) and (2.19), we obtain

$$\begin{aligned}
&|\hat{x}(s_1^{\max}) - x(s_1^{\max})| \\
&\leqslant |g_1(s_1^{\max}, \hat{x}((\hat{t}_1)^-)) - g_1(s_1^{\min}, x(t_1^-))| + |(\mathbf{I}^1_{s_1^{\min}, s_1^{\max}} f)(s_1^{\max}, x)| \\
&\leqslant |g_1(s_1^{\max}, \hat{x}((\hat{t}_1)^-)) - g_1(s_1^{\max}, x(t_1^-))| \\
&\quad + |g_1(s_1^{\max}, x(t_1^-)) - g_1(s_1^{\min}, x(t_1^-))| + M\delta \\
&\leqslant L_{g_1}|\hat{x}((\hat{t}_1)^-) - x(t_1^-)| + |g_1'(\xi)||s_1^{\max} - s_1^{\min}| + M\delta \\
&\leqslant L_{g_1}\delta_{t_1} + |g_1'(\xi)|\delta + M\delta \\
&\leqslant (|g_1'(\xi)| + M)\delta + \delta_{11},
\end{aligned}$$

where $\xi \in (s_1^{\min}, s_1^{\max})$ and let δ_{s_1} be an arbitrary positive number, so if δ and δ_{11} are adequately small, then

$$|\hat{x}(s_1^{\max}) - x(s_1^{\max})| < \delta_{s_1}.$$

From (2.17), we derive the inequality

$$|\hat{x}(t; 0, \hat{x}_0) - x(t; 0, x_0)| \leqslant \delta_{s_1} + L_f(\mathbf{I}^1_{s_1^{\max}, t}|\hat{x} - x|)(t).$$

From [17, theorem 1.1], we obtain

$$|\hat{x}(t; 0, \hat{x}_0) - x(t; 0, x_0)| \leqslant \delta_{s_1} e^{L_f(t_2^{\min} - s_1^{\max})} \leqslant \delta_{s_1} e^{L_f T},$$

if δ_{s_1} is so small, then

$$|\hat{x}(t; 0, \hat{x}_0) - x(t; 0, x_0)| < \delta_{12}, \quad t \in (s_1^{\max}, t_2^{\min}], \tag{2.20}$$

where δ_{12} is an arbitrary positive number.

For $(t_2^{\max}, s_2^{\min}]$, similar to (2.14), we obtain

$$|\hat{x}(t; 0, \hat{x}_0) - x(t; 0, x_0)| \leqslant L_{g_2}|\hat{x}((\hat{t}_2)^-) - x(t_2^-)|. \tag{2.21}$$

Furthermore

$$\begin{aligned}
\hat{x}((\hat{t}_2)^-) &= \hat{x}(s_1^{\max}) + (\mathbf{I}^1_{s_1^{\max}, t_2^{\max}} f)(t_2^{\max}, \hat{x}), \\
x(t_2^-) &= x(s_1^{\min}) + (\mathbf{I}^1_{s_1^{\min}, t_2^{\min}} f)(t_2^{\min}, x),
\end{aligned}$$

then

$$
\begin{aligned}
&|\hat{x}((\hat{t}_2)^-) - x(t_2^-)| \\
&\leqslant |\hat{x}(s_1^{\max}) - x(s_1^{\min})| + |(\mathbf{I}^1_{t_2^{\min},\, t_2^{\max}} f)(t_2^{\max}, \hat{x})| + |(\mathbf{I}^1_{s_1^{\min},\, s_1^{\max}} f)(s_1^{\max}, x)| \\
&\quad + L_f(\mathbf{I}^1_{s_1^{\max},\, t_2^{\min}} |\hat{x} - x|)(t_2^{\min}) \\
&\leqslant |\hat{x}(s_1^{\max}) - x(s_1^{\max})| + |x(s_1^{\max}) - x(s_1^{\min})| + 2M\delta \\
&\quad + L_f(\mathbf{I}^1_{s_1^{\max},\, t_2^{\min}} |\hat{x} - x|)(t_2^{\min}) \\
&\leqslant \delta_{s_1} + \eta + 2M\delta + L_f(t_2^{\min} - s_1^{\max}) e^{L_f(t_2^{\min} - s_1^{\max})} \delta_{s_1}.
\end{aligned}
\tag{2.22}
$$

If δ and δ_{s_1} are sufficiently small, then

$$|\hat{x}((\hat{t}_2)^-) - x(t_2^-)| \leqslant \delta_{t_2}.$$

From (2.21) we have

$$|\hat{x}(t;\, 0,\, \hat{x}_0) - x(t;\, 0,\, x_0)| \leqslant L_{g_2} \delta_{t_2},$$

let δ_{22} be an arbitrary positive number, if δ_{t_2} is small enough, then

$$|\hat{x}(t;\, 0,\, \hat{x}_0) - x(t;\, 0,\, x_0)| < \delta_{22}, \quad t \in (t_2^{\max},\, s_2^{\min}]. \tag{2.23}$$

For $(s_2^{\max},\, t_3^{\min}]$, as for (2.17), we obtain

$$|\hat{x}(t;\, 0,\, \hat{x}_0) - x(t;\, 0,\, x_0)| \leqslant |\hat{x}(s_2^{\max}) - x(s_2^{\max})| + L_f(\mathbf{I}^1_{s_2^{\max},\, t} |\hat{x} - x|)(t).$$

For $(s_2^{\min},\, t_3^{\min}]$ we derive the equality

$$x(t;\, 0,\, x_0) = g_2(s_2^{\min},\, x(t_2^-)) + (\mathbf{I}^1_{s_2^{\min},\, t} f)(t,\, x),$$

and with $t = s_2^{\max}$, we obtain

$$x(s_2^{\max};\, 0,\, x_0) = g_2(s_2^{\min},\, x(t_2^-)) + (\mathbf{I}^1_{s_2^{\min},\, s_2^{\max}} f)(s_2^{\max},\, x).$$

In addition

$$\hat{x}(s_2^{\max};\, 0,\, \hat{x}_0) = g_2(s_2^{\max},\, \hat{x}((\hat{t}_2)^-)),$$

consequently

$$
\begin{aligned}
&|\hat{x}(s_2^{\max}) - x(s_2^{\max})| \\
&\leqslant |g_2(s_2^{\max},\, \hat{x}((\hat{t}_2)^-)) - g_2(s_2^{\min},\, x(t_2^-))| + |(\mathbf{I}^1_{s_2^{\min},\, s_2^{\max}} f)(s_2^{\max},\, x)| \\
&\leqslant |g_2(s_2^{\max},\, \hat{x}((\hat{t}_2)^-)) - g_2(s_2^{\max},\, x(t_2^-))| + |g_2(s_2^{\max},\, x(t_2^-)) - g_2(s_2^{\min},\, x(t_2^-))| + M\delta \\
&\leqslant L_{g_2} |\hat{x}((\hat{t}_2)^-) - x(t_2^-)| + |g_2'(\xi)||s_2^{\max} - s_2^{\min}| + M\delta \\
&\leqslant L_{g_2} \delta_{t_2} + |g_2'(\xi)|\delta + M\delta \\
&\leqslant (|g_2'(\xi)| + M)\delta + \delta_{22},
\end{aligned}
$$

where $\xi \in (s_2^{\min}, s_2^{\max})$, so if δ and δ_{22} are sufficiently small, then

$$|\hat{x}(s_2^{\max}) - x(s_2^{\max})| < \delta_{s_2},$$

where δ_{s_2} is an arbitrary positive number.

Hence we obtain

$$|\hat{x}(t; 0, \hat{x}_0) - x(t; 0, x_0)| \leqslant \delta_{s_2} + L_f(\mathbf{I}^1_{s_2^{\max}, t}|\hat{x} - x|)(t),$$

applying [17, theorem 1.1],

$$|\hat{x}(t; 0, \hat{x}_0) - x(t; 0, x_0)| \leqslant \delta_{s_2} e^{L_f(t_3^{\min} - s_2^{\max})} \leqslant \delta_{s_2} e^{L_f T}.$$

If δ_{s_2} is so small, then

$$|\hat{x}(t; 0, \hat{x}_0) - x(t; 0, x_0)| < \delta_{23}, \quad t \in (s_2^{\max}, t_3^{\min}], \tag{2.24}$$

where δ_{23} denotes an arbitrary positive number.

Repeating the procedure of (2.16) and (2.23), we arrive at

$$\begin{aligned} &\forall\ \delta_{ii} > 0,\ \exists\ \delta > 0,\ \exists\ \delta_{01} > 0,\ \exists\ \delta_{11} > 0,\ \exists\ \delta_{12} > 0,\ \dots,\\ &\qquad \exists\, \delta_{i-1, i} > 0,\ \text{then}\\ &|\hat{x}(t; 0, \hat{x}_0) - x(t; 0, x_0)| \leqslant \delta_{ii}, \quad t \in (t_i^{\max}, s_i^{\min}],\\ &\qquad\qquad i = 1, 2, \dots, k. \end{aligned} \tag{2.25}$$

In the same way, in view of (2.20) and (2.24), we can draw a conclusion

$$\begin{aligned} &\forall\, \delta_{i, i+1} > 0,\ \exists\ \delta > 0,\ \exists\ \delta_{01} > 0,\ \exists\ \delta_{11} > 0,\ \exists\ \delta_{12} > 0,\ \dots,\ \exists\\ &\qquad \delta_{ii} > 0,\ \text{then}\\ &|\hat{x}(t; 0, \hat{x}_0) - x(t; 0, x_0)| \leqslant \delta_{i, i+1}, \quad t \in (s_i^{\max}, t_{i+1}^{\min}],\\ &\qquad\qquad i = 1, 2, \dots, k. \end{aligned} \tag{2.26}$$

Let arbitrary number $\varepsilon > 0$ and put $\delta_{k, k+1} = \varepsilon$. As a result of (2.26), $\exists\ \delta > 0,\ \delta_{01} > 0,\ \delta_{11} > 0,\ \dots,\ \delta_{kk} > 0$ with

$$|\hat{x}(t; 0, \hat{x}_0) - x(t; 0, x_0)| \leqslant \varepsilon, \quad t \in (s_k^{\max}, t_{k+1}^{\min}].$$

If $i = 1$, from (2.25), then $\exists\ \delta > 0,\ \delta_{01} > 0$ such that

$$|\hat{x}(t; 0, \hat{x}_0) - x(t; 0, x_0)| \leqslant \varepsilon, \quad t \in (t_1^{\max}, s_1^{\min}].$$

From (2.13), $\exists\ \delta > 0$ such that

$$|\hat{x}(t; 0, \hat{x}_0) - x(t; 0, x_0)| \leqslant \varepsilon, \quad t \in [0, t_1^{\min}].$$

Therefore, we obtain the conclusion

$$\forall\ \varepsilon > 0,\ \exists\ \delta > 0,\ \text{if } |\hat{x}_0 - x_0| \leqslant \delta,\ |\hat{t}_i - t_i| \leqslant \delta \text{ and } |\hat{s}_i - s_i| \leqslant \delta,\ i = 1, \dots, k,\ \text{then}$$
$$|\hat{x}(t; 0, \hat{x}_0) - x(t; 0, x_0)| \leqslant \varepsilon,\ t \in [0, t_1^{\min}] \bigcup (t_i^{\max}, s_i^{\min}] \bigcup (s_i^{\max}, t_{i+1}^{\min}],$$
$$i = 1, \dots, k.$$

Case 2. Let $t_i^{\min} = t_i, t_i^{\max} = \hat{t}_i, s_i^{\min} = \hat{s}_i, s_i^{\max} = s_i, i = 1, \dots$; see case (c) in figure 2.3 (the case $t_i^{\min} = \hat{t}_i,\ t_i^{\max} = t_i,\ s_i^{\min} = s_i,\ s_i^{\max} = \hat{s}_i,\ i = 1, \dots$ can be considered similarly; see case (d) in figure 2.3).

We still have a similar conclusion (2.11), and if $T^{\min} = T$, then replace the point $t_{k+1}^{\min}$ with T in the rest of case 2 below. Likewise, we need to estimate the $|\hat{x}(t; 0, \hat{x}_0) - x(t; 0, x_0)|$ in each subinterval.

For the intervals $[0, t_1^{\min}]$ and $(t_1^{\max}, s_1^{\min}]$, we have the same result as in (2.13) and (2.16).

For $(s_1^{\max}, t_2^{\min}]$, we also have inequality (2.17). In the interval $(s_1^{\min}, t_2^{\max}]$, we obtain

$$\hat{x}(t; 0, x_0) = g_1(s_1^{\min}, \hat{x}((\hat{t}_1)^-)) + (\mathbf{I}^1_{s_1^{\min}, t} f)(t, \hat{x}),$$

and

$$\hat{x}(s_1^{\max}; 0, x_0) = g_1(s_1^{\min}, \hat{x}((\hat{t}_1)^-)) + (\mathbf{I}^1_{s_1^{\min}, s_1^{\max}} f)(s_1^{\max}, \hat{x}).$$

Also

$$x(s_1^{\max}; 0, x_0) = g_1(s_1^{\max}, x(t_1^-)),$$

and hence

$$\begin{aligned}
&|\hat{x}(s_1^{\max}) - x(s_1^{\max})| \\
&\leqslant |g_1(s_1^{\min}, \hat{x}((\hat{t}_1)^-)) - g_1(s_1^{\max}, x(t_1^-))| + |(\mathbf{I}^1_{s_1^{\min}, s_1^{\max}} f)(s_1^{\max}, \hat{x})| \\
&\leqslant |g_1(s_1^{\min}, \hat{x}((\hat{t}_1)^-)) - g_1(s_1^{\min}, x(t_1^-))| + |g_1(s_1^{\min}, x(t_1^-)) - g_1(s_1^{\max}, x(t_1^-))| + M\delta \\
&\leqslant L_{g_1}|\hat{x}((\hat{t}_1)^-) - x(t_1^-)| + |g_1'(\xi)||s_1^{\min} - s_1^{\max}| + M\delta \\
&\leqslant L_{g_1}\delta_{t_1} + |g_1'(\xi)|\delta + M\delta \\
&\leqslant (|g_1'(\xi)| + M)\delta + \delta_{11},
\end{aligned}$$

where $\xi \in (s_1^{\min}, s_1^{\max})$, δ_{t_1} and δ_{11} come from the procedure in case 1. Let δ_{s_1} be an arbitrary positive number, if δ and δ_{11} are sufficiently small, then

$$|\hat{x}(s_1^{\max}) - x(s_1^{\max})| < \delta_{s_1}.$$

From (2.17) and applying [17, theorem 1.1], we obtain

$$|\hat{x}(t; 0, \hat{x}_0) - x(t; 0, x_0)| \leqslant \delta_{s_1} e^{L_f(t_2^{\min} - s_1^{\max})} \leqslant \delta_{s_1} e^{L_f T},$$

if δ_{s_1} is too small, then

$$|\hat{x}(t; 0, \hat{x}_0) - x(t; 0, x_0)| < \delta_{12}, \quad t \in (s_1^{\max}, t_2^{\min}], \tag{2.27}$$

where δ_{12} denotes an arbitrary positive number.

For $(t_2^{\max}, s_2^{\min}]$, we have the same inequality as in (2.21).

Also

$$\hat{x}((\hat{t}_2)^-) = \hat{x}(s_1^{\min}) + (\mathbf{I}^1_{s_1^{\min}, t_2^{\max}} f)(t_2^{\max}, \hat{x}),$$
$$x(t_2^-) = x(s_1^{\max}) + (\mathbf{I}^1_{s_1^{\max}, t_2^{\min}} f)(t_2^{\min}, x),$$

then

$$\begin{aligned}
&|\hat{x}((\hat{t}_2)^-) - x(t_2^-)| \\
&\leqslant |\hat{x}(s_1^{\min}) - x(s_1^{\max})| + |(\mathbf{I}^1_{s_1^{\min}, s_1^{\max}} f)(s_1^{\max}, \hat{x})| + |(\mathbf{I}^1_{t_2^{\min}, t_2^{\max}} f)(t_2^{\max}, \hat{x})| \\
&\quad + L_f(\mathbf{I}^1_{s_1^{\max}, t_2^{\min}} |\hat{x} - x|)(t_2^{\min}) \\
&\leqslant |\hat{x}(s_1^{\min}) - x(s_1^{\min})| + |x(s_1^{\min}) - x(s_1^{\max})| + 2M\delta + L_f(\mathbf{I}^1_{s_1^{\max}, t_2^{\min}} |\hat{x} - x|)(t_2^{\min}). \\
&\leqslant \delta_{s_1} + \eta + 2M\delta + L_f(t_2^{\min} - s_1^{\max}) e^{L_f(t_2^{\min} - s_1^{\max})} \delta_{s_1}.
\end{aligned}$$

If δ and δ_{s_1} are sufficiently small, then

$$|\hat{x}((\hat{t}_2)^-) - x(t_2^-)| \leqslant \delta_{t_2}.$$

From (2.21), we obtain

$$|\hat{x}(t; 0, \hat{x}_0) - x(t; 0, x_0)| \leqslant L_{g_2} \delta_{t_2},$$

if δ_{t_2} is small enough, then

$$|\hat{x}(t; 0, \hat{x}_0) - x(t; 0, x_0)| < \delta_{22}, \quad t \in (t_2^{\max}, s_2^{\min}], \tag{2.28}$$

where δ_{22} be an arbitrary positive number.

For $(s_2^{\max}, t_3^{\min}]$, consider the process in case 1, we derive

$$|\hat{x}(t; 0, \hat{x}_0) - x(t; 0, x_0)| \leqslant |\hat{x}(s_2^{\max}) - x(s_2^{\max})| + L_f(\mathbf{I}^1_{s_2^{\max}, t} |\hat{x} - x|)(t).$$

For $(s_2^{\min}, t_3^{\max}]$, we can obtain

$$\hat{x}(t; 0, \hat{x}_0) = g_2(s_2^{\min}, \hat{x}((\hat{t}_2)^-)) + (\mathbf{I}^1_{s_2^{\min}, t} f)(t, \hat{x}).$$

Set $t = s_2^{\max}$, then

$$\hat{x}(s_2^{\max}; 0, \hat{x}_0) = g_2(s_2^{\min}, \hat{x}((\hat{t}_2)^-)) + (\mathbf{I}^1_{s_2^{\min}, s_2^{\max}} f)(s_2^{\max}, \hat{x}).$$

Furthermore

$$x(s_2^{\max}; 0, x_0) = g_2(s_2^{\max}, x(t_2^-)),$$

hence

$$\begin{aligned}
&|\hat{x}(s_2^{\max}) - x(s_2^{\max})| \\
&\leqslant |g_2(s_2^{\min}, \hat{x}((\hat{t}_2)^-)) - g_2(s_2^{\max}, x(t_2^-))| + |(\mathbf{I}^1_{s_2^{\min}, s_2^{\max}} f)(s_2^{\max}, \hat{x})| \\
&\leqslant |g_2(s_2^{\min}, \hat{x}((\hat{t}_2)^-)) - g_2(s_2^{\min}, x(t_2^-))| \\
&\quad + |g_2(s_2^{\min}, x(t_2^-)) - g_2(s_2^{\max}, x(t_2^-))| + M\delta \\
&\leqslant L_{g_2}|\hat{x}((\hat{t}_2)^-) - x(t_2^-)| + |g_2'(\xi)||s_2^{\min} - s_2^{\max}| + M\delta \\
&\leqslant L_{g_2}\delta_{t_2} + |g_2'(\xi)|\delta + M\delta \\
&\leqslant (|g_2'(\xi)| + M)\delta + \delta_{22},
\end{aligned}$$

where $\xi \in (s_2^{\min}, s_2^{\max})$, if δ and δ_{22} are sufficiently small, then

$$|\hat{x}(s_2^{\max}) - x(s_2^{\max})| < \delta_{s_2},$$

where δ_{s_2} is an arbitrary positive number.

From [17, theorem 1.1], we can obtain

$$|\hat{x}(t; 0, \hat{x}_0) - x(t; 0, x_0)| \leqslant \delta_{s_2} e^{L_f(t_3^{\min} - s_2^{\max})} \leqslant \delta_{s_2} e^{L_f T},$$

let $\delta_{23} > 0$ be an arbitrary number, so if δ_{s_2} is sufficiently small, then

$$|\hat{x}(t; 0, \hat{x}_0) - x(t; 0, x_0)| < \delta_{23}, \quad t \in (s_2^{\max}, t_3^{\min}].$$

Similarly, from the above procedure we obtain the same conclusions as in (2.25) and (2.26). Let $\varepsilon > 0$ be an arbitrary number and let $\delta_{k, k+1} = \varepsilon$, so we can deduce the conclusion

$$|\hat{x}(t; 0, \hat{x}_0) - x(t; 0, x_0)| \leqslant \varepsilon, \quad t \in (s_k^{\max}, t_{k+1}^{\min}].$$

If $i = 1$, then

$$|\hat{x}(t; 0, \hat{x}_0) - x(t; 0, x_0)| \leqslant \varepsilon, \quad t \in (t_1^{\max}, s_1^{\min}],$$

and

$$|\hat{x}(t; 0, \hat{x}_0) - x(t; 0, x_0)| \leqslant \varepsilon, \quad t \in [0, t_1^{\min}].$$

Therefore

$$\begin{aligned}
&\forall\, \varepsilon > 0, \;\; \exists\, \delta > 0, \;\; \text{if } |\hat{x}_0 - x_0| \leqslant \delta, \;\; |\hat{t}_i - t_i| \leqslant \delta \text{ and} \\
&|\hat{s}_i - s_i| \leqslant \delta, \;\; i = 1, 2, \ldots, k, \;\; \text{then } |\hat{x}(t; 0, \hat{x}_0) - x(t; 0, x_0)| \leqslant \varepsilon, \\
&t \in [0, t_1^{\min}] \bigcup (t_i^{\max}, s_i^{\min}] \bigcup (s_i^{\max}, t_{i+1}^{\min}], \\
&\quad i = 1, 2, \ldots, k.
\end{aligned}$$

Case 3. Let $t_i^{\min} = t_i$, $s_i^{\min} = s_i$, $t_i^{\max} = \hat{t}_i$, $s_i^{\max} = \hat{s}_i$, $i = 1, \ldots$; see case (e) in figure 2.4 (the case $t_i^{\min} = \hat{t}_i$, $s_i^{\min} = \hat{s}_i$, $t_i^{\max} = t_i$, $s_i^{\max} = s_i$, $i = 1, \ldots$ can be considered similarly; see case (f) in figure 2.4).

Since $t_i \to \infty$ $(i \to \infty)$, then $\exists\, k \in \mathbb{N}$ such that $s_k < T \leqslant t_{k+1}$. Assume that $|\hat{x}_0 - x_0| < \delta$, $|\hat{t}_i - t_i| < \delta$, $|\hat{s}_i - s_i| < \delta$, $i = 1, 2, \ldots$. If $\delta > 0$ is sufficiently small, then we can deduce that: $T < s_{k+1}^{\min}$, $t_i^{\min} < s_i^{\min} < t_i^{\max} < s_i^{\max} < t_{i+1}^{\min}$, $i = 1, 2, \dots, k+1$.

Moreover, we have

$$
\begin{aligned}
&[0, T]/\left[\left(\bigcup_{i=1,2,\dots} \langle \hat{t}_i, t_i\rangle\right)\bigcup\left(\bigcup_{i=1,2,\dots} \langle \hat{s}_i, s_i\rangle\right)\right] \\
&= [0, T]/\left[\left(\bigcup_{i=1,2,\dots} (t_i^{\min}, t_i^{\max}]\right)\bigcup\left(\bigcup_{i=1,2,\dots} (s_i^{\min}, s_i^{\max}]\right)\right] \\
&= [0, t_1^{\min}]\bigcup\left(\bigcup_{i=1,2,\dots,k-1} (s_i^{\max}, t_{i+1}^{\min}]\right)\bigcup(s_k^{\max}, T^{\min}] \\
&= [0, t_1^{\min}]\bigcup\left(\bigcup_{i=1,2,\dots,k} (s_i^{\max}, t_{i+1}^{\min}]\right),
\end{aligned}
\tag{2.29}
$$

where $T^{\min} = \min\{T, t_{k+1}^{\min}\}$. For simplicity, we assume that $T^{\min} = t_{k+1}^{\min}$. If $T^{\min} = T$, then replace $t_{k+1}^{\min}$ with T in the rest of case 3 below.

For the interval $[0, t_1^{\min}]$, from the same procedure we have (2.13).

For $(s_1^{\min}, t_2^{\min}]$,

$$x(t; 0, x_0) = g_1(s_1^{\min}, x(t_1^-)) + (\mathbf{I}^1_{s_1^{\min}, t} f)(t, x).$$

In particular, if $t = s_1^{\max}$, then

$$x(s_1^{\max}; 0, x_0) = g_1(s_1^{\min}, x(t_1^-)) + (\mathbf{I}^1_{s_1^{\min}, s_1^{\max}} f)(s_1^{\max}, x),$$

and

$$\hat{x}(s_1^{\max}; 0, \hat{x}_0) = g_1(s_1^{\max}, \hat{x}((\hat{t}_1)^-)).$$

Also, we obtain

$$
\begin{aligned}
&|\hat{x}(s_1^{\max}) - x(s_1^{\max})| \\
&\leqslant |g_1(s_1^{\max}, \hat{x}((\hat{t}_1)^-)) - g_1(s_1^{\min}, x(t_1^-))| + |(\mathbf{I}^1_{s_1^{\min}, s_1^{\max}} f)(s_1^{\max}, x)| \\
&\leqslant |g_1(s_1^{\max}, \hat{x}((\hat{t}_1)^-)) - g_1(s_1^{\max}, x(t_1^-))| \\
&\quad + |g_1(s_1^{\max}, x(t_1^-)) - g_1(s_1^{\min}, x(t_1^-))| + M\delta \\
&\leqslant L_{g_1}|\hat{x}((\hat{t}_1)^-) - x(t_1^-)| + |g_1'(\xi)||s_1^{\max} - s_1^{\min}| + M\delta \\
&\leqslant L_{g_1}|\hat{x}((\hat{t}_1)^-) - x(t_1^-)| + |g_1'(\xi)|\delta + M\delta,
\end{aligned}
\tag{2.30}
$$

where $\xi \in (s_1^{\min}, s_1^{\max})$.

Note that

$$\hat{x}((\hat{t}_1)^-) = \hat{x}_0 + (\mathbf{I}^1_{0,\, t_1^{\max}} f)(t_1^{\max}, \hat{x}) \quad \text{and} \quad x(t_1^-) = x_0 + (\mathbf{I}^1_{0,\, t_1^{\min}} f)(t_1^{\min}, x).$$

Therefore

$$\begin{aligned}
&|\hat{x}((\hat{t}_1)^-) - x(t_1^-)| \\
&\leqslant |\hat{x}_0 - x_0| + |(\mathbf{I}^1_{0,\, t_1^{\min}} f)(t_1^{\min}, \hat{x}) - (\mathbf{I}^1_{0,\, t_1^{\min}} f)(t_1^{\min}, x)| \\
&\quad + |(\mathbf{I}^1_{t_1^{\min},\, t_1^{\max}} f)(t_1^{\max}, \hat{x})| \\
&\leqslant \delta + M\delta + L_f(\mathbf{I}^1_{0,\, t_1^{\min}} |\hat{x} - x|)(t_1^{\min}) \\
&\leqslant (1 + M + L_f t_1 e^{L_f t_1})\delta.
\end{aligned}$$

From (2.30), we have

$$|\hat{x}(s_1^{\max}) - x(s_1^{\max})| \leqslant \left(|g_1'(\xi)| + L_{g_1}(1 + M + L_f t_1 e^{L_f t_1}) + M\right)\delta.$$

Let δ_{s_1} be an arbitrary positive number, so if δ is sufficiently small, then

$$|\hat{x}(s_1^{\max}) - x(s_1^{\max})| < \delta_{s_1}.$$

For $(s_1^{\max}, t_2^{\min}]$, consider as in case 1 and apply [17, theorem 1.1],

$$\begin{aligned}
|\hat{x}(t; 0, \hat{x}_0) - x(t; 0, x_0)| &\leqslant |\hat{x}(s_1^{\max}) - x(s_1^{\max})| + L_f(\mathbf{I}^1_{s_1^{\max},\, t} |\hat{x} - x|)(t) \\
&\leqslant \delta_{s_1} + L_f(\mathbf{I}^1_{s_1^{\max},\, t} |\hat{x} - x|)(t) \\
&\leqslant \delta_{s_1} e^{L_f(t_2^{\min} - s_1^{\max})} \leqslant \delta_{s_1} e^{L_f T}.
\end{aligned}$$

If δ_{s_1} is sufficiently small, then

$$|\hat{x}(t; 0, \hat{x}_0) - x(t; 0, x_0)| < \delta_{12}, \quad t \in (s_1^{\max}, t_2^{\min}], \tag{2.31}$$

where δ_{12} is an arbitrary positive number.

Similar to (2.22), we derive that

$$\begin{aligned}
&|\hat{x}((\hat{t}_2)^-) - x(t_2^-)| \\
&\leqslant |\hat{x}(s_1^{\max}) - x(s_1^{\min})| + |(\mathbf{I}^1_{t_2^{\min},\, t_2^{\max}} f)(t_2^{\max}, \hat{x})| + |(\mathbf{I}^1_{s_1^{\min},\, s_1^{\max}} f)(s_1^{\max}, x)| \\
&\quad + L_f(\mathbf{I}^1_{s_1^{\max},\, t_2^{\min}} |\hat{x} - x|)(t_2^{\min}) \\
&\leqslant |\hat{x}(s_1^{\max}) - x(s_1^{\max})| + |x(s_1^{\max}) - x(s_1^{\min})| + 2M\delta + L_f(\mathbf{I}^1_{s_1^{\max},\, t_2^{\min}} |\hat{x} - x|)(t_2^{\min}). \\
&\leqslant \delta_{s_1} + \eta + 2M\delta + L_f(t_2^{\min} - s_1^{\max}) e^{L_f(t_2^{\min} - s_1^{\max})} \delta_{s_1}.
\end{aligned}$$

For $(s_2^{\min}, t_3^{\min}]$, we obtain the formula

$$x(t; 0, x_0) = g_1(s_2^{\min}, x(t_2^-)) + (\mathbf{I}^1_{s_2^{\min},\, t} f)(t, x),$$

and

$$x(s_2^{\max}; 0, x_0) = g_2(s_2^{\min}, x(t_2^-)) + (\mathbf{I}^1_{s_2^{\min},\, s_2^{\max}} f)(s_2^{\max}, x).$$

Now

$$\hat{x}(s_2^{\max}; 0, \hat{x}_0) = g_2(s_2^{\max}, \hat{x}((\hat{t}_2)^-)),$$

then

$$\begin{aligned}
&|\hat{x}(s_2^{\max}) - x(s_2^{\max})| \\
&\leqslant |g_2(s_2^{\max}, \hat{x}((\hat{t}_2)^-)) - g_2(s_2^{\min}, x(t_2^-))| + |(\mathbf{I}^1_{s_2^{\min}, s_2^{\max}} f)(s_2^{\max}, x)| \\
&\leqslant L_{g_2}|\hat{x}((\hat{t}_2)^-) - x(t_2^-)| + |g_2'(\xi)||s_2^{\max} - s_2^{\min}| + M\delta \\
&\leqslant L_{g_2}\left(\delta_{s_1} + \eta + 2M\delta + L_f(t_2^{\min} - s_1^{\max})e^{L_f(t_2^{\min} - s_1^{\max})}\delta_{s_1}\right) + |g_2'(\xi)|\delta + M\delta \\
&\leqslant (|g_2'(\xi)| + 2ML_{g_2} + M)\delta + L_{g_2}\eta + (L_{g_2} + L_{g_2}L_f t_2 e^{L_f t_2})\delta_{s_1},
\end{aligned}$$

if δ and δ_{s_1} are sufficiently small, then

$$|\hat{x}(s_2^{\max}) - x(s_2^{\max})| < \delta_{s_2},$$

where δ_{s_2} is an arbitrary positive number.

For $(s_2^{\max}, t_3^{\min}]$,

$$\begin{aligned}
|\hat{x}(t; 0, \hat{x}_0) - x(t; 0, x_0)| &\leqslant |\hat{x}(s_2^{\max}) - x(s_2^{\max})| + L_f(\mathbf{I}^1_{s_2^{\max}, t}|\hat{x} - x|)(t) \\
&\leqslant \delta_{s_2} + L_f(\mathbf{I}^1_{s_2^{\max}, t}|\hat{x} - x|)(t).
\end{aligned}$$

Using [17, theorem 1.1], deduce that

$$|\hat{x}(t; 0, \hat{x}_0) - x(t; 0, x_0)| \leqslant \delta_{s_2}e^{L_f(t_3^{\min} - s_2^{\max})} \leqslant \delta_{s_2}e^{L_f T}.$$

Let δ_{23} be an arbitrary positive number, if δ_{s_2} is adequately small, then

$$|\hat{x}(t; 0, \hat{x}_0) - x(t; 0, x_0)| < \delta_{23}, \quad t \in (s_2^{\max}, t_3^{\min}]. \tag{2.32}$$

Repeating the procedure in (2.31) and (2.32), we arrive at

$$\begin{aligned}
&\forall\, \delta_{i,\, i+1} > 0, \quad \exists\, \delta > 0, \quad \exists\, \delta_{s_1} > 0, \quad \exists\, \delta_{s_2} > 0, \quad \dots, \quad \exists\, \delta_{s_i} > 0, \\
&\qquad \text{then} \\
&|\hat{x}(t; 0, \hat{x}_0) - x(t; 0, x_0)| \leqslant \delta_{i,\, i+1}, \quad t \in (s_i^{\max}, t_{i+1}^{\min}], \\
&\qquad\qquad i = 1, 2, \dots, k,
\end{aligned} \tag{2.33}$$

where δ_{s_i} originates from $|\hat{x}(s_i^{\max}) - x(s_i^{\max})| < \delta_{s_i}$, $i = 1, 2, \dots, k$.

Set $\delta_{k,\, k+1} = \varepsilon$, we derive that

$$|\hat{x}(t; 0, \hat{x}_0) - x(t; 0, x_0)| \leqslant \varepsilon, \quad t \in (s_k^{\max}, t_{k+1}^{\min}],$$

where $\varepsilon > 0$ be an arbitrary number.

If $i = 1$, from (2.33), $\exists\, \delta > 0$, $\delta_{s_1} > 0$ such that

$$|\hat{x}(t; 0, \hat{x}_0) - x(t; 0, x_0)| \leqslant \varepsilon, \quad t \in (s_1^{\max}, t_2^{\min}],$$

for $[0, t_1^{\min}]$, $\exists\, \delta > 0$ such that

$$|\hat{x}(t;\, 0,\, \hat{x}_0) - x(t;\, 0,\, x_0)| \leqslant \varepsilon, \quad t \in [0,\, t_1^{\min}].$$

Consequently,

$$\begin{aligned}&\forall\, \varepsilon > 0, \;\; \exists\, \delta > 0, \;\; \text{if } |\hat{x}_0 - x_0| \leqslant \delta, \;\; |\hat{t}_i - t_i| \leqslant \delta \text{ and}\\ &|\hat{s}_i - s_i| \leqslant \delta, \;\; i = 1, \ldots, k, \;\; \text{then } |\hat{x}(t;\, 0,\, \hat{x}_0) - x(t;\, 0,\, x_0)| \leqslant \varepsilon,\\ &t \in [0,\, t_1^{\min}] \bigcup (s_i^{\max},\, t_{i+1}^{\min}],\\ &i = 1, \ldots, k.\end{aligned}$$

The proof is complete. □

Next, we present the continuous dependence on initial point $(0, x_0)$, impulsive points t_1, t_2, ... and junction points s_1, s_2, ... of solutions to (2.2).

From the continuity of $x(\cdot)$, if we set η_α to be an arbitrary positive number, we will suppose $|x(s_1^{\max}) - x(s_1^{\min})| < \eta_\alpha$, if η_α is a sufficiently small constant, i.e. if δ is small enough.

Theorem 2.1.6. *Assume $[H_1]$–$[H_4]$ hold. Then, the solution of (2.2) depends continuously on the initial point, the impulsive points t_i, $i = 1, 2, \ldots$ and the junction points s_i, $i = 1, 2, \ldots$.*

Proof. We divide our proof into several cases.

Case 1. Let $t_i^{\min} = t_i$, $t_i^{\max} = \hat{t}_i$, $s_i^{\min} = s_i$, $s_i^{\max} = \hat{s}_i$, $i = 1, \ldots$; see case (a) in figure 2.2 (the case $t_i^{\min} = \hat{t}_i$, $t_i^{\max} = t_i$, $s_i^{\min} = \hat{s}_i$, $s_i^{\max} = s_i$, $i = 1, \ldots$ can be considered similarly; see case (b) in figure 2.2). In this case, we have the same result as in (2.11).

Now we estimate the $|\hat{x}(t;\, 0,\, \hat{x}_0) - x(t;\, 0,\, x_0)|$ in each subinterval, where $x(t;\, 0,\, x_0)$ and $\hat{x}(t;\, 0,\, \hat{x}_0)$ are given in (2.7) and (2.8), respectively.

For $[0, t_1^{\min}]$, the representation of the solutions of problems (2.2) and (2.4) are given by

$$x(t;\, 0,\, x_0) = x_0 + (\mathbf{I}^{\alpha}_{0,\, t} f)(t,\, x), \tag{2.34}$$

$$\hat{x}(t;\, 0,\, \hat{x}_0) = \hat{x}_0 + (\mathbf{I}^{\alpha}_{0,\, t} f)(t,\, \hat{x}). \tag{2.35}$$

From (2.34) and (2.35), we have

$$\begin{aligned}|\hat{x}(t;\, 0,\, \hat{x}_0) - x(t;\, 0,\, x_0)| &\leqslant |\hat{x}_0 - x_0| + L_f(\mathbf{I}^{\alpha}_{0,\, t}|\hat{x} - x|)(t)\\ &\leqslant \delta + L_f(\mathbf{I}^{\alpha}_{0,\, t}|\hat{x} - x|)(t).\end{aligned}$$

From [18, corollary 2],

$$|\hat{x}(t;0,\hat{x}_0)-x(t;0,x_0)|\leqslant \delta E_\alpha(L_f t_1^{\alpha})\leqslant \delta E_\alpha(L_f T^\alpha), \tag{2.36}$$

so if δ is fully small, then

$$|\hat{x}(t;0,\hat{x}_0)-x(t;0,x_0)|<\delta_{01},\quad t\in[0,t_1^{\min}], \tag{2.37}$$

where δ_{01} denotes an arbitrary positive number.

For $(t_1^{\max}, s_1^{\min}]$, we have the expression of the solutions of problems (2.2) and (2.4), respectively,

$$x(t;0,x_0)=g_1(t,x(t_1^-))\ \text{ and }\ \hat{x}(t;0,\hat{x}_0)=g_1(t,\hat{x}((\hat{t}_1)^-)),$$

then

$$|\hat{x}(t;0,\hat{x}_0)-x(t;0,x_0)|\leqslant L_{g_1}|\hat{x}((\hat{t}_1)^-)-x(t_1^-)|. \tag{2.38}$$

Note that

$$\hat{x}((\hat{t}_1)^-)=\hat{x}_0+(\mathbf{I}^{\alpha}_{0,\,t_1^{\max}}f)(t_1^{\max},\hat{x}),$$
$$x(t_1^-)=x_0+(\mathbf{I}^{\alpha}_{0,\,t_1^{\min}}f)(t_1^{\min},x).$$

Therefore

$$\begin{aligned}
&|\hat{x}((\hat{t}_1)^-)-x(t_1^-)|\\
&\leqslant|\hat{x}_0-x_0|+\frac{1}{\Gamma(\alpha)}\int_0^{t_1^{\min}}(t_1^{\max}-s)^{\alpha-1}|f(s,\hat{x}(s))-f(s,x(s))|ds\\
&+\frac{1}{\Gamma(\alpha)}\int_0^{t_1^{\min}}|(t_1^{\max}-s)^{\alpha-1}-(t_1^{\min}-s)^{\alpha-1}||f(s,x(s))|ds\\
&+|(\mathbf{I}^{\alpha}_{t_1^{\min},\,t_1^{\max}}f)(t_1^{\max},\hat{x})|\\
&\leqslant\delta+\frac{L_f}{\Gamma(\alpha)}\int_0^{t_1^{\min}}(t_1^{\max}-s)^{\alpha-1}|\hat{x}-x(s)|ds+\frac{3M}{\Gamma(\alpha+1)}\delta^\alpha\\
&\leqslant\delta+\frac{3M}{\Gamma(\alpha+1)}\delta^\alpha+\frac{L_f t_1^{\alpha}}{\Gamma(\alpha+1)}E_\alpha(L_f t_1^{\alpha})\delta.
\end{aligned}$$

Let δ_{t_1} be an arbitrary positive number, so if δ is sufficiently small, then

$$|\hat{x}((\hat{t}_1)^-)-x(t_1^-)|<\delta_{t_1}.$$

We also obtain

$$|\hat{x}(t;0,\hat{x}_0)-x(t;0,x_0)|\leqslant L_{g_1}\delta_{t_1},$$

if δ_{t_1} is adequately small, then

$$|\hat{x}(t;0,\hat{x}_0)-x(t;0,x_0)|<\delta_{11},\quad t\in(t_1^{\max},s_1^{\min}], \tag{2.39}$$

where δ_{11} is an arbitrary positive number.

For $(s_1^{\max}, t_2^{\min}]$,

$$x(t;0,x_0)=x(s_1^{\max})+(\mathbf{I}^{\alpha}_{s_1^{\max},\,t}f)(t,x),$$

$$\hat{x}(t;\, 0,\, \hat{x}_0) = \hat{x}(s_1^{\max}) + (\mathbf{I}^{\alpha}_{s_1^{\max},\, t}f)(t,\, \hat{x}),$$

and then

$$|\hat{x}(t;\, 0,\, \hat{x}_0) - x(t;\, 0,\, x_0)| \leqslant |\hat{x}(s_1^{\max}) - x(s_1^{\max})| + L_f(\mathbf{I}^{\alpha}_{s_1^{\max},\, t}|\hat{x} - x|)(t). \tag{2.40}$$

For $(s_1^{\min},\, t_2^{\min}]$, by using the form of expression of the solution and applying the impulsive condition,

$$x(t;\, 0,\, x_0) = g_1(s_1^{\min},\, x(t_1^-)) + (\mathbf{I}^{\alpha}_{s_1^{\min},\, t}f)(t,\, x),$$

from which we find

$$x(s_1^{\max};\, 0,\, x_0) = g_1(s_1^{\min},\, x(t_1^-)) + (\mathbf{I}^{\alpha}_{s_1^{\min},\, s_1^{\max}}f)(s_1^{\max},\, x).$$

Also

$$\hat{x}(s_1^{\max};\, 0,\, \hat{x}_0) = g_1(s_1^{\max},\, \hat{x}((\hat{t}_1)^-)),$$

thus

$$\begin{aligned} &|\hat{x}(s_1^{\max}) - x(s_1^{\max})| \\ &\leqslant |g_1(s_1^{\max},\, \hat{x}((\hat{t}_1)^-)) - g_1(s_1^{\min},\, x(t_1^-))| + |(\mathbf{I}^{\alpha}_{s_1^{\min},\, s_1^{\max}}f)(s_1^{\max},\, x)| \\ &\leqslant L_{g_1}|\hat{x}((\hat{t}_1)^-) - x(t_1^-)| + |g_1'(\zeta)||s_1^{\max} - s_1^{\min}| + \frac{M}{\Gamma(\alpha + 1)}\delta^{\alpha} \\ &\leqslant L_{g_1}\delta_{t_1} + |g_1'(\zeta)|\delta + \frac{M}{\Gamma(\alpha + 1)}\delta^{\alpha} \\ &\leqslant |g_1'(\zeta)|\delta + \frac{M}{\Gamma(\alpha + 1)}\delta^{\alpha} + \delta_{11}, \end{aligned} \tag{2.41}$$

where $\zeta \in (s_1^{\min},\, s_1^{\max})$. Let δ_{s_1} be an arbitrary positive number, if δ, δ_{11} are fully small, then

$$|\hat{x}(s_1^{\max}) - x(s_1^{\max})| < \delta_{s_1}.$$

From (2.40) we draw the conclusion

$$|\hat{x}(t;\, 0,\, \hat{x}_0) - x(t;\, 0,\, x_0)| \leqslant \delta_{s_1} + L_f(\mathbf{I}^{\alpha}_{s_1^{\max},\, t}|\hat{x} - x|)(t).$$

Apply [18, corollary 2] again,

$$|\hat{x}(t;\, 0,\, \hat{x}_0) - x(t;\, 0,\, x_0)| \leqslant \delta_{s_1}E_{\alpha}(L_f(t_2^{\min} - s_1^{\max})^{\alpha}) \leqslant \delta_{s_1}E_{\alpha}(L_fT^{\alpha}).$$

If δ_{s_1} is fully small, then

$$|\hat{x}(t;\, 0,\, \hat{x}_0) - x(t;\, 0,\, x_0)| < \delta_{12}, \quad t \in (s_1^{\max},\, t_2^{\min}],$$

where δ_{12} denotes an arbitrary positive number.

For $(t_2^{\max},\, s_2^{\min}]$, similar to (2.38), we obtain

$$|\hat{x}(t;0,\hat{x}_0) - x(t;0,x_0)| \leqslant L_{g_2}|\hat{x}((\hat{t}_2)^-) - x(t_2^-)|. \tag{2.42}$$

Moreover

$$\hat{x}((\hat{t}_2)^-) = \hat{x}(s_1^{\max}) + (\mathbf{I}^{\alpha}_{s_1^{\max},\, t_2^{\max}} f)(t_2^{\max}, \hat{x}),$$
$$x(t_2^-) = x(s_1^{\min}) + (\mathbf{I}^{\alpha}_{s_1^{\min},\, t_2^{\min}} f)(t_2^{\min}, x),$$

and then

$$\begin{aligned}
&|\hat{x}((\hat{t}_2)^-) - x(t_2^-)| \\
&\leqslant |\hat{x}(s_1^{\max}) - x(s_1^{\min})| + |(\mathbf{I}^{\alpha}_{t_2^{\min},\, t_2^{\max}} f)(t_2^{\max}, \hat{x})| \\
&+ \frac{1}{\Gamma(\alpha)}\int_{s_1^{\max}}^{t_2^{\min}} |(t_2^{\max} - s)^{\alpha-1} - (t_2^{\min} - s)^{\alpha-1}||f(s, x(s))|ds \\
&+ \frac{1}{\Gamma(\alpha)}\int_{s_1^{\min}}^{s_1^{\max}} (t_2^{\min} - s)^{\alpha-1}|f(s, x(s))|ds \\
&+ \frac{L_f}{\Gamma(\alpha)}\int_{s_1^{\max}}^{t_2^{\min}} (t_2^{\max} - s)^{\alpha-1}|\hat{x}(s) - x(s)|ds \\
&\leqslant |\hat{x}(s_1^{\max}) - x(s_1^{\max})| + |x(s_1^{\max}) - x(s_1^{\min})| + \frac{4M}{\Gamma(\alpha+1)}\delta^{\alpha} \\
&+ \frac{L_f}{\Gamma(\alpha)}\int_{s_1^{\max}}^{t_2^{\min}} (t_2^{\max} - s)^{\alpha-1}|\hat{x}(s) - x(s)|ds \\
&\leqslant \delta_{s_1} + \eta_{\alpha} + \frac{4M}{\Gamma(\alpha+1)}\delta^{\alpha} + \frac{L_f(t_2^{\min} - s_1^{\max})^{\alpha}}{\Gamma(\alpha+1)}E_{\alpha}(L_f(t_2^{\min} - s_1^{\max})^{\alpha})\delta_{s_1}.
\end{aligned} \tag{2.43}$$

If δ and δ_{s_1} are sufficiently small, then

$$|\hat{x}((\hat{t}_2)^-) - x(t_2^-)| \leqslant \delta_{t_2}.$$

From (2.42),

$$rcl|\hat{x}(t;0,\hat{x}_0) - x(t;0,x_0)| \leqslant L_{g_2}\delta_{t_2},$$

and let δ_{22} be an arbitrary positive number, if δ_{t_2} is adequately small, then

$$|\hat{x}(t;0,\hat{x}_0) - x(t;0,x_0)| < \delta_{22}, \quad t \in (t_2^{\max}, s_2^{\min}].$$

For $(s_2^{\max}, t_3^{\min}]$, according to (2.40), we come to the conclusion

$$|\hat{x}(t;0,\hat{x}_0) - x(t;0,x_0)| \leqslant |\hat{x}(s_2^{\max}) - x(s_2^{\max})| + L_f(\mathbf{I}^{\alpha}_{s_2^{\max},\, t}|\hat{x} - x|)(t).$$

For $(s_2^{\min}, t_3^{\min}]$, we have

$$x(t; 0, x_0) = g_2(s_2^{\min}, x(t_2^-)) + (\mathbf{I}^{\alpha}_{s_2^{\min}, t} f)(t, x).$$

Let $t = s_2^{\max}$, so we obtain

$$x(s_2^{\max}; 0, x_0) = g_2(s_2^{\min}, x(t_2^-)) + (\mathbf{I}^{\alpha}_{s_2^{\min}, s_2^{\max}} f)(s_2^{\max}, x).$$

In addition

$$\hat{x}(s_2^{\max}; 0, \hat{x}_0) = g_2(s_2^{\max}, \hat{x}((\hat{t}_2)^-)),$$

consequently,

$$\begin{aligned}
&|\hat{x}(s_2^{\max}) - x(s_2^{\max})| \\
&\leqslant |g_2(s_2^{\max}, \hat{x}((\hat{t}_2)^-)) - g_2(s_2^{\min}, x(t_2^-))| + |(\mathbf{I}^{\alpha}_{s_2^{\min}, s_2^{\max}} f)(s_2^{\max}, x)| \\
&\leqslant L_{g_2}|\hat{x}((\hat{t}_2)^-) - x(t_2^-)| + |g_2'(\zeta)||s_2^{\max} - s_2^{\min}| + \frac{M}{\Gamma(\alpha+1)}\delta^{\alpha} \\
&\leqslant L_{g_2}\delta_{t_2} + |g_2'(\zeta)|\delta + \frac{M}{\Gamma(\alpha+1)}\delta^{\alpha} \\
&\leqslant |g_2'(\zeta)|\delta + \frac{M}{\Gamma(\alpha+1)}\delta^{\alpha} + \delta_{22},
\end{aligned}$$

where $\zeta \in (s_2^{\min}, s_2^{\max})$, so if δ and δ_{22} are sufficiently small, then

$$|\hat{x}(s_2^{\max}) - x(s_2^{\max})| < \delta_{s_2},$$

where δ_{s_2} is an arbitrary positive number.

Hence,

$$|\hat{x}(t; 0, \hat{x}_0) - x(t; 0, x_0)| \leqslant \delta_{s_2} + L_f(\mathbf{I}^{\alpha}_{s_2^{\max}, t}|\hat{x} - x|)(t).$$

From [18, corollary 2],

$$|\hat{x}(t; 0, \hat{x}_0) - x(t; 0, x_0)| \leqslant \delta_{s_2} E_{\alpha}(L_f(t_3^{\min} - s_2^{\max})^{\alpha}) \leqslant \delta_{s_2} E_{\alpha}(L_f T^{\alpha}),$$

if δ_{s_2} is sufficiently small, then

$$|\hat{x}(t; 0, \hat{x}_0) - x(t; 0, x_0)| < \delta_{23}, \quad t \in (s_2^{\max}, t_3^{\min}],$$

where δ_{23} denotes an arbitrary positive number.

Similar to case 1 in theorem 2.1.5, we have

$$\begin{aligned}
&\forall\, \varepsilon > 0,\ \exists\, \delta > 0,\ \text{ if } |\hat{x}_0 - x_0| \leqslant \delta,\ |\hat{t}_i - t_i| \leqslant \delta \text{ and} \\
&|\hat{s}_i - s_i| \leqslant \delta,\ i = 1, \dots, k,\ \text{ then } |\hat{x}(t; 0, \hat{x}_0) - x(t; 0, x_0)| \leqslant \varepsilon, \\
&\quad t \in [0, t_1^{\min}] \bigcup (t_i^{\max}, s_i^{\min}] \bigcup (s_i^{\max}, t_{i+1}^{\min}], \\
&i = 1, \dots, k.
\end{aligned}$$

Case 2. Let $t_i^{\min} = t_i$, $t_i^{\max} = \hat{t}_i$, $s_i^{\min} = \hat{s}_i$, $s_i^{\max} = s_i$, $i = 1, \ldots$; see case (c) in figure 2.3 (the case $t_i^{\min} = \hat{t}_i$, $t_i^{\max} = t_i$, $s_i^{\min} = s_i$, $s_i^{\max} = \hat{s}_i$, $i = 1, \ldots$can be considered similarly; see case (d) in figure 2.3).

In the following, we obtain an estimate for $|\hat{x}(t; 0, \hat{x}_0) - x(t; 0, x_0)|$ in each subinterval.

For the intervals $[0, t_1^{\min}]$ and $(t_1^{\max}, s_1^{\min}]$, we have the same result as in (2.37) and (2.39).

For $(s_1^{\max}, t_2^{\min}]$, we also have inequality (2.40).

In the interval $(s_1^{\min}, t_2^{\max}]$, we have

$$\hat{x}(t; 0, x_0) = g_1(s_1^{\min}, \hat{x}((\hat{t}_1)^-)) + (\mathbf{I}^{\alpha}_{s_1^{\min}, t} f)(t, \hat{x}),$$

then

$$\hat{x}(s_1^{\max}; 0, x_0) = g_1(s_1^{\min}, \hat{x}((\hat{t}_1)^-)) + (\mathbf{I}^{\alpha}_{s_1^{\min}, s_1^{\max}} f)(s_1^{\max}, \hat{x}).$$

Also

$$x(s_1^{\max}; 0, x_0) = g_1(s_1^{\max}, x((t_1)^-)),$$

and hence

$$\begin{aligned}
&|\hat{x}(s_1^{\max}) - x(s_1^{\max})| \\
&\leqslant |g_1(s_1^{\min}, \hat{x}((\hat{t}_1)^-)) - g_1(s_1^{\max}, x(t_1^-))| + (\mathbf{I}^{\alpha}_{s_1^{\min}, s_1^{\max}} f)(s_1^{\max}, \hat{x}) \\
&\leqslant L_{g_1}|\hat{x}((\hat{t}_1)^-) - x(t_1^-)| + |g_1'(\zeta)||s_1^{\min} - s_1^{\max}| + \frac{M}{\Gamma(\alpha + 1)}\delta^{\alpha} \\
&\leqslant L_{g_1}\delta_{t_1} + |g_1'(\zeta)|\delta + \frac{M}{\Gamma(\alpha + 1)}\delta^{\alpha} \\
&\leqslant |g_1'(\zeta)|\delta + \frac{M}{\Gamma(\alpha + 1)}\delta^{\alpha} + \delta_{11},
\end{aligned}$$

where $\zeta \in (s_1^{\min}, s_1^{\max})$, and δ_{t_1} and δ_{11} come from the procedure in case 1. Let δ_{s_1} be an arbitrary positive number, if δ, δ_{11} are small enough, then

$$|\hat{x}(s_1^{\max}) - x(s_1^{\max})| < \delta_{s_1}.$$

From (2.40) and applying [18, corollary 2],

$$|\hat{x}(t; 0, \hat{x}_0) - x(t; 0, x_0)| \leqslant \delta_{s_1} E_{\alpha}(L_f(t_2^{\min} - s_1^{\max})^{\alpha}) \leqslant \delta_{s_1} E_{\alpha}(L_f T^{\alpha}),$$

if δ_{s_1} is sufficiently small, then

$$|\hat{x}(t; 0, \hat{x}_0) - x(t; 0, x_0)| < \delta_{12}, \quad t \in (s_1^{\max}, t_2^{\min}], \tag{2.44}$$

where δ_{12} is an arbitrary positive number.

For $(t_2^{\max}, s_2^{\min}]$, we have the same inequality as in (2.42).

Note that

$$\hat{x}((\hat{t}_2)^-) = \hat{x}(s_1^{\min}) + (\mathbf{I}^{\alpha}_{s_1^{\min},\, t_2^{\max}} f)(t_2^{\max}, \hat{x}),$$
$$x(t_2^-) = 7x(s_1^{\max}) + (\mathbf{I}^{\alpha}_{s_1^{\max},\, t_2^{\min}} f)(t_2^{\min}, x),$$

and then

$$\begin{aligned}
&|\hat{x}((\hat{t}_2)^-) - x(t_2^-)| \\
&\leqslant |\hat{x}(s_1^{\min}) - x(s_1^{\max})| + |(\mathbf{I}^{\alpha}_{t_2^{\min},\, t_2^{\max}} f)(t_2^{\max}, \hat{x})| \\
&+ \frac{1}{\Gamma(\alpha)} \int_{s_1^{\max}}^{t_2^{\min}} |(t_2^{\max} - s)^{\alpha-1} - (t_2^{\min} - s)^{\alpha-1})||f(s, x(s))|ds \\
&+ \frac{1}{\Gamma(\alpha)} \int_{s_1^{\min}}^{s_1^{\max}} (t_2^{\max} - s)^{\alpha-1}|f(s, \hat{x}(s)) \\
&|ds + \frac{L_f}{\Gamma(\alpha)} \int_{s_1^{\max}}^{t_2^{\min}} (t_2^{\max} - s)^{\alpha-1}|\hat{x}(s) - x(s)|ds \\
&\leqslant |\hat{x}(s_1^{\min}) - x(s_1^{\min})| + |x(s_1^{\min}) - x(s_1^{\max})| + \frac{4M}{\Gamma(\alpha+1)}\delta^{\alpha} \\
&+ \frac{L_f}{\Gamma(\alpha)} \int_{s_1^{\max}}^{t_2^{\min}} (t_2^{\max} - s)^{\alpha-1}|\hat{x}(s) - x(s)|ds \\
&\leqslant \delta_{s_1} + \eta_{\alpha} + \frac{4M}{\Gamma(\alpha+1)}\delta^{\alpha} + \frac{L_f}{\Gamma(\alpha+1)}(t_2^{\min} - s_1^{\max})^{\alpha} E_{\alpha}(L_f(t_2^{\min} - s_1^{\max})^{\alpha})\delta_{s_1}.
\end{aligned}$$

If δ and δ_{s_1} are sufficiently small, then

$$|\hat{x}((\hat{t}_2)^-) - x(t_2^-)| \leqslant \delta_{t_2}.$$

From (2.42), derive that

$$|\hat{x}(t; 0, \hat{x}_0) - x(t; 0, x_0)| \leqslant L_{g_2}\delta_{t_2},$$

if δ_{t_2} is fully small, then

$$|\hat{x}(t; 0, \hat{x}_0) - x(t; 0, x_0)| < \delta_{22}, \quad t \in (t_2^{\max}, s_2^{\min}], \tag{2.45}$$

where δ_{22} is an arbitrary positive number.

For $(s_2^{\max}, t_3^{\min}]$, we obtain the formula

$$|\hat{x}(t; 0, \hat{x}_0) - x(t; 0, x_0)| \leqslant |\hat{x}(s_2^{\max}) - x(s_2^{\max})| + L_f(\mathbf{I}^{\alpha}_{s_2^{\max},\, t}|\hat{x} - x|)(t).$$

For $(s_2^{\min}, t_3^{\max}]$, we have

$$\hat{x}(t; 0, \hat{x}_0) = g_2(s_2^{\min}, \hat{x}((\hat{t}_2)^-)) + (\mathbf{I}^{\alpha}_{s_2^{\min},\, t} f)(t, \hat{x}).$$

Let $t = s_2^{\max}$, and we obtain

$$\hat{x}(s_2^{\max}; 0, \hat{x}_0) = g_2(s_2^{\min}, \hat{x}((\hat{t}_2)^-)) + (\mathbf{I}^{\alpha}_{s_2^{\min},\, s_2^{\max}} f)(s_2^{\max}, \hat{x}).$$

Moreover

$$x(s_2^{\max}; 0, x_0) = g_2(s_2^{\max}, x(t_2^-)),$$

and hence

$$\begin{aligned}
&|\hat{x}(s_2^{\max}) - x(s_2^{\max})| \\
&\leqslant |g_2(s_2^{\min}, \hat{x}((\hat{t}_2)^-)) - g_2(s_2^{\max}, x(t_2^-))| + |(\mathbf{I}^{\alpha}_{s_2^{\min}, s_2^{\max}} f)(s_2^{\max}, \hat{x})| \\
&\leqslant L_{g_2}|\hat{x}((\hat{t}_2)^-) - x(t_2^-)| + |g_2'(\zeta)||s_2^{\min} - s_2^{\max}| + \frac{M}{\Gamma(\alpha+1)}\delta^{\alpha} \\
&\leqslant L_{g_2}\delta_{t_2} + |g_2'(\zeta)|\delta + \frac{M}{\Gamma(\alpha+1)}\delta^{\alpha} \\
&\leqslant |g_2'(\zeta)|\delta + \frac{M}{\Gamma(\alpha+1)}\delta^{\alpha} + \delta_{22},
\end{aligned}$$

where $\zeta \in (s_2^{\min}, s_2^{\max})$, if δ and δ_{22} are sufficiently small, then

$$|\hat{x}(s_2^{\max}) - x(s_2^{\max})| < \delta_{s_2},$$

where δ_{s_2} denotes an arbitrary positive number.

Consequently, using [18, corollary 2],

$$|\hat{x}(t; 0, \hat{x}_0) - x(t; 0, x_0)| \leqslant \delta_{s_2} E_{\alpha}(L_f(t_3^{\min} - s_2^{\max})^{\alpha}) \leqslant \delta_{s_2} E_{\alpha}(L_f T^{\alpha}),$$

let $\delta_{23} > 0$ be an arbitrary number, so if δ_{s_2} is sufficiently small, then

$$|\hat{x}(t; 0, \hat{x}_0) - x(t; 0, x_0)| < \delta_{23}, \quad t \in (s_2^{\max}, t_3^{\min}].$$

Therefore, we draw the conclusion

$$\begin{aligned}
&\forall\ \varepsilon > 0,\ \ \exists\ \delta > 0,\ \text{ if } |\hat{x}_0 - x_0| \leqslant \delta,\ \ |\hat{t}_i - t_i| \leqslant \delta \text{ and} \\
&|\hat{s}_i - s_i| \leqslant \delta,\ \ i = 1, \ldots, k,\ \text{ then } |\hat{x}(t; 0, \hat{x}_0) - x(t; 0, x_0)| \leqslant \varepsilon, \\
&t \in [0, t_1^{\min}] \bigcup (t_i^{\max}, s_i^{\min}] \bigcup (s_i^{\max}, t_{i+1}^{\min}], \\
&i = 1, \ldots, k.
\end{aligned}$$

Case 3. Let $t_i^{\min} = t_i$, $s_i^{\min} = s_i$, $t_i^{\max} = \hat{t}_i$, $s_i^{\max} = \hat{s}_i$, $i = 1, \ldots$; see case (e) in figure 2.4 (the case $t_i^{\min} = \hat{t}_i$, $s_i^{\min} = \hat{s}_i$, $t_i^{\max} = t_i$, $s_i^{\max} = s_i$, $i = 1, \ldots$ can be considered similarly; see case (f) in figure 2.4). In this case, we have (2.29).

Now we estimate the $|\hat{x}(t; 0, \hat{x}_0) - x(t; 0, x_0)|$ in each subinterval.

For the interval $[0, t_1^{\min}]$, we have the same result as in (2.37).

For $(s_1^{\min}, t_2^{\min}]$, we have

$$x(t; 0, x_0) = g_1(s_1^{\min}, x(t_1^-)) + (\mathbf{I}^{\alpha}_{s_1^{\min}, t} f)(t, x).$$

If $t = s_1^{\max}$, then

$$x(s_1^{\max}; 0, x_0) = g_1(s_1^{\min}, x(t_1^-)) + (\mathbf{I}^{\alpha}_{s_1^{\min}, s_1^{\max}} f)(s_1^{\max}, x).$$

Also

$$\hat{x}(s_1^{\max}; 0, \hat{x}_0) = g_1(s_1^{\max}, \hat{x}((\hat{t}_1)^-)),$$

so

$$\begin{aligned}
&|\hat{x}(s_1^{\max}) - x(s_1^{\max})| \\
&\leqslant |g_1(s_1^{\max}, \hat{x}((\hat{t}_1)^-)) - g_1(s_1^{\min}, x(t_1^-))| + |(\mathbf{I}^{\alpha}_{s_1^{\min}, s_1^{\max}} f)(s_1^{\max}, x)| \\
&\leqslant L_{g_1}|\hat{x}(\hat{t}_1)^-) - x(t_1^-)| + |g_1'(\zeta)||s_1^{\max} - s_1^{\min}| + \frac{M}{\Gamma(\alpha+1)}\delta^{\alpha} \qquad (2.46)\\
&\leqslant L_{g_1}|\hat{x}(\hat{t}_1)^-) - x(t_1^-)| + |g_1'(\zeta)|\delta + \frac{M}{\Gamma(\alpha+1)}\delta^{\alpha},
\end{aligned}$$

where $\zeta \in (s_1^{\min}, s_1^{\max})$.

Note

$$\hat{x}((\hat{t}_1)^-) = \hat{x}_0 + (\mathbf{I}^{\alpha}_{0, t_1^{\max}} f)(t_1^{\max}, \hat{x}),$$

$$x(t_1^-) = x_0 + (\mathbf{I}^{\alpha}_{0, t_1^{\min}} f)(t_1^{\min}, x).$$

Therefore

$$\begin{aligned}
&|\hat{x}((\hat{t}_1)^-) - x(t_1^-)| \\
&\leqslant |\hat{x}_0 - x_0| + \frac{1}{\Gamma(\alpha)}\int_0^{t_1^{\min}} |(t_1^{\max} - s)^{\alpha-1} - (t_1^{\min} - s)^{\alpha-1}||f(s, x(s))|ds \\
&\quad + \frac{L_f}{\Gamma(\alpha)}\int_0^{t_1^{\min}} (t_1^{\max} - s)^{\alpha-1}|\hat{x}(s) - x(s)|ds + |(\mathbf{I}^{\alpha}_{t_1^{\min}, t_1^{\max}} f)(t_1^{\max}, \hat{x})| \\
&\leqslant \delta + \frac{3M}{\Gamma(\alpha+1)}\delta^{\alpha} + \frac{L_f}{\Gamma(\alpha)}\int_0^{t_1^{\min}} (t_1^{\max} - s)^{\alpha-1}|\hat{x}(s) - x(s)|ds \\
&\leqslant \delta + \frac{3M}{\Gamma(\alpha+1)}\delta^{\alpha} + \frac{L_f t_1^{\alpha}}{\Gamma(\alpha+1)}E_{\alpha}(L_f t_1^{\alpha})\delta.
\end{aligned}$$

From (2.46), we have

$$\begin{aligned}
|\hat{x}(s_1^{\max}) - x(s_1^{\max})| \leqslant &\left(L_{g_1} + |g_1'(\xi)| + \frac{L_{g_1} L_f t_1^{\alpha}}{\Gamma(\alpha+1)}E_{\alpha}(L_f t_1^{\alpha})\right)\delta \\
&+ \frac{M}{\Gamma(\alpha+1)}(1 + 3L_{g_1})\delta^{\alpha}.
\end{aligned}$$

Let δ_{s_1} be an arbitrary positive number, so if δ is sufficiently small, then

$$|\hat{x}(s_1^{\max}) - x(s_1^{\max})| < \delta_{s_1}.$$

For $(s_1^{\max}, t_2^{\min}]$, consider as in case 1 and apply [18, corollary 2],

$$\begin{aligned}|\hat{x}(t; 0, \hat{x}_0) - x(t; 0, x_0)| &\leqslant |\hat{x}(s_1^{\max}) - x(s_1^{\max})| + L_f(\mathbf{I}^{\alpha}_{s_1^{\max}, t}|\hat{x} - x|)(t)\\ &\leqslant \delta_{s_1} + \frac{L_f}{\Gamma(\alpha)}\int_{s_1^{\max}}^{t}(t - s)^{\alpha-1}|\hat{x}(s) - x(s)|ds\\ &\leqslant \delta_{s_1}E_\alpha(L_f(t_2^{\min} - s_1^{\max})^\alpha) \leqslant \delta_{s_1}E_\alpha(L_f T^\alpha).\end{aligned}$$

If δ_{s_1} is adequately small, then

$$|\hat{x}(t; 0, \hat{x}_0) - x(t; 0, x_0)| < \delta_{12}, \quad t \in (s_1^{\max}, t_2^{\min}], \tag{2.47}$$

where δ_{12} is an arbitrary positive number.

Similar to (2.43), the inequality

$$\begin{aligned}&|\hat{x}((\hat{t}_2)^-) - x(t_2^-)|\\ &\leqslant |\hat{x}(s_1^{\max}) - x(s_1^{\min})| + |(\mathbf{I}^{\alpha}_{t_2^{\min}, t_2^{\max}}f)(t_2^{\max}, \hat{x})|\\ &\quad + \frac{1}{\Gamma(\alpha)}\int_{s_1^{\max}}^{t_2^{\min}}|(t_2^{\max} - s)^{\alpha-1} - (t_2^{\min} - s)^{\alpha-1}||f(s, x(s))|ds\\ &\quad + \frac{1}{\Gamma(\alpha)}\int_{s_1^{\min}}^{s_1^{\max}}(t_2^{\min} - s)^{\alpha-1}|f(s, x(s))|ds\\ &\quad + \frac{L_f}{\Gamma(\alpha)}\int_{s_1^{\max}}^{t_2^{\min}}(t_2^{\max} - s)^{\alpha-1}|\hat{x}(s) - x(s)|ds\\ &\leqslant |\hat{x}(s_1^{\max}) - x(s_1^{\max})| + |x(s_1^{\max}) - x(s_1^{\min})| + \frac{4M}{\Gamma(\alpha + 1)}\delta^\alpha\\ &\quad + \frac{L_f}{\Gamma(\alpha)}\int_{s_1^{\max}}^{t_2^{\min}}(t_2^{\max} - s)^{\alpha-1}|\hat{x}(s) - x(s)|ds\\ &\leqslant \delta_{s_1} + \eta_\alpha + \frac{4M}{\Gamma(\alpha + 1)}\delta^\alpha + \frac{L_f}{\Gamma(\alpha + 1)}(t_2^{\min} - s_1^{\max})^\alpha E_\alpha(L_f(t_2^{\min} - s_1^{\max})^\alpha)\delta_{s_1}.\end{aligned}$$

For $(s_2^{\min}, t_3^{\min}]$, we have

$$x(t; 0, x_0) = g_1(s_2^{\min}, x(t_2^-)) + (\mathbf{I}^{\alpha}_{s_2^{\min}, t}f)(t, x),$$

with $t = s_2^{\max}$, we obtain

$$x(s_2^{\max}; 0, x_0) = g_2(s_2^{\min}, x(t_2^-)) + (\mathbf{I}^{\alpha}_{s_2^{\min}, s_2^{\max}}f)(s_2^{\max}, x).$$

Also

$$\hat{x}(s_2^{\max}; 0, \hat{x}_0) = g_2(s_2^{\max}, \hat{x}((\hat{t}_2)^-)),$$

and hence

$$
\begin{aligned}
&|\hat{x}(s_2^{\max}) - x(s_2^{\max})| \\
&\leqslant |g_2(s_2^{\max}, \hat{x}((\hat{t}_2)^-)) - g_2(s_2^{\min}, x(t_2^-))| + |(\mathbf{I}^{\alpha}_{s_2^{\min}, s_2^{\max}} f)(s_2^{\max}, x)| \\
&\leqslant L_{g_2}|\hat{x}((\hat{t}_2)^-) - x(t_2^-)| + |g_2'(\zeta)||s_2^{\max} - s_2^{\min}| + \frac{M}{\Gamma(\alpha+1)}\delta^{\alpha} \\
&\leqslant L_{g_2}\left(\delta_{s_1} + \eta_{\alpha} + \frac{4M\delta^{\alpha}}{\Gamma(\alpha+1)} + \frac{L_f(t_2^{\min} - s_1^{\max})^{\alpha}}{\Gamma(\alpha+1)}E_{\alpha}(L_f(t_2^{\min} - s_1^{\max})^{\alpha})\delta_{s_1}\right) \\
&\quad + |g_2'(\zeta)|\delta + \frac{M\delta^{\alpha}}{\Gamma(\alpha+1)} \\
&\leqslant |g_2'(\zeta)|\delta + \frac{M}{\Gamma(\alpha+1)}(1 + 4L_{g_2})\delta^{\alpha} + L_{g_2}\left(1 + \frac{L_f t_2^{\alpha}}{\Gamma(\alpha+1)}E_{\alpha}(L_f t_2^{\alpha})\right)\delta_{s_1} + L_{g_2}\eta_{\alpha},
\end{aligned}
$$

where $\zeta \in (s_2^{\min}, s_2^{\max})$, if δ and δ_{s_1} are sufficiently small, then

$$|\hat{x}(s_2^{\max}) - x(s_2^{\max})| < \delta_{s_2},$$

where δ_{s_2} is an arbitrary positive number.

For $(s_2^{\max}, t_3^{\min}]$,

$$
\begin{aligned}
|\hat{x}(t; 0, \hat{x}_0) - x(t; 0, x_0)| &\leqslant |\hat{x}(s_2^{\max}) - x(s_2^{\max})| + L_f(\mathbf{I}^{\alpha}_{s_2^{\max}, t}|\hat{x} - x|)(t) \\
&\leqslant \delta_{s_2} + L_f(\mathbf{I}^{\alpha}_{s_2^{\max}, t}|\hat{x} - x|)(t).
\end{aligned}
$$

Using [18, corollary 2], the formula is

$$|\hat{x}(t; 0, \hat{x}_0) - x(t; 0, x_0)| \leqslant \delta_{s_2}E_{\alpha}(L_f(t_3^{\min} - s_2^{\max})^{\alpha}) \leqslant \delta_{s_2}E_{\alpha}(L_f T^{\alpha}).$$

Let δ_{23} be an arbitrary positive number, if δ_{s_2} is sufficiently small, then

$$|\hat{x}(t; 0, \hat{x}_0) - x(t; 0, x_0)| < \delta_{23}, \quad t \in (s_2^{\max}, t_3^{\min}]. \tag{2.48}$$

Repeating the procedure as in (2.47) and (2.48), we arrive at

$$
\begin{aligned}
&\forall\, \delta_{i,\, i+1} > 0, \ \exists\, \delta > 0, \ \exists\, \delta_{s_1} > 0, \ \exists\, \delta_{s_2} > 0, \ \ldots, \ \exists\, \delta_{s_i} > 0, \\
&\text{then} \\
&|\hat{x}(t; 0, \hat{x}_0) - x(t; 0, x_0)| \leqslant \delta_{i,\, i+1}, \quad t \in (s_i^{\max}, t_{i+1}^{\min}], \\
&i = 1, 2, \ldots, k,
\end{aligned} \tag{2.49}
$$

where δ_{s_i} originates from $|\hat{x}(s_i^{\max}) - x(s_i^{\max})| < \delta_{s_i}$, $i = 1, 2, \ldots, k$.

Let $\varepsilon > 0$ be an arbitrary number and put $\delta_{k,\, k+1} = \varepsilon$, we can deduce the consequence

$$|\hat{x}(t; 0, \hat{x}_0) - x(t; 0, x_0)| \leqslant \varepsilon, \quad t \in (s_k^{\max}, t_{k+1}^{\min}].$$

If $i = 1$, from (2.49), $\exists\, \delta > 0$, $\delta_{s_1} > 0$ such that

$$|\hat{x}(t; 0, \hat{x}_0) - x(t; 0, x_0)| \leqslant \varepsilon, \quad t \in (s_1^{\max}, t_2^{\min}].$$

Also $\exists\, \delta > 0$ such that

$$|\hat{x}(t; 0, \hat{x}_0) - x(t; 0, x_0)| \leqslant \varepsilon, \quad t \in [0, t_1^{\min}].$$

Consequently, we have

$$\begin{aligned}
&\forall\, \varepsilon > 0, \quad \exists\, \delta > 0, \quad \text{if } |\hat{x}_0 - x_0| \leqslant \delta, \quad |\hat{t}_i - t_i| \leqslant \delta \text{ and} \\
&|\hat{s}_i - s_i| \leqslant \delta, \quad i = 1, \ldots, k, \quad \text{then } |\hat{x}(t; 0, \hat{x}_0) - x(t; 0, x_0)| \leqslant \varepsilon, \\
&t \in [0, t_1^{\min}] \bigcup (s_i^{\max}, t_{i+1}^{\min}], \quad i = 1, \ldots, k.
\end{aligned}$$

The proof is complete. □

2.1.4 The stability of solutions

Now we use definition 2.1.2 to present stability results for the problem (2.1), (2.2) under mild conditions. In the proof of the next two theorems we only consider the case $t_i^{\min} = t_i$, $t_i^{\max} = \hat{t}_i$, $s_i^{\min} = s_i$, $s_i^{\max} = \hat{s}_i$, $i = 1, \ldots$ (see case (a) in figure 2.2). The other possible cases can be considered similarly (see case (b) in figure 2.2, cases (c) and (d) in figure 2.3, and cases (e) and (f) in figure 2.4) with minor adjustments (so as a result we omit the details for these cases).

Theorem 2.1.7. *Assume [H_1]–[H_4] are satisfied and $\prod_{i=1}^{\infty} L_{g_i}$ is finite. If the system (2.9) is gravitating with a constant κ, then the solution of IDE (2.1) is stable about the initial point, the impulsive points t_i, $i = 1, 2, \ldots$ and the junction points s_i, $i = 1, 2, \ldots$.*

Proof. Suppose that there exist positive constants $\delta_0, \delta_1, \delta_2, \ldots,$ such that $|\hat{x}_0 - x_0| \leqslant \delta_0, |\hat{t}_i - t_i| \leqslant \delta_i, |\hat{s}_i - s_i| \leqslant \delta_i, i = 1, 2, \ldots$. From the procedure in theorem 2.1.5, we have

$$\begin{aligned}
&[0, \infty)/\left[\left(\bigcup_{i=1,2,\ldots} \langle \hat{t}_i, t_i \rangle\right) \bigcup \left(\bigcup_{i=1,2,\ldots} \langle \hat{s}_i, s_i \rangle\right)\right] \\
&= [0, \infty)/\left[\left(\bigcup_{i=1,2,\ldots} (t_i^{\min}, t_i^{\max}]\right) \bigcup \left(\bigcup_{i=1,2,\ldots} (s_i^{\min}, s_i^{\max}]\right)\right] \\
&= [0, t_1^{\min}] \bigcup \left(\bigcup_{i=1,2,\ldots} (t_i^{\max}, s_i^{\min}]\right) \bigcup \left(\bigcup_{i=1,2,\ldots} (s_i^{\max}, t_{i+1}^{\min}]\right).
\end{aligned}$$

For $[0, t_1^{\min}]$, from the procedure in [14, theorem 1.2, p 6] and the system (2.9) is gravitating with a constant κ, we obtain

$$|\hat{x}(t; 0, \hat{x}_0) - x(t; 0, x_0)| \leqslant \kappa\delta_0.$$

Let ε be an arbitrary positive number. If δ_0 is fully small, then

$$|\hat{x}(t; 0, \hat{x}_0) - x(t; 0, x_0)| \leqslant \varepsilon, \quad t \in [0, t_1^{\min}]. \tag{2.50}$$

For $(t_1^{\max}, s_1^{\min}]$, from an inequality in (2.15),

$$\begin{aligned}
&|\hat{x}((\hat{t}_1)^-) - x(t_1^-)| \\
&\leqslant |\hat{x}_0 - x_0| + |(\mathbf{I}^1_{0,\, t_1^{\min}} f)(t_1^{\min}, \hat{x}) - (\mathbf{I}^1_{0,\, t_1^{\min}} f)(t_1^{\min}, x)| \\
&\quad + |(\mathbf{I}^1_{t_1^{\min},\, t_1^{\max}} f)(t_1^{\max}, \hat{x})| \\
&\leqslant (1 + \kappa L_f t_1)\delta_0 + M\delta_1.
\end{aligned}$$

From (2.14), obtain the formula

$$|\hat{x}(t; 0, \hat{x}_0) - x(t; 0, x_0)| \leqslant L_{g_1}(1 + \kappa L_f t_1)\delta_0 + ML_{g_1}\delta_1.$$

If δ_0, δ_1 are adequately small, then

$$|\hat{x}(t; 0, \hat{x}_0) - x(t; 0, x_0)| \leqslant \varepsilon, \quad t \in (t_1^{\max}, s_1^{\min}], \tag{2.51}$$

where ε is an arbitrary positive number.

For $(s_1^{\max}, t_2^{\min}]$, from (2.18) and (2.19), we obtain

$$\begin{aligned}
|\hat{x}(s_1^{\max}) - x(s_1^{\max})| &\leqslant |g_1(s_1^{\max}, \hat{x}((\hat{t}_1)^-)) - g_1(s_1^{\min}, x(t_1^-))| \\
&\quad + |(\mathbf{I}^1_{s_1^{\min},\, s_1^{\max}} f)(s_1^{\max}, x)| \\
&\leqslant L_{g_1}|\hat{x}((\hat{t}_1)^-) - x(t_1^-)| + |g_1'(\xi)||s_1^{\max} - s_1^{\min}| + M\delta_1 \\
&\leqslant L_{g_1}(\delta_0 + \kappa L_f t_1\delta_0 + M\delta_1) + |g_1'(\xi)|\delta_1 + M\delta_1 \\
&\leqslant L_{g_1}(1 + \kappa L_f t_1)\delta_0 + (|g_1'(\xi)| + ML_{g_1} + M)\delta_1.
\end{aligned}$$

Since the system (2.9) is gravitating, then

$$|\hat{x}(t; 0, \hat{x}_0) - x(t; 0, x_0)| \leqslant \kappa L_{g_1}(1 + \kappa L_f t_1)\delta_0 + \kappa(|g_1'(\xi)| + ML_{g_1} + M)\delta_1.$$

Let ε be an arbitrary positive number. If δ_0, δ_1 are fully small,

$$|\hat{x}(t; 0, \hat{x}_0) - x(t; 0, x_0)| \leqslant \varepsilon, \quad t \in (s_1^{\max}, t_2^{\min}]. \tag{2.52}$$

For $(t_2^{\max}, s_2^{\min}]$,

$$|\hat{x}(t; 0, \hat{x}_0) - x(t; 0, x_0)| \leqslant L_{g_2}|\hat{x}((\hat{t}_2)^-) - x(t_2^-)|.$$

From inequality in (2.22), we obtain the formula

$$
\begin{aligned}
&|\hat{x}((\hat{t}_2)^-) - x(t_2^-)| \\
&\leqslant |\hat{x}(s_1^{\max}) - x(s_1^{\min})| + |(\mathbf{I}^1_{t_2^{\min}, t_2^{\max}} f)(t_2^{\max}, \hat{x})| + |(\mathbf{I}^1_{s_1^{\min}, s_1^{\max}} f)(s_1^{\max}, x)| \\
&\quad + L_f(\mathbf{I}^1_{s_1^{\max}, t_2^{\min}} |\hat{x} - x|)(t_2^{\min}) \\
&\leqslant L_{g_1}(1 + \kappa L_f t_1)\delta_0 + (|g_1'(\xi)| + ML_{g_1} + M)\delta_1 + \eta + M\delta_2 + M\delta_1 \\
&\quad + \big(\kappa L_{g_1}(1 + \kappa L_f t_1)\delta_0 + \kappa(|g_1'(\xi)| + ML_{g_1} + M)\delta_1 + \eta\big)\kappa L_f(t_2^{\min} - s_1^{\max}) \\
&\leqslant L_{g_1}(1 + \kappa L_f t_1)(1 + \kappa L_f t_2)\delta_0 + M\delta_2 + \kappa L_f t_2 \eta \\
&\quad + \big((|g_1'(\xi)| + ML_{g_1} + M)(1 + \kappa L_f t_2) + M\big)\delta_1.
\end{aligned}
$$

Then

$$
\begin{aligned}
&|\hat{x}(t; 0, \hat{x}_0) - x(t; 0, x_0)| \\
&\leqslant \kappa L_{g_2} L_f t_2 \eta + L_{g_2} L_{g_1}(1 + \kappa L_f t_1)(1 + \kappa L_f t_2)\delta_0 \\
&\quad + L_{g_2}\big((|g_1'(\xi)| + ML_{g_1} + M)(1 + \kappa L_f t_2) + M\big)\delta_1 + ML_{g_2}\delta_2.
\end{aligned}
$$

If δ, δ_0, δ_1 and δ_2 are sufficiently small, then

$$
|\hat{x}(t; 0, \hat{x}_0) - x(t; 0, x_0)| \leqslant \varepsilon, \quad t \in (t_2^{\max}, s_2^{\min}], \tag{2.53}
$$

where ε denotes an arbitrary positive number.

For $(s_2^{\max}, t_3^{\min}]$, from the procedure in case 1 of theorem 2.1.5, we obtain

$$
\begin{aligned}
&|\hat{x}(s_2^{\max}) - x(s_2^{\max})| \\
&\leqslant |g_2(s_2^{\max}, \hat{x}((\hat{t}_2)^-)) - g_2(s_2^{\min}, x(t_2^-))| + |(\mathbf{I}^1_{s_2^{\min}, s_2^{\max}} f)(s_2^{\max}, x)| \\
&\leqslant L_{g_2}|\hat{x}((\hat{t}_2)^-) - x(t_2^-)| + |g_2'(\xi)||s_2^{\max} - s_2^{\min}| + M\delta_2 \\
&\leqslant |g_2'(\xi)|\delta_2 + M\delta_2 + L_{g_2} L_{g_1}(1 + \kappa L_f t_1)(1 + \kappa L_f t_2)\delta_0 + \kappa L_{g_2} L_f t_2 \eta \\
&\quad + L_{g_2}\big((|g_1'(\xi)| + ML_{g_1} + M)(1 + \kappa L_f t_2) + M\big)\delta_1 + ML_{g_2}\delta_2 \\
&= L_{g_2} L_{g_1}(1 + \kappa L_f t_1)(1 + \kappa L_f t_2)\delta_0 + (|g_2'(\xi)| + ML_{g_2} + M)\delta_2 \\
&\quad + L_{g_2}\big((|g_1'(\xi)| + ML_{g_1} + M)(1 + \kappa L_f t_2) + M\big)\delta_1 + \kappa L_{g_2} L_f t_2 \eta.
\end{aligned}
$$

Since the system (2.9) is gravitating with a constant κ, then

$$
\begin{aligned}
&|\hat{x}(t; 0, \hat{x}_0) - x(t; 0, x_0)| \\
&\leqslant \kappa L_{g_2} L_{g_1}(1 + \kappa L_f t_1)(1 + \kappa L_f t_2)\delta_0 + \kappa(|g_2'(\xi)| + ML_{g_2} + M)\delta_2 \\
&\quad + \kappa L_{g_2}\big((|g_1'(\xi)| + ML_{g_1} + M)(1 + \kappa L_f t_2) + M\big)\delta_1 + \kappa^2 L_{g_2} L_f t_2 \eta.
\end{aligned}
$$

Set $\varepsilon > 0$ denotes an arbitrary number. If δ, δ_0, δ_1 and δ_2 are sufficiently small, then

$$
|\hat{x}(t; 0, \hat{x}_0) - x(t; 0, x_0)| \leqslant \varepsilon, \quad t \in (s_2^{\max}, t_3^{\min}]. \tag{2.54}
$$

From (2.50), (2.51), (2.52), (2.53) and (2.54), if δ, δ_i, $i = 1, 2, \ldots$ are sufficiently small, we can come to the conclusion

$$|\hat{x}(t; 0, \hat{x}_0) - x(t; 0, x_0)| \leqslant \varepsilon, \quad t \in [0, t_1^{\min}] \bigcup (t_i^{\max}, s_i^{\min}] \bigcup (s_i^{\max}, t_{i+1}^{\min}], \quad i = 1, 2, \ldots.$$

The proof is complete. □

Next consider the stability of the fractional order IDE (2.2).

Theorem 2.1.8. *Assume [H_1]–[H_4] are satisfied and $\prod_{i=1}^{\infty} L_{g_i}$ is finite. If system (2.10) is gravitating with a constant κ_α, then the solution of IDE (2.2) is stable about the initial point, the impulsive points t_i, $i = 1, 2, \ldots$ and the junction points s_i, $i = 1, 2, \ldots$.*

Proof. Parallel with theorem 2.1.7, we have

$$\begin{aligned}
&[0, \infty)/\left[\left(\bigcup_{i=1,2,\ldots} \langle \hat{t}_i, t_i \rangle\right) \bigcup \left(\bigcup_{i=1,2,\ldots} \langle \hat{s}_i, s_i \rangle\right)\right] \\
&= [0, t_1^{\min}] \bigcup \left(\bigcup_{i=1,2,\ldots} (t_i^{\max}, s_i^{\min}]\right) \bigcup \left(\bigcup_{i=1,2,\ldots} (s_i^{\max}, t_{i+1}^{\min}]\right).
\end{aligned}$$

For $[0, t_1^{\min}]$, the system (2.10) is gravitating, then

$$|\hat{x}(t; 0, \hat{x}_0) - x(t; 0, x_0)| \leqslant \kappa_\alpha \delta_0.$$

Let ε be an arbitrary positive number. If δ_0 is small enough, then

$$|\hat{x}(t; 0, \hat{x}_0) - x(t; 0, x_0)| \leqslant \varepsilon, \quad t \in [0, t_1^{\min}]. \tag{2.55}$$

For $(t_1^{\max}, s_1^{\min}]$, we have

$$\begin{aligned}
&|\hat{x}((\hat{t}_1)^-) - x(t_1^-)| \\
&\leqslant |\hat{x}_0 - x_0| + \frac{1}{\Gamma(\alpha)} \int_0^{t_1^{\min}} (t_1^{\max} - s)^{\alpha-1} |f(s, \hat{x}(s)) - f(s, x(s))| ds \\
&\quad + \frac{1}{\Gamma(\alpha)} \int_0^{t_1^{\min}} |(t_1^{\max} - s)^{\alpha-1} - (t_1^{\min} - s)^{\alpha-1}| |f(s, x(s))| ds \\
&\quad + |(\mathbf{I}^\alpha_{t_1^{\min}, t_1^{\max}} f)(t_1^{\max}, \hat{x})| \\
&\leqslant \delta_0 + \frac{L_f}{\Gamma(\alpha)} \int_0^{t_1^{\min}} (t_1^{\max} - s)^{\alpha-1} |\hat{x}(s) - x(s)| ds + \frac{3M}{\Gamma(\alpha+1)} \delta_1^\alpha \\
&\leqslant \left(1 + \frac{\kappa_\alpha L_f t_1^\alpha}{\Gamma(\alpha+1)}\right) \delta_0 + \frac{3M}{\Gamma(\alpha+1)} \delta_1^\alpha.
\end{aligned}$$

From (2.38), we obtain

$$|\hat{x}(t;0,\hat{x}_0)-x(t;0,x_0)| \leqslant L_{g_1}\left(1+\frac{\kappa_\alpha L_f t_1^\alpha}{\Gamma(\alpha+1)}\right)\delta_0+\frac{3ML_{g_1}}{\Gamma(\alpha+1)}\delta_1^\alpha.$$

If δ_0, δ_1 are fully small, then

$$|\hat{x}(t;0,\hat{x}_0)-x(t;0,x_0)| < \varepsilon, \quad t\in(t_1^{\max}, s_1^{\min}], \tag{2.56}$$

where ε is an arbitrary positive number.

For $(s_1^{\max}, t_2^{\min}]$, from an inequality in (2.41), it follows that

$$\begin{aligned}
&|\hat{x}(s_1^{\max})-x(s_1^{\max})|\\
&\leqslant |g_1(s_1^{\max},\hat{x}((\hat{t}_1)^-))-g_1(s_1^{\min},x(t_1^-))|+|(\mathbf{I}^\alpha_{s_1^{\min},s_1^{\max}}f)(s_1^{\max},x)|\\
&\leqslant L_{g_1}|\hat{x}((\hat{t}_1)^-)-x(t_1^-)|+|g_1'(\xi)||s_1^{\max}-s_1^{\min}|+\frac{M}{\Gamma(\alpha+1)}\delta_1^\alpha\\
&\leqslant L_{g_1}\left(1+\frac{\kappa_\alpha L_f t_1^\alpha}{\Gamma(\alpha+1)}\right)\delta_0+\frac{3ML_{g_1}}{\Gamma(\alpha+1)}\delta_1^\alpha+|g_1'(\xi)|\delta_1+\frac{M}{\Gamma(\alpha+1)}\delta_1^\alpha\\
&= L_{g_1}\left(1+\frac{\kappa_\alpha L_f t_1^\alpha}{\Gamma(\alpha+1)}\right)\delta_0+|g_1'(\xi)|\delta_1+\frac{M}{\Gamma(\alpha+1)}(1+3L_{g_1})\delta_1^\alpha.
\end{aligned}$$

Since system (2.10) is gravitating with a constant κ_α, then

$$\begin{aligned}
&|\hat{x}(t;0,\hat{x}_0)-x(t;0,x_0)|\\
&\leqslant \kappa_\alpha L_{g_1}\left(1+\frac{\kappa_\alpha L_f t_1^\alpha}{\Gamma(\alpha+1)}\right)\delta_0+\kappa_\alpha|g_1'(\xi)|\delta_1+\frac{M\kappa_\alpha}{\Gamma(\alpha+1)}(1+3L_{g_1})\delta_1^\alpha.
\end{aligned}$$

If δ_0 and δ_1 are adequately small, then

$$|\hat{x}(t;0,\hat{x}_0)-x(t;0,x_0)| < \varepsilon, \quad t\in(s_1^{\max}, t_2^{\min}], \tag{2.57}$$

where ε denotes an arbitrary positive number.

For $(t_2^{\max}, s_2^{\min}]$, obtain the formula

$$|\hat{x}(t;0,\hat{x}_0)-x(t;0,x_0)| \leqslant L_{g_2}|\hat{x}((\hat{t}_2)^-)-x(t_2^-)|.$$

From inequality in (2.43), we deduce that

$$
\begin{aligned}
&|\hat{x}((\hat{t}_2)^-) - x(t_2^-)| \\
&\leqslant |\hat{x}(s_1^{\max}) - x(s_1^{\min})| + |(\mathbf{I}^{\alpha}_{t_2^{\min},\, t_2^{\max}} f)(t_2^{\max}, \hat{x})| \\
&\quad + \frac{1}{\Gamma(\alpha)} \int_{s_1^{\max}}^{t_2^{\min}} |(t_2^{\max} - s)^{\alpha-1} - (t_2^{\min} - s)^{\alpha-1}||f(s, x(s))|ds \\
&\quad + \frac{1}{\Gamma(\alpha)} \int_{s_1^{\min}}^{s_1^{\max}} (t_2^{\min} - s)^{\alpha-1}|f(s, x(s))|ds \\
&\quad + \frac{L_f}{\Gamma(\alpha)} \int_{s_1^{\max}}^{t_2^{\min}} (t_2^{\max} - s)^{\alpha-1}|\hat{x}(s) - x(s)|ds \\
&\leqslant L_{g_1}\left(1 + \frac{\kappa_\alpha L_f t_1^\alpha}{\Gamma(\alpha+1)}\right)\delta_0 + |g_1'(\xi)|\delta_1 + \frac{M}{\Gamma(\alpha+1)}(1 + 3L_{g_1})\delta_1^\alpha \\
&\quad + \eta_\alpha + \frac{3M}{\Gamma(\alpha+1)}\delta_2^\alpha + \frac{M}{\Gamma(\alpha+1)}\delta_1^\alpha \\
&\quad + \left[L_{g_1}\left(1 + \frac{\kappa_\alpha L_f t_1^\alpha}{\Gamma(\alpha+1)}\right)\delta_0 + |g_1'(\xi)|\delta_1 + \frac{M}{\Gamma(\alpha+1)}(1 + 3L_{g_1})\delta_1^\alpha + \eta_\alpha\right]\frac{\kappa_\alpha L_f t_2^\alpha}{\Gamma(\alpha+1)} \\
&= L_{g_1}\left(1 + \frac{\kappa_\alpha L_f t_1^\alpha}{\Gamma(\alpha+1)}\right)\left(1 + \frac{\kappa_\alpha L_f t_2^\alpha}{\Gamma(\alpha+1)}\right)\delta_0 + |g_1'(\xi)|\left(1 + \frac{\kappa_\alpha L_f t_2^\alpha}{\Gamma(\alpha+1)}\right)\delta_1 \\
&\quad + \frac{\kappa_\alpha L_f t_2^\alpha}{\Gamma(\alpha+1)}\eta_\alpha + \frac{M}{\Gamma(\alpha+1)}\left[(1 + 3L_{g_1})\left(1 + \frac{\kappa_\alpha L_f t_2^\alpha}{\Gamma(\alpha+1)}\right) + 1\right]\delta_1^\alpha + \frac{3M}{\Gamma(\alpha+1)}\delta_2^\alpha.
\end{aligned}
$$

Then

$$
\begin{aligned}
&|\hat{x}(t; 0, \hat{x}_0) - x(t; 0, x_0)| \\
&\leqslant L_{g_2}L_{g_1}\left(1 + \frac{\kappa_\alpha L_f t_1^\alpha}{\Gamma(\alpha+1)}\right)\left(1 + \frac{\kappa_\alpha L_f t_2^\alpha}{\Gamma(\alpha+1)}\right)\delta_0 + L_{g_2}|g_1'(\xi)|\left(1 + \frac{\kappa_\alpha L_f t_2^\alpha}{\Gamma(\alpha+1)}\right)\delta_1 \\
&\quad + \frac{ML_{g_2}}{\Gamma(\alpha+1)}\left[(1 + 3L_{g_1})\left(1 + \frac{\kappa_\alpha L_f t_2^\alpha}{\Gamma(\alpha+1)}\right) + 1\right]\delta_1^\alpha + \frac{3ML_{g_2}}{\Gamma(\alpha+1)}\delta_2^\alpha + \frac{\kappa_\alpha L_{g_2} L_f t_2^\alpha}{\Gamma(\alpha+1)}\eta_\alpha.
\end{aligned}
$$

If δ_0, δ_1 and δ_2 are fully small, then

$$
|\hat{x}(t; 0, \hat{x}_0) - x(t; 0, x_0)| < \varepsilon, \quad t \in (t_2^{\max}, s_2^{\min}], \tag{2.58}
$$

where $\varepsilon > 0$ is an arbitrary number.

For $(s_2^{\max}, t_3^{\min}]$,

$$
|\hat{x}(t; 0, \hat{x}_0) - x(t; 0, x_0)| \leqslant |\hat{x}(s_2^{\max}) - x(s_2^{\max})| + L_f(\mathbf{I}^{\alpha}_{s_2^{\max},\, t}|\hat{x} - x|)(t).
$$

Note that

$$
\begin{aligned}
&|\hat{x}(s_2^{\max}) - x(s_2^{\max})| \\
&\leqslant |g_2(s_2^{\max}, \hat{x}((\hat{t}_2)^-)) - g_2(s_2^{\min}, x(t_2^-))| + |(\mathbf{I}^\alpha_{s_2^{\min}, s_2^{\max}} f)(s_2^{\max}, x)| \\
&\leqslant |g_2'(\xi)|\delta_2 + L_{g_2}L_{g_1}\left(1 + \frac{\kappa_\alpha L_f t_1^\alpha}{\Gamma(\alpha+1)}\right)\left(1 + \frac{\kappa_\alpha L_f t_2^\alpha}{\Gamma(\alpha+1)}\right)\delta_0 \\
&\quad + L_{g_2}|g_1'(\xi)|\left(1 + \frac{\kappa_\alpha L_f t_2^\alpha}{\Gamma(\alpha+1)}\right)\delta_1 + \frac{3ML_{g_2}}{\Gamma(\alpha+1)}\delta_2^\alpha + \frac{M}{\Gamma(\alpha+1)}\delta_2^\alpha \\
&\quad + \frac{ML_{g_2}}{\Gamma(\alpha+1)}\left[(1 + 3L_{g_1})\left(1 + \frac{\kappa_\alpha L_f t_2^\alpha}{\Gamma(\alpha+1)}\right) + 1\right]\delta_1^\alpha + \frac{\kappa_\alpha L_{g_2} L_f t_2^\alpha}{\Gamma(\alpha+1)}\eta_\alpha \\
&= L_{g_2}L_{g_1}\left(1 + \frac{\kappa_\alpha L_f t_1^\alpha}{\Gamma(\alpha+1)}\right)\left(1 + \frac{\kappa_\alpha L_f t_2^\alpha}{\Gamma(\alpha+1)}\right)\delta_0 + L_{g_2}|g_1'(\xi)|\left(1 + \frac{\kappa_\alpha L_f t_2^\alpha}{\Gamma(\alpha+1)}\right)\delta_1 \\
&\quad + \frac{\kappa_\alpha L_{g_2} L_f t_2^\alpha}{\Gamma(\alpha+1)}\eta_\alpha + \frac{ML_{g_2}}{\Gamma(\alpha+1)}\left[(1 + 3L_{g_1})\left(1 + \frac{\kappa_\alpha L_f t_2^\alpha}{\Gamma(\alpha+1)}\right) + 1\right]\delta_1^\alpha \\
&\quad + |g_2'(\xi)|\delta_2 + \frac{M}{\Gamma(\alpha+1)}(1 + 3L_{g_2})\delta_2^\alpha.
\end{aligned}
$$

Since system (2.10) is gravitating, we draw a conclusion

$$
\begin{aligned}
&|\hat{x}(t; 0, \hat{x}_0) - x(t; 0, x_0)| \\
&\leqslant \kappa_\alpha L_{g_2}L_{g_1}\left(1 + \frac{\kappa_\alpha L_f t_1^\alpha}{\Gamma(\alpha+1)}\right)\left(1 + \frac{\kappa_\alpha L_f t_2^\alpha}{\Gamma(\alpha+1)}\right)\delta_0 + \kappa_\alpha L_{g_2}|g_1'(\xi)|\left(1 + \frac{\kappa_\alpha L_f t_2^\alpha}{\Gamma(\alpha+1)}\right)\delta_1 \\
&\quad + \frac{\kappa_\alpha ML_{g_2}}{\Gamma(\alpha+1)}\left[(1 + 3L_{g_1})\left(1 + \frac{\kappa_\alpha L_f t_2^\alpha}{\Gamma(\alpha+1)}\right) + 1\right]\delta_1^\alpha + \kappa_\alpha|g_2'(\xi)|\delta_2 \\
&\quad + \frac{M\kappa_\alpha}{\Gamma(\alpha+1)}(1 + 3L_{g_2})\delta_2^\alpha + \frac{\kappa_\alpha^2 L_{g_2} L_f t_2^\alpha}{\Gamma(\alpha+1)}\eta_\alpha.
\end{aligned}
$$

Set $\varepsilon > 0$ denotes an arbitrary number. If δ, δ_0, δ_1 and δ_2 are sufficiently small, then

$$
|\hat{x}(t; 0, \hat{x}_0) - x(t; 0, x_0)| \leqslant \varepsilon, \quad t \in (s_2^{\max}, t_3^{\min}]. \tag{2.59}
$$

From (2.55), (2.56), (2.57), (2.58) and (2.59), for adequately small values of δ, δ_i, $i = 1, 2, \ldots$, the result is as follows

$$
|\hat{x}(t; 0, \hat{x}_0) - x(t; 0, x_0)| \leqslant \varepsilon, \quad t \in [0, t_1^{\min}] \bigcup (t_i^{\max}, s_i^{\min}] \bigcup (s_i^{\max}, t_{i+1}^{\min}],
$$

$$
i = 1, 2, \ldots.
$$

The proof is complete. □

2.1.5 Examples

Set $t_0 = s_0 = 0$, $t_i = 2i - 1$ and $s_i = 2i$, $i \in \{0\} \cup \mathbb{N}$. Obviously, $s_i < t_{i+1} \to \infty$ $(i \to \infty)$.

Example 2.1.9. *Consider the following linear problem*

$$\begin{cases} x'(t) = Px(t), \ t \in (2i, 2i+1], \ i \in \{0\} \bigcup \mathbb{N}, \ P \geqslant 0, \\ x((2i-1)^+) = \rho_i \sin(2i-1)x((2i-1)^-), \ i \in \mathbb{N}, \\ x(t) = \rho_i(\sin t)x((2i-1)^-), \ t \in (2i-1, 2i], \ i \in \mathbb{N}, \\ x(0) = x_0. \end{cases} \tag{2.60}$$

Then one can derive the solution of (2.60), *namely*

$$x(t) = \begin{cases} e^{Pt}x_0, \ \text{for } t \in (0, 1], \\ \rho_1(\sin t)x(1^-), \ \text{for } t \in (1, 2], \\ \rho_1(\sin 2)x_0 e^P e^{P(t-2)}, \ \text{for } t \in (2, 3], \\ \vdots \\ \rho_\gamma(\sin t)x((2\gamma-1)^-), \ \text{for } t \in (2\gamma-1, 2\gamma], \\ \prod_{i=1}^{\gamma} \rho_i(\sin 2i)e^{\gamma P}e^{P(t-2\gamma)}x_0, \ \text{for } t \in (2\gamma, 2\gamma+1], \\ \vdots \end{cases} \tag{2.61}$$

Let $f(t, x) = Px$, $g_i(t, x) = \rho_i(\sin t)x$. *Note* $g_i \in C^1([2i-1, 2i] \times \mathbb{R}, \mathbb{R})$, $i = 1, 2, \ldots$.

Let $t \in (2i, 2i+1]$, $|f(t, x)| \leqslant M := P\max_{u_i, v_j \in \mathbb{R}}|u_i - v_j|$, $\forall\, x \in \mathbb{R}$. *Moreover,* $|f(t, x) - f(t, y)| \leqslant P|x - y|$ *and* $|g_i(t, x) - g_i(t, y)| \leqslant \rho_i|x - y|$, $\forall\, x, y \in \mathbb{R}$. *Choose* $L_f = P$, $L_{g_i} = \rho_i$. *Thus, [H_1]–[H_4] hold. Therefore, the listed hypotheses of theorem 2.1.5 are fulfilled.*

From definition 2.1.3, we can see that (2.60) *without impulses is gravitating with constant which is related to the function* e^{Pt} *by a direct computation via [17, theorem 1.1].*

Concerning the formula of the solution (2.61) on each subinterval $t \in (s_i, t_{i+1}]$, *i.e.* $t \in (2i, 2i+1]$, $i \in \mathbb{N}$, *we obtain* $T_i = t_{i+1} - s_i = 1$, $i = 1, 2, \ldots$. *Then, we obtain another possible more explicit gravitational constant, namely* $\kappa := e^P$. *In particular, we choose* $\rho_i = e^{\frac{1}{i^2}}$. *Now, the listed hypotheses of theorem 2.1.7 are checked.*

Summarizing, theorems 2.1.5 and 2.1.7 can be used to solve (2.60).

Example 2.1.10. *Consider the following nonlinear problem*

$$\begin{cases} x'(t) = P\arctan x(t), \ t \in (2i, 2i+1], \ i \in \{0\} \bigcup \mathbb{N}, \ P \geqslant 0, \\ x((2i-1)^+) = \dfrac{\rho_i|x((2i-1)^-)|}{1 + |x((2i-1)^-)|}, \ i \in \mathbb{N}, \\ x(t) = \dfrac{\rho_i|x((2i-1)^-)|}{1 + |x((2i-1)^-)|}, \ t \in (2i-1, 2i], \ i \in \mathbb{N}, \\ x(0) = x_0. \end{cases} \tag{2.62}$$

Then one can derive the solution of (2.62), namely,

$$x(t) = \begin{cases} x_0 + P\int_0^t \arctan x(s)ds, \ \text{for } t \in (0, 1], \\ \frac{\rho_1|x(1^-)|}{1 + |x(1^-)|}, \ \text{for } t \in (1, 2], \\ \frac{\rho_1|x(1^-)|}{1 + |x(1^-)|} + P\int_2^t \arctan x(s)ds, \ \text{for } t \in (2, 3], \\ \vdots \\ \frac{\rho_\gamma|x((2\gamma - 1)^-)|}{1 + |x((2\gamma - 1)^-)|}, \ \text{for } t \in (2\gamma - 1, 2\gamma], \\ \frac{\rho_\gamma|x((2\gamma - 1)^-)|}{1 + |x((2\gamma - 1)^-)|} + P\int_{2\gamma}^t \arctan x(s)ds, \ \text{for } t \in (2\gamma, 2\gamma + 1], \\ \vdots \end{cases} \tag{2.63}$$

Let $f(t, x) = P\arctan x$, $g_i(t, x) = \frac{\rho_i|x|}{1+|x|}$. *Note* $g_i \in C^1([2i - 1, 2i] \times \mathbb{R}, \mathbb{R})$, $i = 1, 2, \ldots$.

Let $t \in (2i, 2i + 1]$, $|f(t, x)| \leqslant M := \frac{P\pi}{2}$, $\forall\, x \in \mathbb{R}$. *Moreover,* $|f(t, x) - f(t, y)| \leqslant P|\arctan x - \arctan y| \leqslant P|x - y|$ *and* $|g_i(t, x) - g_i(t, y)| \leqslant \rho_i|x - y|$, $\forall\, x, y \in \mathbb{R}$. *Choose* $L_f = P$, $L_{g_i} = \rho_i := e^{\frac{1}{i^2}}$.

Concerning the formula of the solution (2.63) and $T_i = 1$, $i = 1, 2, \ldots$, *one can derive the gravitational constant* $\kappa := e^P$.

Summarizing, theorems 2.1.5 and 2.1.7 can be used to solve (2.62).

Example 2.1.11. *Consider the following linear problem of fractional order*

$$\begin{cases} {}^c\mathbf{D}_{2i,\,t}^{\frac{1}{2}}x(t) = Px(t), \ t \in (2i, 2i + 1], \ i \in \{0\} \cup \mathbb{N}, \ P \geqslant 0, \\ x((2i - 1)^+) = \rho_i \sin(2i - 1)x((2i - 1)^-), \ i \in \mathbb{N}, \\ x(t) = \rho_i(\sin t)x((2i - 1)^-), \ t \in (2i - 1, 2i], \ i \in \mathbb{N}, \\ x(0) = x_0. \end{cases} \tag{2.64}$$

Obviously, $f(t, x) = Px$, $g_i(t, x) = \rho_i(\sin t)x$, *which are the same as in example 2.1.9. Thus,* $[H_1]$–$[H_4]$ *hold. Next, one can derive the solution of (2.64), namely*

$$x(t) = \begin{cases} E_\alpha(Pt^\alpha)x_0, \ \text{for } t \in (0, 1], \\ \rho_1(\sin t)x(1^-), \ \text{for } t \in (1, 2], \\ \rho_1(\sin 2)x_0E_\alpha(P)E_\alpha(P(t - 2)^\alpha), \ \text{for } t \in (2, 3], \\ \vdots \\ \rho_\gamma(\sin t)x((2\gamma - 1)^-), \ \text{for } t \in (2\gamma - 1, 2\gamma], \\ \prod_{i=1}^{\gamma} \rho_i(\sin 2i)(E_\alpha(P))^\gamma E_\alpha(P(t - 2\gamma)^\alpha)x_0, \ \text{for } t \in (2\gamma, 2\gamma + 1], \\ \vdots \end{cases} \tag{2.65}$$

From definition 2.1.4, then we can see that (2.64) without impulses is gravitating with constant which is related to the function $E_{\frac{1}{2}}(Pt^{\frac{1}{2}})$ *by a direct computation via [18, corollary 2].*

Concerning the formula of the solution (2.65) on each sub-interval $t \in (s_i, t_{i+1}]$ *and* $T_i = 1$, $i = 1, 2, \ldots$. *Then, we obtain another possible more explicit gravitational constant, namely* $\kappa_\alpha := E_{\frac{1}{2}}(P)$. *In particular, we choose* $\rho_i = e^{\frac{1}{i^2}}$. *Now, the listed hypotheses of theorem 2.1.7 are checked.*

Summarizing, theorems 2.1.6 and 2.1.8 can be used to solve (2.64).

Example 2.1.12. *Consider the following nonlinear problem of fractional order*

$$\begin{cases} {}^c\mathbf{D}^{\frac{1}{2}}_{2i,\,t}x(t) = P\arctan x(t), \ t \in (2i, 2i+1], \ i \in \{0\} \cup \mathbb{N}, \ P \geqslant 0, \\ x((2i-1)^+) = \dfrac{\rho_i|x((2i-1)^-)|}{1+|x((2i-1)^-)|}, \ i \in \mathbb{N}, \\ x(t) = \dfrac{\rho_i|x((2i-1)^-)|}{1+|x((2i-1)^-)|}, \ t \in (2i-1, 2i], \ i \in \mathbb{N}, \\ x(0) = x_0. \end{cases} \tag{2.66}$$

Obviously, $f(t, x) = P\arctan x$, $g_i(t, x) = \frac{\rho_i|x|}{1+|x|}$, *which are the same as example 2.1.10. Next, one can derive the solution of (2.66), namely*

$$x(t) = \begin{cases} x_0 + \dfrac{P}{\sqrt{\pi}}\displaystyle\int_0^t (t-s)^{-\frac{1}{2}}\arctan x(s)ds, \ t \in (0, 1], \\ \dfrac{\rho_1|x(1^-)|}{1+|x(1^-)|}, \ \text{for } t \in (1, 2], \\ \dfrac{\rho_1|x(1^-)|}{1+|x(1^-)|} + \dfrac{P}{\sqrt{\pi}}\displaystyle\int_2^t (t-s)^{-\frac{1}{2}}\arctan x(s)ds, \ \text{for } t \in (2, 3], \\ \vdots \\ \dfrac{\rho_\gamma|x((2\gamma-1)^-)|}{1+|x((2\gamma-1)^-)|}, \ \text{for } t \in (2\gamma-1, 2\gamma], \\ \dfrac{\rho_\gamma|x((2\gamma-1)^-)|}{1+|x((2\gamma-1)^-)|} \\ +\dfrac{P}{\sqrt{\pi}}\displaystyle\int_{2\gamma}^t (t-s)^{-\frac{1}{2}}\arctan x(s)ds, \ \text{for } t \in (2\gamma, 2\gamma+1], \\ \vdots \end{cases} \tag{2.67}$$

Concerning the formula of the solution (2.67) and $T_i = 1$, $i = 1, 2, \ldots$, *one can obtain another possible more explicit gravitational constant, namely* $\kappa_\alpha := E_{\frac{1}{2}}(P)$ *and we choose* $\rho_i = e^{\frac{1}{i^2}}$.

Summarizing, theorems 2.1.6 and 2.1.8 can be used to solve (2.66).

2.1.6 Notes and remarks

This section uses the ideas in [14] to investigate the asymptotic properties of the solutions, continuous dependence and stability of integer order and fractional order nonlinear non-instantaneous impulsive differential equations. We introduce the concept of continuous dependence and stability of solutions to integer order and fractional order non-instantaneous impulsive Cauchy problems, and establish a framework to seek sufficient conditions to guarantee that the solutions of both the original equations and the perturbed non-instantaneous equations are close to each other in a certain sense.

The results in section 2.1 are motivated by [22].

2.2 Orbital Hausdorff dependence of the solutions

2.2.1 Introduction

The concept of orbital Hausdorff dependence of the solutions to integer order instantaneous impulsive differential equations was introduced in the monograph [21], where the measure between their respective trajectories is given in the whole domain using the Hausdorff distance. In [22], we study asymptotic properties of solutions, continuous dependence and stability, of integer order and fractional order NIDEs. Sufficient conditions are presented to guarantee that the solutions of both the original and the perturbed problems are close to each other in sense of the uniform metric.

We extend the ideas in [21, 22] to investigate orbital Hausdorff dependence of the solutions of the integer order NIDEs:

$$\begin{cases} \chi'(\tau) = f(\tau, \chi(\tau)), \quad \tau \in (\varsigma_i, \tau_{i+1}], \ \tau_{i+1} = \varsigma_i + d, \ i \in \{0\} \cup \mathbb{N}, \\ \chi(\tau_i^+) = g_i(\tau_i, \chi(\tau_i^-)), \quad i \in \mathbb{N}, \\ \chi(\tau) = g_i(\tau, \chi(\tau_i^-)), \quad \tau \in (\tau_i, \varsigma_i], \ \varsigma_i = \tau_i + d, \ i \in \mathbb{N}, \\ \chi(0) = \chi_0, \end{cases} \tag{2.68}$$

and the fractional order NIDEs:

$$\begin{cases} {}^c\mathbf{D}^{\alpha}_{\varsigma_i,\tau}\chi(\tau) = f(\tau, \chi(\tau)), \\ \qquad \tau \in (\varsigma_i, \tau_{i+1}], \ \tau_{i+1} = \varsigma_i + d, \ i \in \{0\} \cup \mathbb{N}, \ \alpha \in (0, 1), \\ \chi(\tau_i^+) = g_i(\tau_i, \chi(\tau_i^-)), \quad i \in \mathbb{N}, \\ \chi(\tau) = g_i(\tau, \chi(\tau_i^-)), \quad \tau \in (\tau_i, \varsigma_i], \ \varsigma_i = \tau_i + d, \ i \in \mathbb{N}, \\ \chi(0) = \chi_0, \end{cases} \tag{2.69}$$

where ${}^{c}\mathbf{D}^{\alpha}_{\varsigma_i,t}$ denotes the classical Caputo fractional derivative of order α by changing the lower limit ς_i [1], τ_i acts as an impulsive point and ς_i acts as a junction point satisfying $\varsigma_i < \tau_{i+1} \to \infty$ with $\tau_0 = \varsigma_0 = 0$, the constant $d > 0$ is the difference between the impulsive points and the junction points, Now $\chi(\tau_i^+) = \lim_{\varepsilon\to 0^+}\chi(\tau_i + \varepsilon)$ and $\chi(\tau_i^-) = \lim_{\varepsilon\to 0^+}\chi(\tau_i - \varepsilon) := \chi(\tau_i)$. The function $f \in C([0, \infty) \times \mathbb{D}, \mathbb{R}^n)$, $\varnothing \neq \mathbb{D} \subset \mathbb{R}^n$ and $g_i \in C([\tau_i, \varsigma_i] \times \mathbb{D}, \mathbb{R}^n)$, $i \in \mathbb{N}$.

Consider the corresponding perturbation problems of the form:

$$\begin{cases} \tilde{\chi}'(\tau) = f(\tau, \tilde{\chi}(\tau)), \quad \tau \in (\tilde{\varsigma}_i, \tilde{\tau}_{i+1}],\ \tilde{\tau}_{i+1} = \tilde{\varsigma}_i + \tilde{d}_{\tau_{i+1}}, \quad i \in \{0\} \bigcup \mathbb{N}, \\ \tilde{\chi}((\tilde{\tau}_i)^+) = g_i(\tilde{\tau}_i, \tilde{\chi}((\tilde{\tau}_i)^-)), \quad i \in \mathbb{N}, \\ \tilde{\chi}(\tau) = g_i(\tau, \tilde{\chi}((\tilde{\tau}_i)^-)), \quad \tau \in (\tilde{\tau}_i, \tilde{\varsigma}_i],\ \tilde{\varsigma}_i = \tilde{\tau}_i + \tilde{d}_{\varsigma_i}, \quad i \in \mathbb{N}, \\ \tilde{\chi}(0) = \tilde{\chi}_0, \end{cases} \tag{2.70}$$

and

$$\begin{cases} {}^{c}\mathbf{D}^{\alpha}_{\tilde{\varsigma}_i,\tau}\tilde{\chi}(\tau) = f(\tau, \tilde{\chi}(\tau)), \quad \tau \in (\tilde{\varsigma}_i, \tilde{\tau}_{i+1}], \\ \qquad \tilde{\tau}_{i+1} = \tilde{\varsigma}_i + \tilde{d}_{\tau_{i+1}}, \quad i \in \{0\} \bigcup \mathbb{N}, \quad \alpha \in (0, 1), \\ \tilde{\chi}((\tilde{\tau}_i)^+) = g_i(\tilde{\tau}_i, \tilde{\chi}((\tilde{\tau}_i)^-)), \quad i \in \mathbb{N}, \\ \tilde{\chi}(\tau) = g_i(\tau, \tilde{\chi}((\tilde{\tau}_i)^-)), \quad \tau \in (\tilde{\tau}_i, \tilde{\varsigma}_i],\ \tilde{\varsigma}_i = \tilde{\tau}_i + \tilde{d}_{\varsigma_i}, \quad i \in \mathbb{N}, \\ \tilde{\chi}(0) = \tilde{\chi}_0, \end{cases} \tag{2.71}$$

where $\tilde{\tau}_0 = \tilde{\varsigma}_0 = 0$, $\tilde{\varsigma}_i < \tilde{\tau}_{i+1} \to \infty$, the constants $\tilde{d}_{\tau_{i+1}}, \tilde{d}_{\varsigma_i} > 0$ denote the differences between the impulsive points and the junction points.

The representation of piecewise continuous solutions to problems (2.68) and (2.70) which we denote respectively by $\chi(\cdot\,; 0, \chi_0) \in PC([0, \infty), \mathbb{R}^n)$ and $\tilde{\chi}(\cdot\,; 0, \tilde{\chi}_0) \in PC([0, \infty), \mathbb{R}^n)$ is as follows:

$$\chi(\tau; 0, \chi_0) = \begin{cases} \chi_0 + (\mathbf{I}^1_{0,\tau}f)(\tau, \chi), \quad \tau \in [0, \tau_1], \\ g_i(\tau, \chi(\tau_i^-)), \quad \tau \in (\tau_i, \varsigma_i],\ i \in \mathbb{N}, \\ g_i(\varsigma_i, \chi(\tau_i^-)) + (\mathbf{I}^1_{\varsigma_i,\tau}f)(\tau, \chi), \quad \tau \in (\varsigma_i, \tau_{i+1}],\ i \in \mathbb{N}, \end{cases} \tag{2.72}$$

and

$$\tilde{\chi}(\tau; 0, \tilde{\chi}_0) = \begin{cases} \tilde{\chi}_0 + (\mathbf{I}^1_{0,\tau}f)(\tau, \tilde{\chi}), \quad \tau \in [0, \tilde{\tau}_1], \\ g_i(\tau, \tilde{\chi}((\tilde{\tau}_i)^-)), \quad \tau \in (\tilde{\tau}_i, \tilde{\varsigma}_i],\ i \in \mathbb{N}, \\ g_i(\tilde{\varsigma}_i, \tilde{\chi}((\tilde{\tau}_i)^-)) + (\mathbf{I}^1_{\tilde{\varsigma}_i,\tau}f)(\tau, \tilde{\chi}), \quad t \in (\tilde{\varsigma}_i, \tilde{\tau}_{i+1}],\ i \in \mathbb{N}, \end{cases} \tag{2.73}$$

where

$$(\mathbf{I}^p_{a,\tau}f)(\tau, \chi) := \frac{1}{\Gamma(p)} \int_a^\tau \frac{f(\sigma, \chi(\sigma))}{(\tau - \sigma)^{1-p}} d\sigma, \ p > 0.$$

Similarly, we get the solutions of problems (2.69) and (2.71), namely:

$$\chi(\tau; 0, \chi_0) = \begin{cases} \chi_0 + (\mathbf{I}^{\alpha}_{0,\tau}f)(\tau, \chi), & \tau \in [0, \tau_1], \\ g_i(\tau, \chi(\tau_i^-)), & \tau \in (\tau_i, \varsigma_i], \ i \in \mathbb{N}, \\ g_i(\varsigma_i, \chi(\tau_i^-)) + (\mathbf{I}^{\alpha}_{\varsigma_i,\tau}f)(\tau, \chi) & \tau \in (\varsigma_i, \tau_{i+1}], \ i \in \mathbb{N}, \end{cases} \tag{2.74}$$

and

$$\tilde{\chi}(\tau; 0, \tilde{\chi}_0) = \begin{cases} \tilde{\chi}_0 + (\mathbf{I}^{\alpha}_{0,\tau}f)(\tau, \tilde{\chi}), & \tau \in [0, \tilde{\tau}_1], \\ g_i(\tau, \tilde{\chi}((\tilde{\tau}_i)^-)), & \tau \in (\tilde{\tau}_i, \tilde{\varsigma}_i], \ i \in \mathbb{N}, \\ g_i(\tilde{\varsigma}_i, \tilde{\chi}((\tilde{\tau}_i)^-)) + (\mathbf{I}^{\alpha}_{\tilde{\varsigma}_i,\tau}f)(\tau, \tilde{\chi}), & \tau \in (\tilde{\varsigma}_i, \tilde{\tau}_{i+1}], \ i \in \mathbb{N}. \end{cases} \tag{2.75}$$

The rest of this part is organized as follows. In section 2.2.2, we introduce the definition of orbital Hausdorff continuous dependence of solutions for our problems. In section 2.2.3, we establish sufficient conditions to guarantee the Hausdorff continuous dependence of solutions. Two examples are given in section 2.2.4 to illustrate our results.

2.2.2 Preliminaries

Let $J = [0, \infty)$. Consider the piecewise continuous function space $PC(J, \mathbb{R}^n) := \{v: J \to \mathbb{R}^n: v \in C((t_k, t_{k+1}], \mathbb{R}^n),\ k = 0, 1, \dots$ and $\exists\ v(t_k^+),\ v(t_k^-),\ k = 1, 2, \dots$ with $v(t_k^-) = v(t_k)\}$ with the norm $\|v\|_{PC} := \sup_{t\in J}\|v(t)\|$, where $C(J, \mathbb{R}^n) = \{v: J \to \mathbb{R}^n$ is continuous$\}$.

Next we recall some concepts from [23].

With the points $x = (x_1, x_2, \dots, x_n)$, $y = (y_1, y_2, \dots, y_n) \in \mathbb{R}^n$, the Euclidean distance and Euclidean norm are defined as: $\rho(x, y) = \sqrt{\Sigma_{j=1}^n (x_j - y_j)^2}$ and $\|x\| = \sqrt{\Sigma_{j=1}^n x_j^2}$. Clearly, $\|x - y\| = \rho(x, y)$.

If $\varnothing \neq X$, $Y \subset \mathbb{R}^n$, the Euclidean and Hausdorff distance between them are introduced as:

$$E(X, Y) = \inf\{\inf\{\rho(x, y), x \in X\}, y \in Y\},$$

and

$$H(X, Y) = \max\{\sup\{\inf\{\rho(x, y), x \in X\}, y \in Y\}, \sup\{\inf\{\rho(x, y), y \in Y\}, x \in X\}\}.$$

When $X = \varnothing$ or $Y = \varnothing$, we suppose that $E(X, Y) = 0$ and $H(X, Y) = 0$.

Theorem 2.2.1. *If the sets* $X_1, X_2, \dots, X_k, Y_1, Y_2, \dots, Y_k \subset \mathbb{R}^n$ *are bounded and* $X = \bigcup_{i=1}^k X_i$, $Y = \bigcup_{i=1}^k Y_i$, *then*

$$\begin{aligned} H(X, Y) &= H(X_1 \cup X_2 \cup \cdots \cup X_k, Y_1 \cup Y_2 \cup \cdots \cup Y_k) \\ &\leqslant \max\{H(X_1, Y_1), H(X_2, Y_2), \ldots, H(X_k, Y_k)\}. \end{aligned}$$

Set the functions $h, \tilde{h} \in C(\mathbb{R}^+, \mathbb{R}^n)$ and the constants $t_1, t_2, \tilde{t}_1, \tilde{t}_2 \in \mathbb{R}^+$. We define the notation of the parametric curves:

$$r[t_1, t_2] = \begin{cases} \{h(t);\ t_1 \leqslant t \leqslant t_2\},\ t_1 \leqslant t_2; \\ \varnothing;\ t_1 > t_2, \end{cases}$$

and

$$\tilde{r}[\tilde{t}_1, \tilde{t}_2] = \begin{cases} \{\tilde{h}(t);\ \tilde{t}_1 \leqslant t \leqslant \tilde{t}_2\},\ \tilde{t}_1 \leqslant \tilde{t}_2; \\ \varnothing;\ \tilde{t}_1 > \tilde{t}_2. \end{cases}$$

Similarly, we can also define the parametric curves in half-open and open intervals.

Now we give the Hausdorff distance between continuous parametric curves.

Remark 2.2.2. *[21, remark 1.4] Let $0 \leqslant t_1 \leqslant t_2$, $0 \leqslant \tilde{t}_1 \leqslant \tilde{t}_2$. We give the definition concerning the Euclidean, Hausdorff and uniform distances between the curves $r[t_1, t_2]$ and $\tilde{r}[\tilde{t}_1, \tilde{t}_2]$, respectively:*

$$E(\tilde{r}[\tilde{t}_1, \tilde{t}_2], r[t_1, t_2]) = \inf\left\{\inf\left\{\rho(\tilde{h}(\tilde{t}), h(t)), \tilde{t}_1 \leqslant \tilde{t} \leqslant \tilde{t}_2\right\}, t_1 \leqslant t \leqslant t_2\right\};$$

and

$$\begin{aligned} &H(\tilde{r}[\tilde{t}_1, \tilde{t}_2], r[t_1, t_2]) \\ &= \max\left\{\sup\left\{\inf\left\{\rho(\tilde{h}(\tilde{t}), h(t)), \tilde{t}_1 \leqslant \tilde{t} \leqslant \tilde{t}_2\right\}, t_1 \leqslant t \leqslant t_2\right\},\right. \\ &\quad \left.\sup\left\{\inf\left\{\rho(\tilde{h}(\tilde{t}), h(t)),\right\}, t_1 \leqslant t \leqslant t_2\right\}, \tilde{t}_1 \leqslant \tilde{t} \leqslant \tilde{t}_2\right\}; \end{aligned}$$

and

$$R(\tilde{r}[t_1, t_2], r[t_1, t_2]) = \sup\{\rho(\tilde{h}(t), h(t)), t_1 \leqslant t \leqslant t_2\}.$$

For brevity, we set $t_i^{\min} = \min\{\tilde{t}_i, t_i\}$, $t_i^{\max} = \max\{\tilde{t}_i, t_i\}$, $s_i^{\min} = \min\{\tilde{s}_i, s_i\}$ and $s_i^{\max} = \max\{\tilde{s}_i, s_i\}$, $i = 1, 2, \ldots$.

We consider the following hypothesis:

$[H_1]$ The function $f: J \times \mathbb{D} \to \mathbb{R}^n$ is continuous and $g_i \in C([\tau_i, \varsigma_i] \times \mathbb{D}, \mathbb{R}^n)$, $i \in \mathbb{N}$.

$[H_2]$ There exists a positive constant L_f such that $\|f(\tau, \chi) - f(\tau, \psi)\| \leqslant L_f\|\chi - \psi\|$, for each $\tau \in [\varsigma_i, \tau_{i+1}]$, $i \in \{0\} \cup \mathbb{N}$, for all $\chi, \psi \in \mathbb{R}^n$.

$[H_3]$ There exists a positive constant L_{g_i}, $i \in \mathbb{N}$ such that $\|g_i(\tau_1, \chi) - g_i(\tau_2, \psi)\| \leqslant L_{g_i}(|\tau_1 - \tau_2| + \|\chi - \psi\|)$, for $\tau_1, \tau_2 \in [\tau_i, \varsigma_i],$, $i \in \mathbb{N}$, for all $\chi, \psi \in \mathbb{R}^n$.

As in [20, theorem 4.1], the following theorem is a direct consequence of conditions $[H_1]$, $[H_2]$ and $[H_3]$.

Lemma 2.2.3. *Assume $[H_1]$, $[H_2]$ and $[H_3]$ are satisfied. Then (2.68)(or (2.69)) has a unique solution in $PC(J, \mathbb{R}^n)$.*

As in [21, theorem 2.2], we have

Lemma 2.2.4. *Suppose that the functions $h, \tilde{h}: \mathbb{R}^+ \to \mathbb{R}^n$ are continuous on the left hand side in $\mathbb{R}^+$, if $i = 1$ and $t_1^{\max} \leqslant s_1^{\min}$ (see figure 2.5). Then*

$$
\begin{aligned}
&H(\tilde{r}(\tilde{t}_1, \tilde{s}_1], r(t_1, s_1])\\
&\quad\leqslant \max\Big\{R\Big(\tilde{r}\Big(t_1^{\max}, s_1^{\min}\Big], r\Big(t_1^{\max}, s_1^{\min}\Big]\Big), H(h(t_1+0), \tilde{r}(\tilde{t}_1, t_1]),\\
&\qquad H(\tilde{h}(\tilde{t}_1+0), r(t_1, \tilde{t}_1]), H(h(s_1), \tilde{r}(s_1, \tilde{s}_1]), H(\tilde{h}(\tilde{s}_1), r(\tilde{s}_1, s_1])\Big\}.
\end{aligned}
$$

Definition 2.2.5. *The solution of (2.68)(or (2.69)) is orbital Hausdorff dependent on the initial condition and the differences between the impulsive points and the junction points, if $\forall\,(0, \chi_0) \in [0, T] \times \mathbb{D}$, $\forall\, d > 0$, $\forall\, \varepsilon > 0$, $\forall\, T > 0$, $\exists\, \delta = \delta(\chi_0, d, \varepsilon, T) > 0$, for $\forall\, \tilde{\chi}_0 \in D$, $\|\tilde{\chi}_0 - \chi_0\| < \delta$, $\forall\, \tilde{d}_{\tau_i} > 0$, $\tilde{d}_{\varsigma_i} > 0$, $|\tilde{d}_{\tau_i} - d| < \delta$, $|\tilde{d}_{\varsigma_i} - d| < \delta$, $i = 1, 2, \ldots$, then*

$$H(\tilde{r}[0, T], r[0, T]) < \varepsilon.$$

2.2.3 Orbital Hausdorff continuous dependence

In this section, we investigate orbital Hausdorff continuous dependence of the solutions for our problems.

We need the following condition:

$[H_4]$ There exists a positive constant M such that $\|f(\tau, \chi)\| \leqslant M$, for any $(\tau, \chi) \in J \times \mathbb{R}^n$.

Remark 2.2.6. *In fact, $[H_4]$ could be changed to $\sup_{t\in J}\|f(t, 0)\| < \infty$. Then, one can apply the impulsive Gronwall inequality [15, lemma 1](or [16, lemma 2.8]) to derive a prior estimates of solutions to (2.68)(or (2.69)) under $[H_2]$. Here we keep $[H_4]$ so that the proofs are more straightforward.*

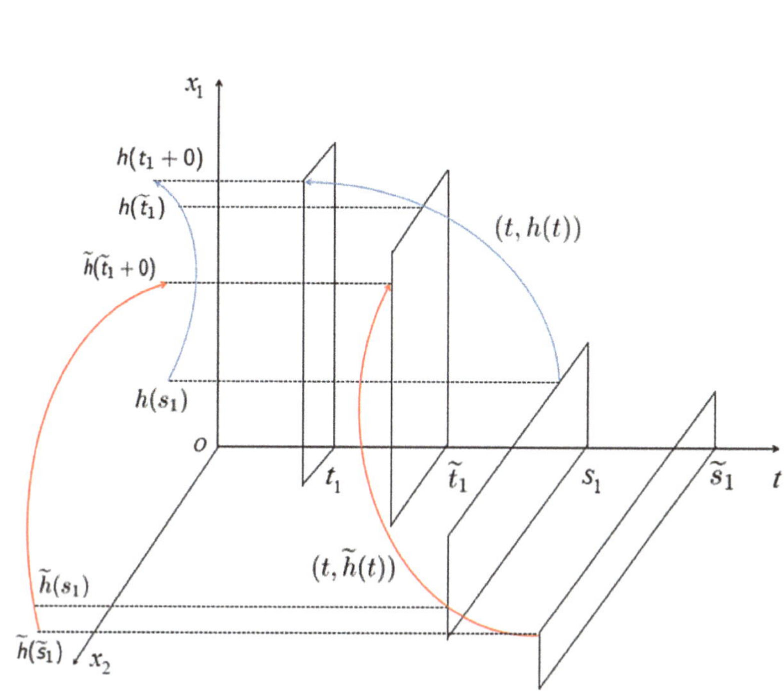

Figure 2.5. The line $(t, h(t))$ denotes the orbital of the solution of original problem and the line $(t, \tilde{h}(t))$ denotes the orbital of the solution of perturbation problem. Reprinted from [22]. Copyright (2017), with permission from Elsevier.

Theorem 2.2.7. *Suppose $[H_1] - [H_4]$ are satisfied. Then, the solution of problem (2.68) is orbital Hausdorff dependent on the initial condition and the difference between the impulsive points τ_i and the junction points ς_i, $i = 1, 2, \ldots$.*

Proof. Consider the possible location of the distribution of the impulsive points τ_i, $\tilde{\tau}_i$ and the junction points ς_i, $\tilde{\varsigma}_i$, so we divide our proof into several cases.

Case 1. Let $\tau_i^{\min} = \tau_i$, $\tau_i^{\max} = \tilde{\tau}_i$, $\varsigma_i^{\min} = \varsigma_i$, $\varsigma_i^{\max} = \tilde{\varsigma}_i$, $i = 1, 2, \ldots$; (the case $\tau_i^{\min} = \tilde{\tau}_i$, $\tau_i^{\max} = \tau_i$, $\varsigma_i^{\min} = \tilde{\varsigma}_i$, $\varsigma_i^{\max} = \varsigma_i$, $i = 1, 2, \ldots$ can be considered similarly).

For the point $(0, \chi_0) \in [0, \infty) \times \mathbb{R}^n$, let ϵ and T be positive constants. Since $\tau_i \to \infty$ $(i \to \infty)$, then $\exists\ k \in \mathbb{N}$ such that $2kd = \varsigma_k < T < \tau_{k+1} = (2k+1)d$. Therefore, we can select a constant $\delta_T = \delta_T(d, T) > 0$, which is sufficiently small, and then $\forall\ \tilde{d}_{\tau_i}, \tilde{d}_{\varsigma_i} > 0$, $|\tilde{d}_{\tau_i} - d| < \delta_T$, $|\tilde{d}_{\varsigma_i} - d| < \delta_T$ and $T < \varsigma_{k+1}^{\min}$, $\varsigma_{i-1}^{\max} < \tau_i^{\min}$, $\tau_i^{\max} < \varsigma_i^{\min}$, $i = 1, 2, \ldots, k+1$.

Furthermore,

$$\tilde{\tau}_1 < \varsigma_1 \Leftrightarrow \tilde{d}_{\tau_1} < 2d \Rightarrow \delta_T < d;$$

$$\tilde{\varsigma}_1 < \tau_2 \Leftrightarrow \tilde{d}_{\tau_1} + \tilde{d}_{\varsigma_1} < 3d \Rightarrow \delta_T < \frac{d}{2};$$

$$\tilde{\tau}_2 < \varsigma_2 \Leftrightarrow \tilde{d}_{\tau_1} + \tilde{d}_{\varsigma_1} + \tilde{d}_{\tau_2} < 4d \Rightarrow \delta_T < \frac{d}{3};$$

$$\tilde{\varsigma}_2 < \tau_3 \Leftrightarrow \tilde{d}_{\tau_1} + \tilde{d}_{\varsigma_1} + \tilde{d}_{\tau_2} + \tilde{d}_{\varsigma_2} < 5d \Rightarrow \delta_T < \frac{d}{4};$$

$$\vdots$$

$$\tilde{\tau}_k < \varsigma_k \Leftrightarrow \tilde{d}_{\tau_1} + \tilde{d}_{\varsigma_1} + \cdots + \tilde{d}_{\varsigma_{k-1}} + \tilde{d}_{\tau_k} < 2kd \Rightarrow \delta_T < \frac{d}{2k-1};$$

$$\tilde{\varsigma}_k < T \Leftrightarrow \tilde{d}_{\tau_1} + \tilde{d}_{\varsigma_1} + \cdots + \tilde{d}_{\tau_k} + \tilde{d}_{\varsigma_k} < T \Rightarrow \delta_T < \frac{T - \varsigma_k}{2k};$$

$$T < \tau_{k+1} \Leftrightarrow T < (2k+1)d \Rightarrow \delta_T < \frac{\tilde{\tau}_{k+1} - T}{2k+1}.$$

From the inequalities, we suppose that $0 < \delta_T < \min\{\frac{d}{2k}, \frac{T-\varsigma_k}{2k}, \frac{\tilde{\tau}_{k+1}-T}{2k+1}\}$. Now we consider the Hausdorff distance between the trajectories on the corresponding subintervas.

The Hausdorff distance of the trajectories on the intervals $[0, \tilde{\tau}_1]$ and $[0, \tau_1]$, according to the property $H(\bar{X}, \bar{Y}) = H(X, Y)$ and lemma 2.2.4, is

$$\begin{aligned} H(\tilde{r}[0, \tilde{\tau}_1], r[0, \tau_1]) &= H(\tilde{r}(0, \tilde{\tau}_1], r(0, \tau_1]) \\ &\leqslant \max\Big\{R\big(\tilde{r}\big(0, \tau_1^{\min}\big], r\big(0, \tau_1^{\min}\big]\big), H\big(\chi(\tau_1; 0, \chi_0), \tilde{r}(\tau_1, \tilde{\tau}_1]\big), \\ &\quad H\big(\tilde{\chi}(\tilde{\tau}_1; 0, \tilde{\chi}_0), r(\tilde{\tau}_1, \tau_1]\big)\Big\}. \end{aligned} \tag{2.76}$$

Since $(\tilde{\tau}_1, \tau_1] = \varnothing$,

$$H(\tilde{\chi}(\tilde{\tau}_1; 0, \tilde{\chi}_0), r(\tilde{\tau}_1, \tau_1]) = 0.$$

(see figure 2.6)

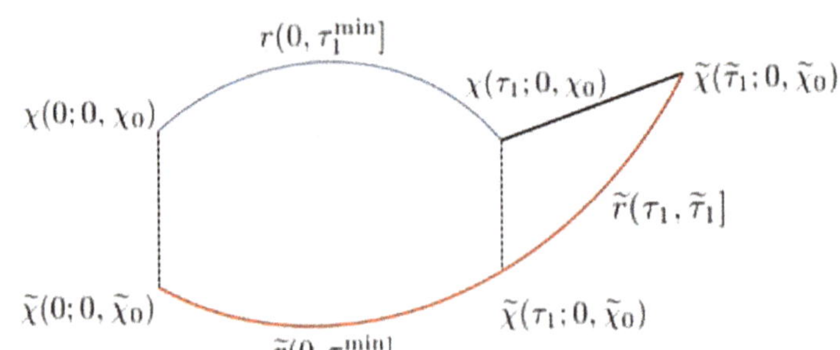

Figure 2.6. The line r denotes the orbital of solution of (2.68) in interval $(0, \tau_1]$ and the line $\tilde{r}$ denotes the orbital of solution of (2.70) in interval $(0, \tilde{\tau}_1]$. Reprinted from [22]. Copyright (2017), with permission from Elsevier.

Motivated from [21, definition 2.1], we introduce the following definition.

We need to evaluate the other two terms in (2.76).

Let $0 < \eta_{01} < \varepsilon$, and we infer that $\exists\ \delta_0 > 0$, $\delta_0 < \min\{\delta_T, \frac{\eta_{01}}{2M}\}$, $\forall\ \tilde{\chi}_0 \in \mathbb{D}$, $\|\tilde{\chi}_0 - \chi_0\| < \delta_0$, $\forall\ \tilde{d}_{\tau_1} > 0$, $|\tilde{d}_{\tau_1} - d| < \delta_0$, then $\|\tilde{\chi}(\tau; 0, \tilde{\chi}_0) - \chi(\tau; 0, \chi_0)\| < \frac{1}{2}\eta_{01}$, $0 < \tau \leqslant \tau_1^{\min}$, that is $R(\tilde{r}(0, \tau_1^{\min}], r(0, \tau_1^{\min}]) < \frac{1}{2}\eta_{01} < \varepsilon$.

Note that $|\tilde{\tau}_1 - \tau_1| = |\tilde{d}_{\tau_1} - d| < \delta_0 < \frac{\eta_{01}}{2M}$. For $\tau_1 < \tau \leqslant \tilde{\tau}_1$,

$$
\begin{aligned}
&\|\chi(\tau_1; 0, \chi_0) - \tilde{\chi}(\tau; 0, \tilde{\chi}_0)\| \\
&\leqslant \|\chi(\tau_1; 0, \chi_0) - \tilde{\chi}(\tau_1; 0, \tilde{\chi}_0)\| + \|\tilde{\chi}(\tau; 0, \tilde{\chi}_0) - \tilde{\chi}(\tau_1; 0, \tilde{\chi}_0)\| \\
&< \frac{1}{2}\eta_{01} + \left\|\left(\mathbf{I}^1_{\tau_1, \tau}f\right)(\tau, \tilde{\chi})\right\| \\
&\leqslant \frac{1}{2}\eta_{01} + M|\tilde{\tau}_1 - \tau_1| \\
&< \frac{1}{2}\eta_{01} + M\delta_0 < \eta_{01} < \varepsilon,
\end{aligned}
$$

i.e. $H(\chi(\tau_1; 0, \chi_0), \tilde{r}(\tau_1, \tilde{\tau}_1]) < \varepsilon$.

Let $\delta_{\tau_1} > 0$ be an arbitrary constant, assume that $\eta_{01} < 2M\delta_{\tau_1}$, and then $|\tilde{\tau}_1 - \tau_1| < \delta_{\tau_1}$. Therefore, $\forall\ \delta_{\tau_1} > 0$, $\exists\ \delta_0 > 0$, $\forall\ \tilde{\chi}_0 \in \mathbb{D}$, $\|\tilde{\chi}_0 - \chi_0\| < \delta_0$, $\forall\ \tilde{d}_{\tau_1} > 0$, $|\tilde{d}_{\tau_1} - d| < \delta_0$, then $H(\tilde{r}[0, \tilde{\tau}_1], r[0, \tau_1]) < \varepsilon$, $|\tilde{\tau}_1 - \tau_1| < \delta_{\tau_1}$.

For the trajectories $\tilde{r}(\tilde{\tau}_1, \tilde{\varsigma}_1]$ and $r(\tau_1, \varsigma_1]$ (see figure 2.7), the Hausdorff distance is

$$
\begin{aligned}
&H(\tilde{r}(\tilde{\tau}_1, \tilde{\varsigma}_1], r(\tau_1, \varsigma_1]) \\
&\leqslant \max\Big\{R(\tilde{r}(\tau_1^{\max}, \varsigma_1^{\min}], r(\tau_1^{\max}, \varsigma_1^{\min}]), H\big(\chi(\tau_1 + 0; 0, \chi_0), \tilde{r}(\tilde{\tau}_1, \tau_1]\big), \\
&H\big(\tilde{\chi}(\tilde{\tau}_1 + 0; 0, \tilde{\chi}_0), r(\tau_1, \tilde{\tau}_1]\big), H\big(\chi(\varsigma_1; 0, \chi_0), \tilde{r}(\varsigma_1, \tilde{\varsigma}_1]\big), \\
&H\big(\tilde{\chi}(\tilde{\varsigma}_1; 0, \tilde{\chi}_0), r(\tilde{\varsigma}_1, \varsigma_1]\big)\Big\}.
\end{aligned}
\tag{2.77}
$$

Since $(\tilde{\tau}_1, \tau_1] = \varnothing$ and $(\tilde{\varsigma}_1, \varsigma_1] = \varnothing$, $H(\chi(\tau_1 + 0; 0, \chi_0), \tilde{r}(\tilde{\tau}_1, \tau_1]) = 0$, $H(\tilde{\chi}(\tilde{\varsigma}_1; 0, \tilde{\chi}_0), r(\tilde{\varsigma}_1, \varsigma_1]) = 0$.

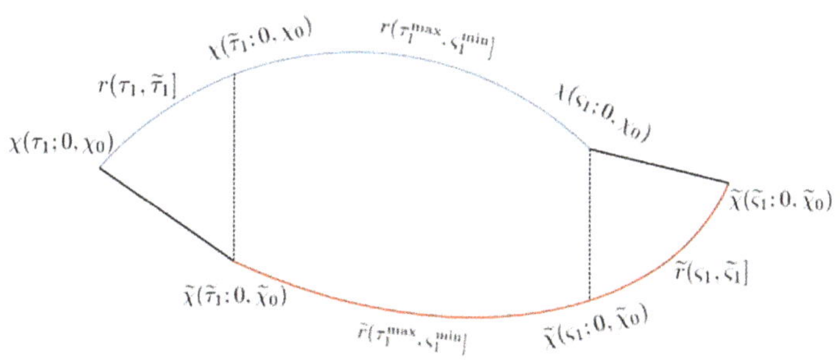

Figure 2.7. The line r denotes the orbital of solution of (2.68) in interval $(\tau_1, \varsigma_1]$ and the line $\tilde{r}$ denotes the orbital of solution of (2.70) in interval $(\tilde{\tau}_1, \tilde{\varsigma}_1]$. Reprinted from [22]. Copyright (2017), with permission from Elsevier.

Set $0 < \eta_{11} < \varepsilon$, and we have that $\exists\ \delta_{\tau_1} > 0,\ \delta_{\tau_1} < \frac{\eta_{11}}{4L_{g_1}},\ \forall\ \tilde{\chi}_0 \in \mathbb{D}, \|\tilde{\chi}_0 - \chi_0\| < \delta_0$, $\forall\ \tilde{d}_{\varsigma_1} > 0, |\tilde{d}_{\varsigma_1} - d| < \delta_{\tau_1}$, then $\|\tilde{\chi}(\tau; 0, \tilde{\chi}_0) - \chi(\tau; 0, \chi_0)\| < \frac{1}{2}\eta_{11}, \tau_1^{\max} < \tau \leqslant \varsigma_1^{\min}$, that is $R(\tilde{r}(\tau_1^{\max}, \varsigma_1^{\min}], r(\tau_1^{\max}, \varsigma_1^{\min}]) < \frac{1}{2}\eta_{11} < \varepsilon$.

For $\tau_1 < \tau \leqslant \tilde{\tau}_1$,

$$
\begin{aligned}
&\left\|\tilde{\chi}(\tilde{\tau}_1 + 0; 0, \tilde{\chi}_0) - \chi(\tau; 0, \chi_0)\right\| \\
&\leqslant \left\|\tilde{\chi}(\tilde{\tau}_1; 0, \tilde{\chi}_0) - \chi(\tilde{\tau}_1; 0, \chi_0)\right\| + \left\|\chi(\tilde{\tau}_1; 0, \chi_0) - \chi(\tau; 0, \chi_0)\right\| \\
&< \frac{1}{2}\eta_{11} + \|g_1(\tilde{\tau}_1, \chi(\tau_1^-)) - g_1(\tau, \chi(\tau_1^-))\| \\
&\leqslant \frac{1}{2}\eta_{11} + L_{g_1}|\tilde{\tau}_1 - \tau_1| \\
&< \frac{1}{2}\eta_{11} + L_{g_1}\delta_{\tau_1} < \eta_{11} < \varepsilon,
\end{aligned}
\tag{2.78}
$$

i.e. $H(\tilde{\chi}(\tilde{\tau}_1 + 0; 0, \tilde{\chi}_0), r(\tau_1, \tilde{\tau}_1]) < \varepsilon$.

Note that $|\tilde{\varsigma}_1 - \varsigma_1| \leqslant |\tilde{\tau}_1 - \tau_1| + |\tilde{d}_{\varsigma_1} - d| < 2\delta_{\tau_1}$. For $\varsigma_1 < \tau \leqslant \tilde{\varsigma}_1$,

$$
\begin{aligned}
&\left\|\chi(\varsigma_1; 0, \chi_0) - \tilde{\chi}(\tau; 0, \tilde{\chi}_0)\right\| \\
&\leqslant \left\|\chi(\varsigma_1; 0, \chi_0) - \tilde{\chi}(\varsigma_1; 0, \tilde{\chi}_0)\right\| + \left\|\tilde{\chi}(\tau; 0, \tilde{\chi}_0) - \tilde{\chi}(\varsigma_1; 0, \tilde{\chi}_0)\right\| \\
&< \frac{1}{2}\eta_{11} + \left\|g_1(\tau, \tilde{\chi}(\tilde{\tau}_1^-)) - g_1(\varsigma_1, \tilde{\chi}(\tilde{\tau}_1^-))\right\| \\
&\leqslant \frac{1}{2}\eta_{11} + L_{g_1}|\tilde{\varsigma}_1 - \varsigma_1| \\
&< \frac{1}{2}\eta_{11} + 2L_{g_1}\delta_{\tau_1} < \eta_{11} < \varepsilon,
\end{aligned}
$$

i.e. $H(\chi(\varsigma_1; 0, \chi_0), \tilde{r}(\varsigma_1, \tilde{\varsigma}_1]) < \varepsilon$.

Set $\eta_{11} < 2L_{g_1}\delta_{\varsigma_1}$, and then $|\tilde{\varsigma}_1 - \varsigma_1| < \delta_{\varsigma_1}$, where δ_{ς_1} is an arbitrary positive constant. Hence, $\forall\ \delta_{\varsigma_1} > 0, \exists\ \delta_{\tau_1} > 0, \forall\ \tilde{\chi}_0 \in \mathbb{D}, \|\tilde{\chi}_0 - \chi_0\| < \delta_0, \forall\ \tilde{d}_{\varsigma_1} > 0,\ |\tilde{d}_{\varsigma_1} - d| < \delta_{\tau_1}$, then $H(\tilde{r}(\tilde{\tau}_1, \tilde{\varsigma}_1], r(\tau_1, \varsigma_1]) < \varepsilon, |\tilde{\varsigma}_1 - \varsigma_1| < \delta_{\varsigma_1}$.

For the Hausdorff distance between the trajectories $\tilde{r}(\tilde{\varsigma}_1, \tilde{\tau}_2]$ and $r(\varsigma_1, \tau_2]$ (see figure 2.8),

$$
\begin{aligned}
&H(\tilde{r}(\tilde{\varsigma}_1, \tilde{\tau}_2], r(\varsigma_1, \tau_2]) \\
&\leqslant \max\Big\{R\big(\tilde{r}\big(\varsigma_1^{\max}, \tau_2^{\min}\big], r\big(\varsigma_1^{\max}, \tau_2^{\min}\big]\big), H\big(\chi(\varsigma_1 + 0; 0, \chi_0), \tilde{r}(\tilde{\varsigma}_1, \varsigma_1]\big), \\
&H\big(\tilde{\chi}(\tilde{\varsigma}_1 + 0; 0, \tilde{\chi}_0), r(\varsigma_1, \tilde{\varsigma}_1]\big), H\big(\chi(\tau_2; 0, \chi_0), \tilde{r}(\tau_2, \tilde{\tau}_2]\big), \\
&H\big(\tilde{\chi}(\tilde{\tau}_2; 0, \tilde{\chi}_0), r(\tilde{\tau}_2, \tau_2]\big)\Big\}.
\end{aligned}
\tag{2.79}
$$

Since $(\tilde{\varsigma}_1, \varsigma_1] = \varnothing$, $(\tilde{\tau}_2, \tau_2] = \varnothing$, then $H(\chi(\varsigma_1 + 0; 0, \chi_0), \tilde{r}(\tilde{\varsigma}_1, \varsigma_1]) = 0$, $H(\tilde{\chi}(\tilde{\tau}_2; 0, \tilde{\chi}_0), r(\tilde{\tau}_2, \tau_2]) = 0$.

Let $0 < \eta_{12} < \varepsilon$, we have that $\exists\ \delta_{\varsigma_1} > 0$, $\delta_{\varsigma_1} < \frac{\eta_{12}}{4M}$, $\forall\ \tilde{\varsigma}_1 \in R^+$, $|\tilde{\varsigma}_1 - \varsigma_1| < \delta_{\varsigma_1}$, $\forall\ \tilde{d}_{\tau_2} > 0$, $|\tilde{d}_{\tau_2} - d| < \delta_{\varsigma_1}$, then $\|\tilde{\chi}(\tau; 0, \tilde{\chi}_0) - \chi(\tau; 0, \chi_0)\| < \frac{1}{2}\eta_{12}$, $\varsigma_1^{\max} < \tau \leqslant \tau_2^{\min}$, that is $R(\tilde{r}(\varsigma_1^{\max}, \tau_2^{\min}], r(\varsigma_1^{\max}, \tau_2^{\min}]) < \frac{1}{2}\eta_{12} < \varepsilon$.

For $\varsigma_1 < \tau \leqslant \tilde{\varsigma}_1$,

$$\begin{aligned}
&\left\|\tilde{\chi}(\tilde{\varsigma}_1 + 0; 0, \tilde{\chi}_0) - \chi(\tau; 0, \chi_0)\right\| \\
\leqslant&\left\|\tilde{\chi}(\tilde{\varsigma}_1; 0, \tilde{\chi}_0) - \chi(\tilde{\varsigma}_1; 0, \chi_0)\right\| + \left\|\chi(\tilde{\varsigma}_1; 0, \chi_0) - \chi(\tau; 0, \chi_0)\right\| \\
<&\frac{1}{2}\eta_{12} + \left\|(\mathbf{I}^1_{\tau,\tilde{\varsigma}_1} f)(\tilde{\varsigma}_1, \chi)\right\| \\
\leqslant&\frac{1}{2}\eta_{12} + M|\tilde{\varsigma}_1 - \varsigma_1| \\
<&\frac{1}{2}\eta_{12} + M\delta_{\varsigma_1} < \eta_{12} < \varepsilon,
\end{aligned}$$

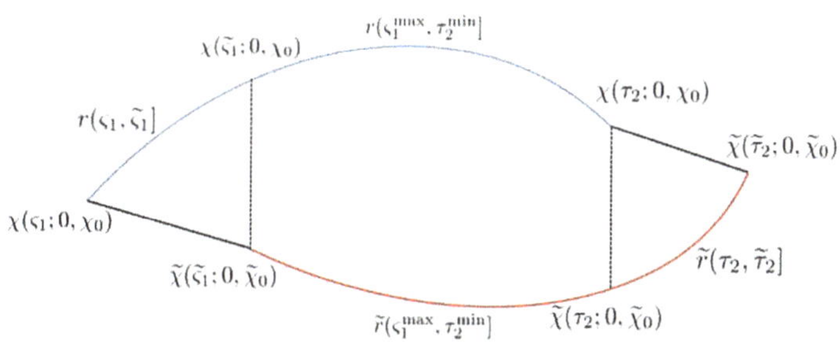

Figure 2.8. The line r denotes the orbital of solution of (2.68) in interval $(\varsigma_1, \tau_2]$ and the line $\tilde{r}$ denotes the orbital of solution of (2.70) in interval $(\tilde{\varsigma}_1, \tilde{\tau}_2]$. Reprinted from [22]. Copyright (2017), with permission from Elsevier.

i.e. $H(\tilde{\chi}(\tilde{\varsigma}_1 + 0; 0, \tilde{\chi}_0), r(\varsigma_1, \tilde{\varsigma}_1]) < \varepsilon$.

Note that $|\tilde{\tau}_2 - \tau_2| \leqslant |\tilde{\varsigma}_1 - \varsigma_1| + |\tilde{d}_{\tau_2} - d| < 2\delta_{\varsigma_1}$. For $\tau_2 < \tau \leqslant \tilde{\tau}_2$,

$$\begin{aligned}
&\left\|\chi(\tau_2; 0, \chi_0) - \tilde{\chi}(\tau; 0, \tilde{\chi}_0)\right\| \\
\leqslant&\left\|\chi(\tau_2; 0, \chi_0) - \tilde{\chi}(\tau_2; 0, \tilde{\chi}_0)\right\| + \left\|\tilde{\chi}(\tau; 0, \tilde{\chi}_0) - \tilde{\chi}(\tau_2; 0, \tilde{\chi}_0)\right\| \\
<&\frac{1}{2}\eta_{12} + \left\|(\mathbf{I}^1_{\tau_2,\tau} f)(\tau, \tilde{\chi})\right\| \\
\leqslant&\frac{1}{2}\eta_{12} + M|\tilde{\tau}_2 - \tau_2| \\
<&\frac{1}{2}\eta_{12} + 2M\delta_{\varsigma_1} < \eta_{12} < \varepsilon,
\end{aligned} \tag{2.80}$$

i.e. $H(\chi(\tau_2; 0, \chi_0), \tilde{r}(\tau_2, \tilde{\tau}_2]) < \varepsilon$.

Let δ_{τ_2} denote an arbitrary positive constant, and we presume that $\eta_{12} < 2M\delta_{\tau_2}$, and then $|\tilde{\tau}_2 - \tau_2| < \delta_{\tau_2}$.

Therefore, $\forall\ \delta_{\tau_2} > 0,\ \exists\ \delta_{\varsigma_1} > 0\ ,\forall\ \tilde{\varsigma}_1 \in R^+,\quad |\tilde{\varsigma}_1 - \varsigma_1| < \delta_{\varsigma_1},\quad \forall\ \tilde{d}_{\tau_2} > 0$, $|\tilde{d}_{\tau_2} - d| < \delta_{\varsigma_1}$, then $H(\tilde{r}(\tilde{\varsigma}_1, \tilde{\tau}_2], r(\varsigma_1, \tau_2]) < \varepsilon$, $|\tilde{\tau}_2 - \tau_2| < \delta_{\tau_2}$.

Consider the trajectories $\tilde{r}(\tilde{\tau}_2, \tilde{\varsigma}_2]$ and $r(\tau_2, \varsigma_2]$ and note

$$\begin{aligned} &H(\tilde{r}(\tilde{\tau}_2, \tilde{\varsigma}_2], r(\tau_2, \varsigma_2]) \\ &\leqslant \max\Big\{R\Big(\tilde{r}\big(\tau_2^{\max}, \varsigma_2^{\min}\big], r\big(\tau_2^{\max}, \varsigma_2^{\min}\big]\Big), H\big(\chi(\tau_2 + 0; 0, \chi_0), \tilde{r}(\tilde{\tau}_2, \tau_2]\big), \\ &\quad H\big(\tilde{\chi}(\tilde{\tau}_2 + 0; 0, \tilde{\chi}_0), r(\tau_2, \tilde{\tau}_2]\big), H\big(\chi(\varsigma_2; 0, \chi_0), \tilde{r}(\varsigma_2, \tilde{\varsigma}_2]\big), \\ &\quad H\Big(\tilde{\chi}(\tilde{\varsigma}_2; 0, \tilde{\chi}_0), r(\tilde{\varsigma}_2, \varsigma_2]\Big)\Big\}. \end{aligned} \tag{2.81}$$

Since $(\tilde{\tau}_2, \tau_2] = \varnothing$, $(\tilde{\varsigma}_2, \varsigma_2] = \varnothing$, then $H(\chi(\tau_2 + 0; 0, \chi_0), \tilde{r}(\tilde{\tau}_2, \tau_2]) = 0$, $H(\tilde{\chi}(\tilde{\varsigma}_2; 0, \tilde{\chi}_0), r(\tilde{\varsigma}_2, \varsigma_2]) = 0$.

Let $0 < \eta_{22} < \varepsilon$, we have that $\exists\ \delta_{\tau_2} > 0$, $\delta_{\tau_2} < \frac{\eta_{22}}{4L_{g_2}}$, $\forall\ \tilde{d}_{\varsigma_2} > 0$, $|\tilde{d}_{\varsigma_2} - d| < \delta_{\tau_2}$, then $\|\tilde{\chi}(\tau; 0, \tilde{\chi}_0) - \chi(\tau; 0, \chi_0)\| < \frac{1}{2}\eta_{22}$, $\tau_2^{\max} < \tau \leqslant \varsigma_2^{\min}$, that is $R(\tilde{r}(\tau_2^{\max}, \varsigma_2^{\min}], r(\tau_2^{\max}, \varsigma_2^{\min}]) < \frac{1}{2}\eta_{22} < \varepsilon$.

For $\tau_2 < \tau \leqslant \tilde{\tau}_2$,

$$\begin{aligned} &\big\|\tilde{\chi}(\tilde{\tau}_2 + 0; 0, \tilde{\chi}_0) - \chi(\tau; 0, \chi_0)\big\| \\ &\leqslant \big\|\tilde{\chi}(\tilde{\tau}_2; 0, \tilde{\chi}_0) - \chi(\tilde{\tau}_2; 0, \chi_0)\big\| + \big\|\chi(\tilde{\tau}_2; 0, \chi_0) - \chi(\tau; 0, \chi_0)\big\| \\ &< \frac{1}{2}\eta_{22} + \Big\|g_2(\tilde{\tau}_2, \chi(\tau_2^-)) - g_2(\tau, \chi(\tau_2^-))\Big\| \\ &\leqslant \frac{1}{2}\eta_{22} + L_{g_2}|\tilde{\tau}_2 - \tau_2| \\ &< \frac{1}{2}\eta_{22} + L_{g_2}\delta_{\tau_2} < \eta_{22} < \varepsilon, \end{aligned}$$

i.e. $H(\tilde{\chi}(\tilde{\tau}_2 + 0; 0, \tilde{\chi}_0), r(\tau_2, \tilde{\tau}_2]) < \varepsilon$.

Note that $|\tilde{\varsigma}_2 - \varsigma_2| \leqslant |\tilde{\tau}_2 - \tau_2| + |\tilde{d}_{\varsigma_2} - d| < 2\delta_{\tau_2}$. For $\varsigma_2 < \tau \leqslant \tilde{\varsigma}_2$,

$$\begin{aligned} &\big\|\chi(\varsigma_2; 0, \chi_0) - \tilde{\chi}(\tau; 0, \tilde{\chi}_0)\big\| \\ &\leqslant \big\|\chi(\varsigma_2; 0, \chi_0) - \tilde{\chi}(\varsigma_2; 0, \tilde{\chi}_0)\big\| + \big\|\tilde{\chi}(\tau; 0, \tilde{\chi}_0) - \tilde{\chi}(\varsigma_2; 0, \tilde{\chi}_0)\big\| \\ &< \frac{1}{2}\eta_{22} + \Big\|g_2(\tau, \tilde{\chi}(\tilde{\tau}_2^-)) - g_2(\varsigma_2, \tilde{\chi}(\tilde{\tau}_2^-))\Big\| \\ &\leqslant \frac{1}{2}\eta_{22} + L_{g_2}|\tilde{\varsigma}_2 - \varsigma_2| \\ &< \frac{1}{2}\eta_{22} + 2L_{g_2}\delta_{\tau_2} < \eta_{22} < \varepsilon, \end{aligned}$$

i.e. $H(\chi(\varsigma_2; 0, \chi_0), \tilde{r}(\varsigma_2, \tilde{\varsigma}_2]) < \varepsilon$.

Let δ_{ς_2} be an arbitrary positive constant, put $\eta_{22} < 2L_{g_2}\delta_{\varsigma_2}$, and then $|\tilde{\xi}_2 - \varsigma_2| < \delta_{\varsigma_2}$. Hence $\forall\ \delta_{\varsigma_2} > 0,\ \exists\ \delta_{\tau_2} > 0, \forall\ \tilde{d}_{\varsigma_2} > 0,\ |\tilde{d}_{\varsigma_2} - d| < \delta_{\tau_2}$, then $H(\tilde{r}(\tilde{\tau}_2, \tilde{\xi}_2], r(\tau_2, \varsigma_2]) < \varepsilon$, $|\tilde{\xi}_2 - \varsigma_2| < \delta_{\varsigma_2}$.

The Hausdorff distance about the trajectories $\tilde{r}(\tilde{\xi}_2, \tilde{\tau}_3]$ and $r(\varsigma_2, \tau_3]$ is

$$\begin{aligned}
&H(\tilde{r}(\tilde{\xi}_2, \tilde{\tau}_3], r(\varsigma_2, \tau_3]) \\
&\leqslant \max\Big\{R(\tilde{r}(\varsigma_2^{\max}, \tau_3^{\min}], r(\varsigma_2^{\max}, \tau_3^{\min}]), H\big(\chi(\varsigma_2 + 0; 0, \chi_0), \tilde{r}(\tilde{\xi}_2, \varsigma_2]\big), \\
&\quad H\big(\tilde{\chi}(\tilde{\xi}_2 + 0; 0, \tilde{\chi}_0), r(\varsigma_2, \tilde{\xi}_2]\big), H\big(\chi(\tau_3; 0, \chi_0), \tilde{r}(\tau_3, \tilde{\tau}_3]\big), \\
&\quad H\big(\tilde{\chi}(\tilde{\tau}_3; 0, \tilde{\chi}_0), r(\tilde{\tau}_3, \tau_3]\big)\Big\}.
\end{aligned} \tag{2.82}$$

Since $(\tilde{\xi}_2, \varsigma_2] = \varnothing$, $(\tilde{\tau}_3, \tau_3] = \varnothing$, then $H(\chi(\varsigma_2 + 0; 0, \chi_0), \tilde{r}(\tilde{\xi}_2, \varsigma_2]) = 0$, $H(\tilde{\chi}(\tilde{\tau}_3; 0, \tilde{\chi}_0), r(\tilde{\tau}_3, \tau_3]) = 0$.

Let $0 < \eta_{23} < \varepsilon$, and we have that $\exists\ \delta_{\varsigma_2} > 0,\ \delta_{\varsigma_2} < \frac{\eta_{23}}{4M},\ \forall\ \tilde{\xi}_2 \in \mathbb{R}^+, |\tilde{\xi}_2 - \varsigma_2| < \delta_{\varsigma_2}$, $\forall\ \tilde{d}_{\tau_3} > 0,\ |\tilde{d}_{\tau_3} - d| < \delta_{\varsigma_2}$, then $\|\tilde{\chi}(\tau; 0, \tilde{\chi}_0) - \chi(\tau; 0, \chi_0)\| < \frac{1}{2}\eta_{23},\ \varsigma_2^{\max} < \tau \leqslant \tau_3^{\min}$, that is $R(\tilde{r}(\varsigma_2^{\max}, \tau_3^{\min}], r(\varsigma_2^{\max}, \tau_3^{\min}]) < \frac{1}{2}\eta_{23} < \varepsilon$.

For $\varsigma_2 < \tau \leqslant \tilde{\xi}_2$,

$$\begin{aligned}
&\left\|\tilde{\chi}(\tilde{\xi}_2 + 0; 0, \tilde{\chi}_0) - \chi(\tau; 0, \chi_0)\right\| \\
&\leqslant \left\|\tilde{\chi}(\tilde{\xi}_2; 0, \tilde{\chi}_0) - \chi(\tilde{\xi}_2; 0, \chi_0)\right\| + \left\|\chi(\tilde{\xi}_2; 0, \chi_0) - \chi(\tau; 0, \chi_0)\right\| \\
&< \frac{1}{2}\eta_{23} + \left\|(\mathbf{I}^1_{\tau,\tilde{\xi}_2} f)(\tilde{\xi}_2, \chi)\right\| \\
&\leqslant \frac{1}{2}\eta_{23} + M|\tilde{\xi}_2 - \varsigma_2| \\
&< \frac{1}{2}\eta_{23} + M\delta_{\varsigma_2} < \eta_{23} < \varepsilon,
\end{aligned}$$

i.e. $H(\tilde{\chi}(\tilde{\xi}_2 + 0; 0, \tilde{\chi}_0), r(\varsigma_2, \tilde{\xi}_2]) < \varepsilon$.

Consider $|\tilde{\tau}_3 - \tau_3| \leqslant |\tilde{\xi}_2 - \varsigma_2| + |\tilde{d}_{\tau_3} - d| < 2\delta_{\varsigma_2}$. For $\tau_3 < \tau \leqslant \tilde{\tau}_3$,

$$\begin{aligned}
&\left\|\chi(\tau_3; 0, \chi_0) - \tilde{\chi}(\tau; 0, \tilde{\chi}_0)\right\| \\
&\leqslant \left\|\chi(\tau_3; 0, \chi_0) - \tilde{\chi}(\tau_3; 0, \tilde{\chi}_0)\right\| + \left\|\tilde{\chi}(\tau; 0, \tilde{\chi}_0) - \tilde{\chi}(\tau_3; 0, \tilde{\chi}_0)\right\| \\
&< \frac{1}{2}\eta_{23} + \left\|(\mathbf{I}^1_{\tau_3,\tau} f)(\tau, \tilde{\chi})\right\| \\
&\leqslant \frac{1}{2}\eta_{23} + M|\tilde{\tau}_3 - \tau_3| \\
&< \frac{1}{2}\eta_{23} + 2M\delta_{\varsigma_2} < \eta_{23} < \varepsilon,
\end{aligned}$$

i.e. $H(\chi(\tau_3; 0, \chi_0), \tilde{r}(\tau_3, \tilde{\tau}_3]) < \varepsilon$.

Set $\eta_{23} < 2M\delta_{\tau_3}$, then $|\tilde{\tau}_3 - \tau_3| < \delta_{\tau_3}$, where δ_{τ_3} denotes an arbitrary positive constant. Therefore, $\forall\ \delta_{\tau_3} > 0,\ \exists\ \delta_{\varsigma_2} > 0,\ \forall\ \tilde{\xi}_2 \in \mathbb{R}^+,\ |\tilde{\xi}_2 - \varsigma_2| < \delta_{\varsigma_2},\ \forall\ \tilde{d}_{\tau_3} > 0,$ $|\tilde{d}_{\tau_3} - d| < \delta_{\varsigma_2}$, then $H(\tilde{r}(\tilde{\xi}_2, \tilde{\tau}_3], r(\varsigma_2, \tau_3]) < \varepsilon$, $|\tilde{\tau}_3 - \tau_3| < \delta_{\tau_3}$.

From the above procedure, we arrive at the conclusion:

$$\begin{aligned}&\forall\ \delta_{\varsigma_i} > 0,\ \exists\ \delta_{\tau_i} > 0,\ \forall\ \tilde{d}_{\varsigma_i} > 0,\ |\tilde{d}_{\varsigma_i} - d| < \delta_{\tau_i},\ \text{then}\\ &H(\tilde{r}(\tilde{\tau}_i, \tilde{\xi}_i], r(\tau_i, \varsigma_i]) < \varepsilon,\quad |\tilde{\xi}_i - \varsigma_i| < \delta_{\varsigma_i},\quad i = 1, 2, \ldots, k.\end{aligned} \tag{2.83}$$

$$\begin{aligned}&\forall\,\delta_{\tau_{i+1}} > 0,\ \exists\ \delta_{\varsigma_i} > 0,\ \forall\ \tilde{\xi}_i \in \mathbb{R}^+,\ |\tilde{\xi}_i - \varsigma_i| < \delta_{\varsigma_i},\ \forall\,\tilde{d}_{\tau_{i+1}} > 0,\ |\tilde{d}_{\tau_{i+1}} - d| < \delta_{\varsigma_i},\\ &\text{then}\quad H(\tilde{r}(\tilde{\xi}_i, \tilde{\tau}_{i+1}], r(\varsigma_i, \tau_{i+1}]) < \varepsilon,\quad |\tilde{\tau}_{i+1} - \tau_{i+1}| < \delta_{\tau_{i+1}},\quad i = 1, 2, \ldots, k-1.\end{aligned} \tag{2.84}$$

Finally, consider the trajectories $\tilde{r}(\tilde{\xi}_k, T]$ and $r(\varsigma_k, T]$,

$$\begin{aligned}&H(\tilde{r}(\tilde{\xi}_k, T], r(\varsigma_k, T])\\ &\leqslant \max\Big\{R(\tilde{r}(\varsigma_k^{\max}, T], r(\varsigma_k^{\max}, T]), H\big(\chi(\varsigma_k + 0; 0, \chi_0), \tilde{r}(\tilde{\xi}_k, \varsigma_k]\big),\\ &H\big(\tilde{\chi}(\tilde{\xi}_k + 0; 0, \tilde{\chi}_0), r(\varsigma_k, \tilde{\xi}_k]\big)\Big\}.\end{aligned} \tag{2.85}$$

Since $(\tilde{\xi}_k, \varsigma_k] = \varnothing$, $H(\chi(\varsigma_k + 0; 0, \chi_0), \tilde{r}(\tilde{\xi}_k, \varsigma_k]) = 0$.

One can deduce that $\forall\ \varepsilon > 0,\ \exists\ \delta_{\varsigma_k},\ 0 < \delta_{\varsigma_k} < \frac{\varepsilon}{2M},\ \forall\ \tilde{\xi}_k \in \mathbb{R}^+,\ |\tilde{\xi}_k - \varsigma_k| < \delta_{\varsigma_k},$ $\forall\,\tilde{d}_{\tau_{k+1}} > 0,\ |\tilde{d}_{\tau_{k+1}} - d| < \delta_{\varsigma_k}$, then $\|\tilde{\chi}(\tau; 0, \tilde{\chi}_0) - \chi(\tau; 0, \chi_0)\| < \frac{\varepsilon}{2}$, $\varsigma_k^{\max} < \tau \leqslant T$, that is $R(\tilde{r}(\varsigma_k^{\max}, T], r(\varsigma_k^{\max}, T]) < \varepsilon$.

For $\varsigma_k < \tau \leqslant \tilde{\xi}_k$,

$$\begin{aligned}&\big\|\tilde{\chi}(\tilde{\xi}_k + 0; 0, \tilde{\chi}_0) - \chi(\tau; 0, \chi_0)\big\|\\ &\leqslant \big\|\tilde{\chi}(\tilde{\xi}_k; 0, \tilde{\chi}_0) - \chi(\tilde{\xi}_k; 0, \chi_0)\big\| + \big\|\chi(\tilde{\xi}_k; 0, \chi_0) - \chi(\tau; 0, \chi_0)\big\|\\ &< \frac{\varepsilon}{2} + \left\|(\mathbf{I}^1_{\tau, \tilde{\xi}_k} f)(\tilde{\xi}_k, \chi)\right\|\\ &\leqslant \frac{\varepsilon}{2} + M|\tilde{\xi}_k - \varsigma_k|\\ &< \frac{\varepsilon}{2} + M\delta_{\varsigma_k} < \varepsilon,\end{aligned}$$

i.e. $H(\tilde{\chi}(\tilde{\xi}_k + 0; 0, \tilde{\chi}_0), r(\varsigma_k, \tilde{\xi}_k]) < \varepsilon$.

Therefore, $\forall\ \varepsilon > 0,\ \exists\ \delta_{\varsigma_k} > 0,\quad \forall\ \tilde{\xi}_k \in R^+,\quad |\tilde{\xi}_k - \varsigma_k| < \delta_{\varsigma_k},\ \forall\,\tilde{d}_{\tau_k+1} > 0,$ $|\tilde{d}_{\tau_k+1} - d| < \delta_{\varsigma_k}$, then $H(\tilde{r}(\tilde{\xi}_k, T], r(\varsigma_k, T]) < \varepsilon$.

Now $\delta_{\varsigma_k} = \delta_{\varsigma_k}(\varepsilon)$, $\delta_{\tau_k} = \delta_{\tau_k}(\delta_{\varsigma_k}, \varepsilon)$, $\delta_{\varsigma_{k-1}} = \delta_{\varsigma_{k-1}}(\delta_{\tau_k}, \varepsilon)$, $\cdots$, $\delta_{\tau_1} = \delta_{\tau_1}(\delta_{\varsigma_1}, \varepsilon)$ and $\delta_0 = \delta_0(\delta_T, \delta_{\tau_1}, \varepsilon)$.

Consequently, one has the conclusion: $\forall\ \varepsilon > 0,\ \exists\ \delta_0 > 0,\ \forall\ \tilde{\chi}_0 \in \mathbb{D}$, $\|\tilde{\chi}_0 - \chi_0\| < \delta_0,\ \forall\ \tilde{d}_{\tau_i} > 0,\ \forall\ \tilde{d}_{\varsigma_i} > 0,\ |\tilde{d}_{\tau_i} - d| < \delta_0,\ |\tilde{d}_{\varsigma_i} - d| < \delta_0,\ i = 1, 2, \ldots, k,$

then $H(\tilde{r}[0, \tilde{\tau}_1], r[0, \tau_1]) < \varepsilon$, $H(\tilde{r}(\tilde{\tau}_i, \tilde{\varsigma}_i], r(\tau_i, \varsigma_i]) < \varepsilon$, $i = 1, 2, \ldots, k$, $H(\tilde{r}(\tilde{\varsigma}_i, \tilde{\tau}_{i+1}], r(\varsigma_i, \tau_{i+1}]) < \varepsilon$, $i = 1, 2, \ldots, k - 1$, $H(\tilde{r}(\tilde{\varsigma}_k, T], r(\varsigma_k, T]) < \varepsilon$.

Note that

$$r[0, T] = r[0, \tau_1] \bigcup \left(\bigcup_{i=1,2,\ldots,k} r(\tau_i, \varsigma_i] \right) \bigcup \left(\bigcup_{i=1,2,\ldots,k-1} r(\varsigma_i, \tau_{i+1}] \right) \bigcup_r (\varsigma_k, T],$$

$$\tilde{r}[0, T] = \tilde{r}[0, \tilde{\tau}_1] \bigcup \left(\bigcup_{i=1,2,\ldots,k} \tilde{r}(\tilde{\tau}_i, \tilde{\varsigma}_i] \right) \bigcup \left(\bigcup_{i=1,2,\ldots,k-1} \tilde{r}(\tilde{\varsigma}_i, \tilde{\tau}_{i+1}] \right) \bigcup \tilde{r}(\tilde{\varsigma}_k, T].$$

Now apply theorem 2.2.1, and then

$$\begin{aligned} & H(\tilde{r}[0, T], r[0, T]) \\ &= H\Bigg(\tilde{r}[0, \tilde{\tau}_1] \bigcup \left(\bigcup_{i=1,2,\ldots,k} \tilde{r}(\tilde{\tau}_i, \tilde{\varsigma}_i] \right) \bigcup \left(\bigcup_{i=1,2,\ldots,k-1} \tilde{r}(\tilde{\varsigma}_i, \tilde{\tau}_{i+1}] \right) \bigcup \tilde{r}(\tilde{\varsigma}_k, T], \\ & \quad r[0, \tau_1] \bigcup \left(\bigcup_{i=1,2,\ldots,k} r(\tau_i, \varsigma_i] \right) \bigcup \left(\bigcup_{i=1,2,\ldots,k-1} r(\varsigma_i, \tau_{i+1}] \right) \bigcup_r (\varsigma_k, T] \Bigg) \\ &\leqslant \max\big\{ H(\tilde{r}[0, \tilde{\tau}_1], r[0, \tau_1]),\ H(\tilde{r}(\tilde{\tau}_i, \tilde{\varsigma}_i], r(\tau_i, \varsigma_i]),\ i = 1, \ldots, k, \\ & \quad H(\tilde{r}(\tilde{\varsigma}_i, \tilde{\tau}_{i+1}], r(\varsigma_i, \tau_{i+1}]),\ i = 1, \ldots, k - 1,\ H(\tilde{r}(\tilde{\varsigma}_k, T], r(\varsigma_k, T])\big\} < \varepsilon. \end{aligned} \tag{2.86}$$

Case 2. Let $\tau_i^{\min} = \tau_i$, $\tau_i^{\max} = \tilde{\tau}_i$, $\varsigma_i^{\min} = \tilde{\varsigma}_i$, $\varsigma_i^{\max} = \varsigma_i$, $i = 1, 2, \ldots$; (the case $\tau_i^{\min} = \tilde{\tau}_i$, $\tau_i^{\max} = \tau_i$, $\varsigma_i^{\min} = \varsigma_i$, $\varsigma_i^{\max} = \tilde{\varsigma}_i$, $i = 1, 2, \ldots$ can be considered analogously).

For the point $(0, \chi_0) \in [0, \infty) \times \mathbb{R}^n$, let ϵ and T be positive constants. Since $\tau_i \to \infty$ $(i \to \infty)$, then $\exists\, k \in \mathbb{N}$ such that $2kd = \varsigma_k < T < \tau_{k+1} = (2k + 1)d$.

In this case, we still have the formula (2.76) (see figure (a)), the Hausdorff distance between the trajectories $\tilde{r}[0, \tilde{\tau}_1]$ and $r[0, \tau_1]$, and we have the same conclusion that $\forall\, \delta_{\tau_1} > 0$, $\exists\, \delta_0 > 0$, $\forall\, \tilde{\chi}_0 \in \mathbb{D}$, $\|\tilde{\chi}_0 - \chi_0\| < \delta_0$, $\forall\, \tilde{d}_{\tau_1} > 0$, $|\tilde{d}_{\tau_1} - d| < \delta_0$, then $H(\tilde{r}[0, \tilde{\tau}_1], r[0, \tau_1]) < \varepsilon$, $|\tilde{\tau}_1 - \tau_1| < \delta_{\tau_1}$.

For the trajectories $\tilde{r}(\tilde{\tau}_1, \tilde{\varsigma}_1]$ and $r(\tau_1, \varsigma_1]$, we have the inequality (2.77). Since $(\tilde{\tau}_1, \tau_1] = \varnothing$ and $(\varsigma_1, \tilde{\varsigma}_1] = \varnothing$, then $H(\chi(\tau_1 + 0; 0, \chi_0), \tilde{r}(\tilde{\tau}_1, \tau_1]) = 0$, $H(\chi(\varsigma_1; 0, \chi_0), \tilde{r}(\varsigma_1, \tilde{\varsigma}_1]) = 0$ (see figure 2.9).

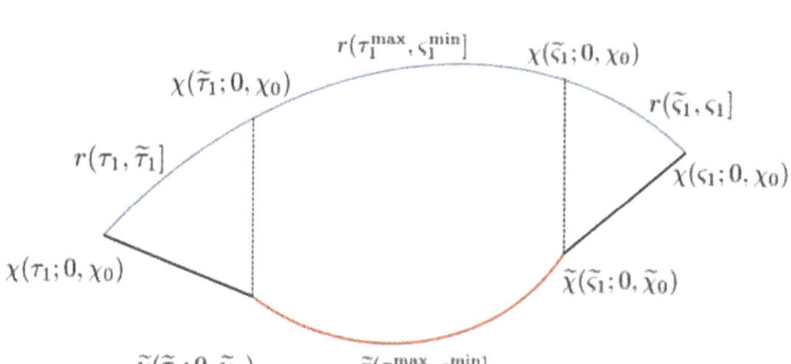

Figure 2.9. The line r denotes the orbital of solution of (2.68) in interval $(\tau_1, \varsigma_1]$ and the line $\tilde{r}$ denotes the orbital of solution of (2.70) in interval $(\tilde{\tau}_1, \tilde{\varsigma}_1]$. Reprinted from [22]. Copyright (2017), with permission from Elsevier.

Let $0 < \eta_{11} < \varepsilon$, we have that $\exists\ \delta_{\tau_1} > 0,\ \delta_{\tau_1} < \frac{\eta_{11}}{4L_{g_1}},\ \forall\ \tilde{d}_{\varsigma_1} > 0,\ |d - \tilde{d}_{\varsigma_1}| < \delta_{\tau_1}$, then $\|\tilde{\chi}(\tau; 0, \tilde{\chi}_0) - \chi(\tau; 0, \chi_0)\| < \frac{1}{2}\eta_{11}$, $\tau_1^{\max} < \tau \leqslant \varsigma_1^{\min}$, that is $R(\tilde{r}(\tau_1^{\max}, \varsigma_1^{\min}], r(\tau_1^{\max}, \varsigma_1^{\min}]) < \frac{1}{2}\eta_{11} < \varepsilon$.

For $\tau_1 < \tau \leqslant \tilde{\tau}_1$, similar to (2.78), we obtain $H(\tilde{\chi}(\tilde{\tau}_1 + 0; 0, \tilde{\chi}_0), r(\tau_1, \tilde{\tau}_1]) < \varepsilon$.

Note that $|\varsigma_1 - \tilde{\varsigma}_1| \leqslant |\tau_1 - \tilde{\tau}_1| + |d - \tilde{d}_{\varsigma_1}| < 2\delta_{\tau_1}$. For $\tilde{\varsigma}_1 < \tau \leqslant \varsigma_1$,

$$\begin{aligned}
&\left\|\tilde{\chi}(\tilde{\varsigma}_1; 0, \tilde{\chi}_0) - \chi(\tau; 0, \chi_0)\right\| \\
\leqslant\ & \left\|\tilde{\chi}(\tilde{\varsigma}_1; 0, \tilde{\chi}_0) - \chi(\tilde{\varsigma}_1; 0, \chi_0)\right\| + \left\|\chi(\tau; 0, \chi_0) - \chi(\tilde{\varsigma}_1; 0, \chi_0)\right\| \\
<\ & \frac{1}{2}\eta_{11} + \left\|g_1(\tau, \chi(\tau_1^-)) - g_1(\tilde{\varsigma}_1, \chi(\tau_1^-))\right\| \\
\leqslant\ & \frac{1}{2}\eta_{11} + L_{g_1}|\varsigma_1 - \tilde{\varsigma}_1| \\
<\ & \frac{1}{2}\eta_{11} + 2L_{g_1}\delta_{\tau_1} < \eta_{11} < \varepsilon,
\end{aligned} \tag{2.87}$$

i.e. $H(\tilde{\chi}(\tilde{\varsigma}_1; 0, \tilde{\chi}_0), r(\tilde{\varsigma}_1, \varsigma_1]) < \varepsilon$.

Let δ_{ς_1} denote an arbitrary positive constant, and assume that $\eta_{11} < 2L_{g_1}\delta_{\varsigma_1}$, and then $|\varsigma_1 - \tilde{\varsigma}_1| < \delta_{\varsigma_1}$.

Hence, $\forall\ \delta_{\varsigma_1} > 0,\ \exists\ \delta_{\tau_1} > 0,\ \forall\ \tilde{d}_{\varsigma_1} > 0,\ |d - \tilde{d}_{\varsigma_1}| < \delta_{\tau_1}$, then $H(\tilde{r}(\tilde{\tau}_1, \tilde{\varsigma}_1], r(\tau_1, \varsigma_1]) < \varepsilon$, $|\varsigma_1 - \tilde{\varsigma}_1| < \delta_{\varsigma_1}$.

For the Hausdorff distance about the trajectories $\tilde{r}(\tilde{\varsigma}_1, \tilde{\tau}_2]$ and $r(\varsigma_1, \tau_2]$ (see figure 2.10),we have the inequality (2.79). Since $(\varsigma_1, \tilde{\varsigma}_1] = \varnothing$ and $(\tilde{\tau}_2, \tau_2] = \varnothing$, then $H(\tilde{\chi}(\tilde{\varsigma}_1 + 0; 0, \tilde{\chi}_0), r(\varsigma_1, \tilde{\varsigma}_1]) = 0$. $H(\tilde{\chi}(\tilde{\tau}_2; 0, \tilde{\chi}_0), r(\tilde{\tau}_2, \tau_2]) = 0$.

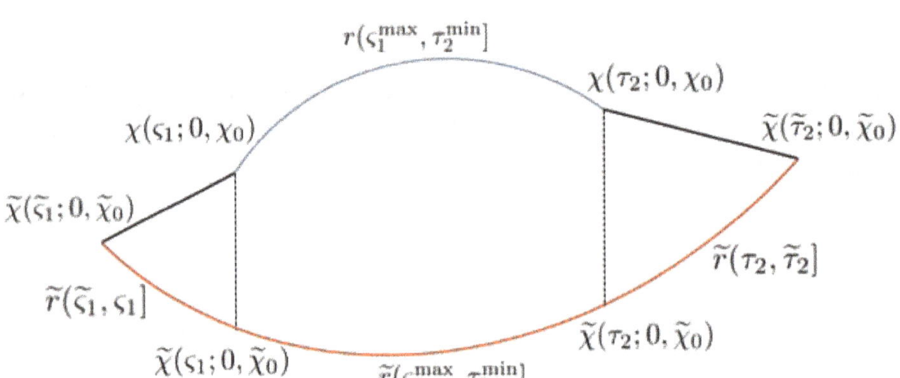

Figure 2.10. The line r denotes the orbital of solution of (2.68) in interval $(\varsigma_1, \tau_2]$ and the line $\tilde{r}$ denotes the orbital of solution of (2.70) in interval $(\tilde{\varsigma}_1, \tilde{\tau}_2]$. Reprinted from [22]. Copyright (2017), with permission from Elsevier.

Set $0 < \eta_{12} < \varepsilon$, and we have that $\exists\ \delta_{\varsigma_1} > 0$, $\delta_{\varsigma_1} < \frac{\eta_{12}}{4M}$, $\forall\ \tilde{\varsigma}_1 \in \mathbb{R}^+$, $|\varsigma_1 - \tilde{\varsigma}_1| < \delta_{\varsigma_1}$, $\forall\ \tilde{d}_{\tau_2} > 0$, $|d - \tilde{d}_{\tau_2}| < \delta_{\varsigma_1}$, then $\|\tilde{\chi}(\tau; 0, \tilde{\chi}_0) - \chi(\tau; 0, \chi_0)\| < \frac{1}{2}\eta_{12}$, $\varsigma_1^{\max} < \tau \leqslant \tau_2^{\min}$, that is $R(\tilde{r}(\varsigma_1^{\max}, \tau_2^{\min}], r(\varsigma_1^{\max}, \tau_2^{\min}]) < \frac{1}{2}\eta_{12} < \varepsilon$.

For $\tilde{\varsigma}_1 < \tau \leqslant \varsigma_1$,

$$\begin{aligned}
&\left\|\chi(\varsigma_1 + 0; 0, \chi_0) - \tilde{\chi}(\tau; 0, \tilde{\chi}_0)\right\| \\
\leqslant\ & \left\|\chi(\varsigma_1; 0, \chi_0) - \tilde{\chi}(\varsigma_1; 0, \tilde{\chi}_0)\right\| + \left\|\tilde{\chi}(\varsigma_1; 0, \tilde{\chi}_0) - \tilde{\chi}(\tau; 0, \tilde{\chi}_0)\right\| \\
<\ & \frac{1}{2}\eta_{12} + \left\|(\mathbf{I}^1_{\tau, \varsigma_1} f)(\varsigma_1, \tilde{\chi})\right\| \\
\leqslant\ & \frac{1}{2}\eta_{12} + M|\varsigma_1 - \tilde{\varsigma}_1| \\
<\ & \frac{1}{2}\eta_{12} + M\delta_{\varsigma_1} < \eta_{12} < \varepsilon,
\end{aligned}$$

i.e. $H(\chi(\varsigma_1 + 0; 0, \chi_0), \tilde{r}(\tilde{\varsigma}_1, \varsigma_1]) < \varepsilon$.

For $\tau_2 < \tau \leqslant \tilde{\tau}_2$, similar to (2.80), we get $H(\chi(\tau_2; 0, \chi_0), \tilde{r}(\tau_2, \tilde{\tau}_2]) < \varepsilon$.

Set $\eta_{12} < 2M\delta_{\tau_2}$, and then $|\tilde{\tau}_2 - \tau_2| < \delta_{\tau_2}$, where δ_{τ_2} is an arbitrary positive constant. Therefore, $\forall\ \delta_{\tau_2} > 0$, $\exists\ \delta_{\varsigma_1} > 0$, $\forall\ \tilde{\varsigma}_1 \in \mathbb{R}^+$, $|\varsigma_1 - \tilde{\varsigma}_1| < \delta_{\varsigma_1}$, $\forall\ \tilde{d}_{\tau_2} > 0$, $|d - \tilde{d}_{\tau_2}| < \delta_{\varsigma_1}$, then $H(\tilde{r}(\tilde{\varsigma}_1, \tilde{\tau}_2], r(\varsigma_1, \tau_2]) < \varepsilon$, $|\tilde{\tau}_2 - \tau_2| < \delta_{\tau_2}$.

Repeating the procedure, we obtain the same conclusion as in (2.83) and (2.84).

Considering the trajectories $\tilde{r}(\tilde{\varsigma}_k, T]$ and $r(\varsigma_k, T]$, we still have the expression (2.85).

Since $(\varsigma_k, \tilde{\varsigma}_k] = \varnothing$, $H(\tilde{\chi}(\tilde{\varsigma}_k + 0; 0, \tilde{\chi}_0), r(\varsigma_k, \tilde{\varsigma}_k]) = 0$.

We have that $\forall\ \varepsilon > 0, \exists\ \delta_{\varsigma_k}$, $0 < \delta_{\varsigma_k} < \frac{\varepsilon}{2M}$, $\forall\ \tilde{\varsigma}_k \in \mathbb{R}^+$, $|\varsigma_k - \tilde{\varsigma}_k| < \delta_{\varsigma_k}$, $\forall\, \tilde{d}_{\tau_{k+1}} > 0$, $|d - \tilde{d}_{\tau_{k+1}}| < \delta_{\varsigma_k}$, then $\|\tilde{\chi}(\tau; 0, \tilde{\chi}_0) - \chi(\tau; 0, \chi_0)\| < \frac{\varepsilon}{2}$, $\varsigma_k^{\max} < \tau \leqslant T$, that is $R(\tilde{r}(\varsigma_k^{\max}, T], r(\varsigma_k^{\max}, T]) < \varepsilon$.

For $\tilde{\varsigma}_k < \tau \leqslant \varsigma_k$,

$$\begin{aligned}
&\left\|\chi(\varsigma_k + 0; 0, \chi_0) - \tilde{\chi}(\tau; 0, \tilde{\chi}_0)\right\| \\
\leqslant\ & \left\|\chi(\varsigma_k; 0, \chi_0) - \tilde{\chi}(\varsigma_k; 0, \tilde{\chi}_0)\right\| + \left\|\tilde{\chi}(\varsigma_k; 0, \tilde{\chi}_0) - \tilde{\chi}(\tau; 0, \tilde{\chi}_0)\right\| \\
<\ & \frac{\varepsilon}{2} + \left\|(\mathbf{I}^1_{\tau, \varsigma_k} f)(\varsigma_k, \tilde{\chi})\right\| \\
\leqslant\ & \frac{\varepsilon}{2} + M|\varsigma_k - \tilde{\varsigma}_k| \\
<\ & \frac{\varepsilon}{2} + M\delta_{\varsigma_k} < \varepsilon,
\end{aligned}$$

i.e. $H(\chi(\varsigma_k + 0; 0, \chi_0), \tilde{r}(\tilde{\varsigma}_k, \varsigma_k]) < \varepsilon$.

Therefore, $\forall\ \varepsilon > 0$, $\exists\ \delta_{\varsigma_k} > 0$, $\forall\ \tilde{\varsigma}_k \in \mathbb{R}^+$, $|\varsigma_k - \tilde{\varsigma}_k| < \delta_{\varsigma_k}$, $\forall\, \tilde{d}_{\tau_k+1} > 0$, $|d - \tilde{d}_{\tau_k+1}| < \delta_{\varsigma_k}$, then $H(\tilde{r}(\tilde{\varsigma}_k, T], r(\varsigma_k, T]) < \varepsilon$.

Now $\delta_{\varsigma_k} = \delta_{\varsigma_k}(\varepsilon)$, $\delta_{\tau_k} = \delta_{\tau_k}(\delta_{\varsigma_k}, \varepsilon)$, $\delta_{\varsigma_{k-1}} = \delta_{\varsigma_{k-1}}(\delta_{\tau_k}, \varepsilon)$, $\cdots$, $\delta_{\tau_1} = \delta_{\tau_1}(\delta_{\varsigma_1}, \varepsilon)$ and $\delta_0 = \delta_0(\delta_{\tau_1}, \varepsilon)$.

Thus we infer that $\forall\, \varepsilon > 0$, $\exists\, \delta_0 > 0$, $\forall\, \tilde{\chi}_0 \in \mathbb{D}$, $\|\tilde{\chi}_0 - \chi_0\| < \delta_0$, $\forall\, \tilde{d}_{\tau_i} > 0$, $\forall\, \tilde{d}_{\varsigma_i} > 0$, $|\tilde{d}_{\tau_i} - d| < \delta_0$, $|\tilde{d}_{\varsigma_i} - d| < \delta_0$, $i = 1, 2, \ldots, k$, then $H(\tilde{r}[0, \tilde{\tau}_1], r[0, \tau_1]) < \varepsilon$, $H(\tilde{r}(\tilde{\tau}_i, \tilde{\varsigma}_i], r(\tau_i, \varsigma_i]) < \varepsilon$, $i = 1, \ldots, k$, $H(\tilde{r}(\tilde{\varsigma}_i, \tilde{\tau}_{i+1}], r(\varsigma_i, \tau_{i+1}]) < \varepsilon$, $i = 1, \ldots, k-1$, $H(\tilde{r}(\tilde{\varsigma}_k, T], r(\varsigma_k, T]) < \varepsilon$.

Note that

$$r[0, T] = r[0, \tau_1] \bigcup \left(\bigcup_{i=1,2,\ldots,k} r(\tau_i, \varsigma_i] \right) \bigcup \left(\bigcup_{i=1,2,\ldots,k-1} r(\varsigma_i, \tau_{i+1}] \right) \bigcup_r (\varsigma_k, T],$$

$$\tilde{r}[0, T] = \tilde{r}[0, \tilde{\tau}_1] \bigcup \left(\bigcup_{i=1,2,\ldots,k} \tilde{r}(\tilde{\tau}_i, \tilde{\varsigma}_i] \right) \bigcup \left(\bigcup_{i=1,2,\ldots,k-1} \tilde{r}(\tilde{\varsigma}_i, \tilde{\tau}_{i+1}] \right) \bigcup \tilde{r}(\tilde{\varsigma}_k, T].$$

Apply theorem 2.2.1 and we have the result (2.86), that is $H(\tilde{r}[0, T], r[0, T]) < \varepsilon$.

Case 3. Let $\tau_i^{\min} = \tau_i$, $\varsigma_i^{\min} = \varsigma_i$, $\tau_i^{\max} = \tilde{\tau}_i$, $\varsigma_i^{\max} = \tilde{\varsigma}_i$, $i = 1, 2, \ldots$; (the case $\tau_i^{\min} = \tilde{\tau}_i$, $\varsigma_i^{\min} = \tilde{\varsigma}_i$, $\tau_i^{\max} = \tau_i$, $\varsigma_i^{\max} = \varsigma_i$, $i = 1, 2, \ldots$ can be considered similarly).

For the point $(0, \chi_0) \in [0, \infty) \times \mathbb{R}^n$, let ϵ and T be positive constants. Since $\tau_i \to \infty$ $(i \to \infty)$, then $\exists\, k \in \mathbb{N}$ such that $2kd = \varsigma_k < T < \tau_{k+1} = (2k+1)d$. Therefore, we can select a constant $\delta_T = \delta_T(d, T) > 0$, which is sufficiently small, and then $\forall\, \tilde{d}_{\tau_i}, \tilde{d}_{\varsigma_i} > 0$, $|\tilde{d}_{\tau_i} - d| < \delta_T$, $|\tilde{d}_{\varsigma_i} - d| < \delta_T$ and $T < \varsigma_{k+1}^{\min}$, $\tau_i^{\min} < \varsigma_i^{\min} < \tau_i^{\max} < \varsigma_i^{\max} < \tau_{i+1}^{\min}$, $i = 1, 2, \ldots, k+1$.

Furthermore,

$$\tilde{\varsigma}_1 < \tau_2 \Leftrightarrow \tilde{d}_{\tau_1} + \tilde{d}_{\varsigma_1} < 3d \Rightarrow \delta_T < \frac{d}{2};$$

$$\tilde{\varsigma}_2 < \tau_3 \Leftrightarrow \tilde{d}_{\tau_1} + \tilde{d}_{\varsigma_1} + \tilde{d}_{\tau_2} + \tilde{d}_{\varsigma_2} < 5d \Rightarrow \delta_T < \frac{d}{4};$$

$$\tilde{\varsigma}_3 < \tau_4 \Leftrightarrow \tilde{d}_{\tau_1} + \tilde{d}_{\varsigma_1} + \cdots + \tilde{d}_{\tau_3} + \tilde{d}_{\varsigma_3} < 7d \Rightarrow \delta_T < \frac{d}{6};$$

$$\vdots$$

$$\tilde{\varsigma}_k < T \Leftrightarrow \tilde{d}_{\tau_1} + \tilde{d}_{\varsigma_1} + \cdots + \tilde{d}_{\tau_k} + \tilde{d}_{\varsigma_k} < T \Rightarrow \delta_T < \frac{T - \varsigma_k}{2k}.$$

From the inequalities, $0 < \delta_T < \min\{\frac{d}{2k}, \frac{T - \varsigma_k}{2k}\}$. With the property $H(\bar{X}, \bar{Y}) = H(X, Y)$ and lemma 2.2.4, we consider the Hausdorff distance between the trajectories on the corresponding subintervals.

For the trajectories $\tilde{r}[0, \tilde{\tau}_1]$ and $r[0, \tau_1]$ (see figure 2.11),

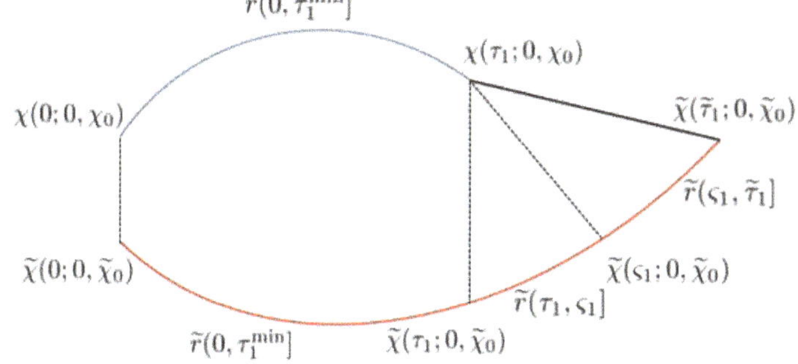

Figure 2.11. The line r denotes the orbital of solution of (2.68) in interval $(0, \tau_1]$ and the line $\tilde{r}$ denotes the orbital of solution of (2.70) in interval $(0, \tilde{\tau}_1]$. Reprinted from [22]. Copyright (2017), with permission from Elsevier.

$$\begin{aligned} &H(\tilde{r}[0, \tilde{\tau}_1], r[0, \tau_1]) = H(\tilde{r}(0, \tilde{\tau}_1], r(0, \tau_1]) \\ &\leqslant \max\Big\{R(\tilde{r}(0, \tau_1^{\min}], r(0, \tau_1^{\min}]), H\big(\chi(\tau_1; 0, \chi_0), \tilde{r}(\tau_1, \varsigma_1]\big), \\ &\quad H(\tilde{r}(\varsigma_1, \tilde{\tau}_1], r(\tilde{\tau}_1, \varsigma_1]), H\big(\tilde{\chi}(\tilde{\tau}_1; 0, \tilde{\chi}_0), r(\varsigma_1, \tau_1]\big)\Big\}. \end{aligned} \tag{2.88}$$

Since $(\tilde{\tau}_1, \varsigma_1] = \varnothing$ and $(\varsigma_1, \tau_1] = \varnothing$, $H(\tilde{r}(\varsigma_1, \tilde{\tau}_1], r(\tilde{\tau}_1, \varsigma_1]) = 0$, $H(\tilde{\chi}(\tilde{\tau}_1; 0, \tilde{\chi}_0), r(\varsigma_1, \tau_1]) = 0$.

Next, we will estimate the other two terms.

Let $0 < \eta_{01} < \varepsilon$, we have that $\exists\, \delta_0 > 0$, $\delta_0 < \min\{\delta_T, \frac{\eta_{01}}{2M}\}$, $\forall\, \tilde{\chi}_0 \in \mathbb{D}$, $\|\tilde{\chi}_0 - \chi_0\| < \delta_0$, $\forall\, \tilde{d}_{\tau_1} > 0$, $|\tilde{d}_{\tau_1} - d| < \delta_0$, then $\|\tilde{\chi}(\tau; 0, \tilde{\chi}_0) - \chi(\tau; 0, \chi_0)\| < \frac{1}{2}\eta_{01}$, $0 < \tau \leqslant \tau_1^{\min}$, that is $R(\tilde{r}(0, \tau_1^{\min}], r(0, \tau_1^{\min}]) < \frac{1}{2}\eta_{01} < \varepsilon$.

Note $|\tilde{\tau}_1 - \tau_1| = |\tilde{d}_{\tau_1} - d| < \delta_0 < \frac{\eta_{01}}{2M}$. For $\tau_1 < \tau \leqslant \varsigma_1$,

$$\begin{aligned} &\left\|\chi(\tau_1; 0, \chi_0) - \tilde{\chi}(\tau; 0, \tilde{\chi}_0)\right\| \\ \leqslant\ &\left\|\chi(\tau_1; 0, \chi_0) - \tilde{\chi}(\tau_1; 0, \tilde{\chi}_0)\right\| + \left\|\tilde{\chi}(\tau; 0, \tilde{\chi}_0) - \tilde{\chi}(\tau_1; 0, \tilde{\chi}_0)\right\| \\ <\ &\frac{1}{2}\eta_{01} + \left\|(\mathbf{I}^1_{\tau_1,\tau}f)(\tau, \tilde{\chi})\right\| \\ \leqslant\ &\frac{1}{2}\eta_{01} + M|\varsigma_1 - \tau_1| < \frac{1}{2}\eta_{01} + M|\tilde{\tau}_1 - \tau_1| \\ <\ &\frac{1}{2}\eta_{01} + M\delta_0 < \eta_{01} < \varepsilon, \end{aligned}$$

i.e. $H(\chi(\tau_1; 0, \chi_0), \tilde{r}(\tau_1, \varsigma_1]) < \varepsilon$.

We assume that $\eta_{01} < 2M\delta_{\tau_1}$, and then $|\tilde{\tau}_1 - \tau_1| < \delta_{\tau_1}$, where δ_{τ_1} is an arbitrary positive constant.

Therefore, $\forall\ \delta_{\tau_1} > 0,\ \exists\ \delta_0 > 0,\ \ \forall\ \tilde{\chi}_0 \in \mathbb{D},\ \ \|\tilde{\chi}_0 - \chi_0\| < \delta_0,\ \ \forall\ \tilde{d}_{\tau_1} > 0$, $|\tilde{d}_{\tau_1} - d| < \delta_0$, then $H(\tilde{r}[0, \tilde{\tau}_1], r[0, \tau_1]) < \varepsilon$, $|\tilde{\tau}_1 - \tau_1| < \delta_{\tau_1}$.

Consider the trajectories $\tilde{r}(\tilde{\varsigma}_1, \tilde{\tau}_2]$ and $r(\varsigma_1, \tau_2]$ (see figure 2.12), and the Hausdorff distance

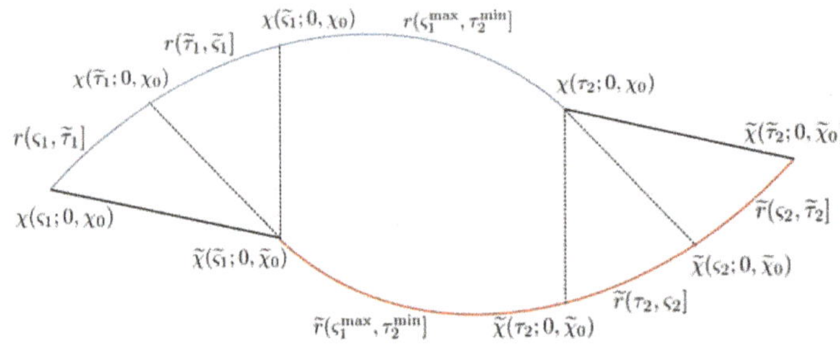

Figure 2.12. The line r denotes the orbital of solution of (2.68) in interval $(\varsigma_1, \tau_2]$ and the line r denotes the orbital of solution of (2.70) in interval $(\tilde{\varsigma}_1, \tilde{\tau}_2]$. Reprinted from [22]. Copyright (2017), with permission from Elsevier.

$$
\begin{aligned}
&H(\tilde{r}(\tilde{\varsigma}_1, \tilde{\tau}_2], r(\varsigma_1, \tau_2]) \\
&\leqslant \max\Big\{R(\tilde{r}(\varsigma_1^{\max}, \tau_2^{\min}], r(\varsigma_1^{\max}, \tau_2^{\min}]), H\big(\chi(\varsigma_1 + 0; 0, \chi_0), \tilde{r}(\tilde{\varsigma}_1, \tilde{\tau}_1]\big), \\
&H(r(\varsigma_1, \tilde{\tau}_1], \tilde{r}(\tilde{\tau}_1, \varsigma_1]), H\big(\tilde{\chi}(\tilde{\varsigma}_1 + 0; 0, \tilde{\chi}_0), r(\tilde{\tau}_1, \tilde{\varsigma}_1]\big), H\big(\chi(\tau_2; 0, \chi_0), \tilde{r}(\tau_2, \varsigma_2]\big), \\
&H(\tilde{r}(\varsigma_2, \tilde{\tau}_2], r(\tilde{\tau}_2, \varsigma_2]), H\big(\tilde{\chi}(\tilde{\tau}_2; 0, \tilde{\chi}_0), r(\varsigma_2, \tau_2]\big)\Big\}.
\end{aligned}
\tag{2.89}
$$

Since $(\tilde{\varsigma}_1, \tilde{\tau}_1] = \varnothing$, $(\tilde{\tau}_1, \varsigma_1] = \varnothing$, $(\tilde{\tau}_2, \varsigma_2] = \varnothing$ and $(\varsigma_2, \tau_2] = \varnothing$, $H(\chi(\varsigma_1 + 0; 0, \chi_0)$, $\tilde{r}(\tilde{\varsigma}_1, \tilde{\tau}_1]) = 0$, $H(r(\varsigma_1, \tilde{\tau}_1], \tilde{r}(\tilde{\tau}_1, \varsigma_1]) = 0$, $H(\tilde{r}(\varsigma_2, \tilde{\tau}_2], r(\tilde{\tau}_2, \varsigma_2]) = 0$ and $H(\tilde{\chi}(\tilde{\tau}_2; 0, \tilde{\chi}_0)$, $r(\varsigma_2, \tau_2]) = 0$.

We need to estimate the other three parts in inequality (2.89).

Set $|\tilde{d}_{\varsigma_1} - d| < \delta_{\tau_1}$, and then $|\tilde{\varsigma}_1 - \varsigma_1| \leqslant |\tilde{\tau}_1 - \tau_1| + |\tilde{d}_{\varsigma_1} - d| < 2\delta_{\tau_1}$, if δ_{τ_1} is sufficiently small, then $|\tilde{\varsigma}_1 - \varsigma_1| < \delta_{\varsigma_1}$, where δ_{ς_1} denotes an arbitrary positive number.

Let $0 < \eta_{12} < \varepsilon$, we have that $\exists\ \delta_{\varsigma_1} > 0$, $\delta_{\varsigma_1} < \frac{\eta_{12}}{4M}$, $\forall\ \tilde{\varsigma}_1 \in \mathbb{R}^+$, $|\tilde{\varsigma}_1 - \varsigma_1| < \delta_{\varsigma_1}$, $\forall\ \tilde{d}_{\tau_2} > 0$, $|\tilde{d}_{\tau_2} - d| < \delta_{\varsigma_1}$, then $\|\tilde{\chi}(\tau; 0, \tilde{\chi}_0) - \chi(\tau; 0, \chi_0)\| < \frac{1}{2}\eta_{12}$, $\varsigma_1^{\max} < \tau \leqslant \tau_2^{\min}$, that is $R(\tilde{r}(\varsigma_1^{\max}, \tau_2^{\min}], r(\varsigma_1^{\max}, \tau_2^{\min}]) < \frac{1}{2}\eta_{12} < \varepsilon$.

For $\tilde{\tau}_1 < \tau \leqslant \tilde{\varsigma}_1$,

$$
\begin{aligned}
& \left\|\tilde{\chi}(\tilde{\varsigma}_1 + 0; 0, \tilde{\chi}_0) - \chi(\tau; 0, \chi_0)\right\| \\
\leqslant\ & \left\|\tilde{\chi}(\tilde{\varsigma}_1; 0, \tilde{\chi}_0) - \chi(\tilde{\varsigma}_1; 0, \chi_0)\right\| + \left\|\chi(\tilde{\varsigma}_1; 0, \chi_0) - \chi(\tau; 0, \chi_0)\right\| \\
<\ & \frac{1}{2}\eta_{12} + \left\|(\mathbf{I}^1_{\tau, \tilde{\varsigma}_1} f)(\tilde{\varsigma}_1, \chi)\right\| \\
\leqslant\ & \frac{1}{2}\eta_{12} + M|\tilde{\varsigma}_1 - \tilde{\tau}_1| < \frac{1}{2}\eta_{12} + M|\tilde{\varsigma}_1 - \varsigma_1| \\
<\ & \frac{1}{2}\eta_{12} + M\delta_{\varsigma_1} < \eta_{12} < \varepsilon,
\end{aligned}
$$

i.e. $H(\tilde{\chi}(\tilde{\varsigma}_1 + 0; 0, \tilde{\chi}_0), r(\tilde{\tau}_1, \tilde{\varsigma}_1]) < \varepsilon$.

Now $|\tilde{\tau}_2 - \tau_2| \leqslant |\tilde{\varsigma}_1 - \varsigma_1| + |\tilde{d}_{\tau_2} - d| < 2\delta_{\varsigma_1}$. For $\tau_2 < \tau \leqslant \varsigma_2$,

$$
\begin{aligned}
& \left\|\chi(\tau_2; 0, \chi_0) - \tilde{\chi}(\tau; 0, \tilde{\chi}_0)\right\| \\
\leqslant\ & \left\|\chi(\tau_2; 0, \chi_0) - \tilde{\chi}(\tau_2; 0, \tilde{\chi}_0)\right\| + \left\|\tilde{\chi}(\tau; 0, \tilde{\chi}_0) - \tilde{\chi}(\tau_2; 0, \tilde{\chi}_0)\right\| \\
<\ & \frac{1}{2}\eta_{12} + \left\|(\mathbf{I}^1_{\tau_2, \tau} f)(\tau, \tilde{\chi})\right\| \\
\leqslant\ & \frac{1}{2}\eta_{12} + M|\varsigma_2 - \tau_2| < \frac{1}{2}\eta_{12} + M|\tilde{\tau}_2 - \tau_2| \\
<\ & \frac{1}{2}\eta_{12} + 2M\delta_{\varsigma_1} < \eta_{12} < \varepsilon,
\end{aligned}
$$

i.e. $H(\chi(\tau_2; 0, \chi_0), \tilde{r}(\tau_2, \varsigma_2]) < \varepsilon$.

Let $\delta_{\tau_2} > 0$ be an arbitrary number, set $\eta_{12} < 2M\delta_{\tau_2}$, and then $|\tilde{\tau}_2 - \tau_2| < \delta_{\tau_2}$.

Therefore, $\forall\ \delta_{\tau_2} > 0,\ \exists\ \delta_{\varsigma_1} > 0, \forall\ \tilde{\varsigma}_1 \in \mathbb{R}^+, |\tilde{\varsigma}_1 - \varsigma_1| < \delta_{\varsigma_1}, \forall\ \tilde{d}_{\tau_2} > 0, |\tilde{d}_{\tau_2} - d| < \delta_{\varsigma_1}$, then $H(\tilde{r}(\tilde{\varsigma}_1, \tilde{\tau}_2], r(\varsigma_1, \tau_2]) < \varepsilon,\ |\tilde{\tau}_2 - \tau_2| < \delta_{\tau_2}$.

From the process above we have the conclusion:

$$
\begin{aligned}
&\forall\,\delta_{\tau_{i+1}} > 0,\ \exists\ \delta_{\varsigma_i} > 0,\ \forall\ \tilde{\varsigma}_i \in \mathbb{R}^+,\ |\tilde{\varsigma}_i - \varsigma_i| < \delta_{\varsigma_i},\ \forall\,\tilde{d}_{\tau_{i+1}} > 0,\ |\tilde{d}_{\tau_{i+1}} - d| < \delta_{\varsigma_i}, \\
&\text{then } H(\tilde{r}(\tilde{\varsigma}_i, \tilde{\tau}_{i+1}], r(\varsigma_i, \tau_{i+1}]) < \varepsilon, \quad |\tilde{\tau}_{i+1} - \tau_{i+1}| < \delta_{\tau_{i+1}},\quad i = 1, 2, \ldots, k-1.
\end{aligned} \tag{2.90}
$$

For the trajectories $\tilde{r}(\tilde{\varsigma}_k, T]$ and $r(\varsigma_k, T]$,

$$
\begin{aligned}
& H(\tilde{r}(\tilde{\varsigma}_k, T], r(\varsigma_k, T]) \\
\leqslant\ & \max\Big\{R(\tilde{r}(\varsigma_k^{\max}, T], r(\varsigma_k^{\max}, T]), H\big(\chi(\varsigma_k + 0; 0, \chi_0), \tilde{r}(\tilde{\varsigma}_k, \tilde{\tau}_k]\big), \\
& H(r(\varsigma_k, \tilde{\tau}_k], \tilde{r}(\tilde{\tau}_k, \varsigma_k]), H\big(\tilde{\chi}(\tilde{\varsigma}_k + 0; 0, \tilde{\chi}_0), r(\tilde{\tau}_k, \tilde{\varsigma}_k]\big)\Big\}.
\end{aligned} \tag{2.91}
$$

Since $(\tilde{\varsigma}_k, \tilde{\tau}_k] = \varnothing$, $(\tilde{\tau}_k, \varsigma_k] = \varnothing$, then $H(\chi(\varsigma_k + 0; 0, \chi_0), \tilde{r}(\tilde{\varsigma}_k, \tilde{\tau}_k] = 0$, $H(r(\varsigma_k, \tilde{\tau}_k], \tilde{r}(\tilde{\tau}_k, \varsigma_k] = 0$.

Set $|\tilde{d}_{\varsigma_k} - d| < \delta_{\tau_k}$, $|\tilde{\varsigma}_k - \varsigma_k| \leqslant |\tilde{\tau}_k - \tau_k| + |\tilde{d}_{\varsigma_k} - d| < 2\delta_{\tau_k}$, let $\delta_{\varsigma_k} > 0$ denote an arbitrary number, so if δ_{τ_k} is sufficiently small, then $|\tilde{\varsigma}_k - \varsigma_k| < \delta_{\varsigma_k}$.

We have that $\forall\ \varepsilon > 0$, $\exists\ \delta_{\varsigma_k}$, $0 < \delta_{\varsigma_k} < \frac{\varepsilon}{2M}$, $\forall\ \tilde{\varsigma}_k \in R^+$, $|\tilde{\varsigma}_k - \varsigma_k| < \delta_{\varsigma_k}$, $\forall\, \tilde{d}_{\tau_{k+1}} > 0$, $|\tilde{d}_{\tau_{k+1}} - d| < \delta_{\varsigma_k}$, then $\|\tilde{\chi}(\tau; 0, \tilde{\chi}_0) - \chi(\tau; 0, \chi_0)\| < \frac{\varepsilon}{2}$, $\varsigma_k^{\max} < \tau \leqslant T$, that is $R(\tilde{r}(\varsigma_k^{\max}, T], r(\varsigma_k^{\max}, T]) < \varepsilon$.

For $\tilde{\tau}_k < \tau \leqslant \tilde{\varsigma}_k$,

$$
\begin{aligned}
&\left\|\tilde{\chi}(\tilde{\varsigma}_k + 0; 0, \tilde{\chi}_0) - \chi(\tau; 0, \chi_0)\right\| \\
&\leqslant \left\|\tilde{\chi}(\tilde{\varsigma}_k; 0, \tilde{\chi}_0) - \chi(\tilde{\varsigma}_k; 0, \chi_0)\right\| + \left\|\chi(\tilde{\varsigma}_k; 0, \chi_0) - \chi(\tau; 0, \chi_0)\right\| \\
&< \frac{\varepsilon}{2} + \left\|(\mathbf{I}^1_{\tau, \tilde{\varsigma}_k} f)(\tilde{\varsigma}_k, \chi)\right\| \\
&< \frac{\varepsilon}{2} + M|\tilde{\varsigma}_k - \varsigma_k| \\
&< \frac{\varepsilon}{2} + M\delta_{\varsigma_k} < \varepsilon,
\end{aligned}
$$

i.e. $H(\tilde{\chi}(\tilde{\varsigma}_k + 0; 0, \tilde{\chi}_0), r(\tilde{\tau}_k, \tilde{\varsigma}_k]) < \varepsilon$.

Therefore, $\forall\ \varepsilon > 0$, $\exists\ \delta_{\varsigma_k} > 0$, $\forall\ \tilde{\varsigma}_k \in \mathbb{R}^+$, $|\tilde{\varsigma}_k - \varsigma_k| < \delta_{\varsigma_k}$, $\forall\, \tilde{d}_{\tau_k+1} > 0$, $|\tilde{d}_{\tau_k+1} - d| < \delta_{\varsigma_k}$, then $H(\tilde{r}(\tilde{\varsigma}_k, T], r(\varsigma_k, T]) < \varepsilon$.

Now $\delta_{\varsigma_k} = \delta_{\varsigma_k}(\varepsilon)$, $\delta_{\tau_k} = \delta_{\tau_k}(\delta_{\varsigma_k}, \varepsilon)$, $\delta_{\varsigma_{k-1}} = \delta_{\varsigma_{k-1}}(\delta_{\tau_k}, \varepsilon)$, $\cdots$, $\delta_{\tau_1} = \delta_{\tau_1}(\delta_{\varsigma_1}, \varepsilon)$ and $\delta_0 = \delta_0(\delta_T, \delta_{\tau_1}, \varepsilon)$.

Consequently, we have the conclusion: $\forall\ \varepsilon > 0$, $\exists\ \delta_0 > 0$, $\forall\ \tilde{\chi}_0 \in \mathbb{D}$, $\|\tilde{\chi}_0 - \chi_0\| < \delta_0$, $\forall\ \tilde{d}_{\tau_i} > 0$, $\forall\ \tilde{d}_{\varsigma_i} > 0$, $|\tilde{d}_{\tau_i} - d| < \delta_0$, $|\tilde{d}_{\varsigma_i} - d| < \delta_0$, $i = 1, 2, \ldots, k$, then $H(\tilde{r}[0, \tilde{\tau}_1], r[0, \tau_1]) < \varepsilon$, $H(\tilde{r}(\tilde{\varsigma}_i, \tilde{\tau}_{i+1}], r(\varsigma_i, \tau_{i+1}]) < \varepsilon$, $i = 1, 2, \ldots, k-1$, $H(\tilde{r}(\tilde{\varsigma}_k, T], r(\varsigma_k, T]) < \varepsilon$.

Consider that

$$
r[0, T] = r[0, \tau_1] \bigcup \left(\bigcup_{i=1,2,\ldots,k} r(\tau_i, \varsigma_i] \right) \bigcup \left(\bigcup_{i=1,2,\ldots,k-1} r(\varsigma_i, \tau_{i+1}] \right) \bigcup r(\varsigma_k, T],
$$

$$
\tilde{r}[0, T] = \tilde{r}[0, \tilde{\tau}_1] \bigcup \left(\bigcup_{i=1,2,\ldots,k} \tilde{r}(\tilde{\tau}_i, \tilde{\varsigma}_i] \right) \bigcup \left(\bigcup_{i=1,2,\ldots,k-1} \tilde{r}(\tilde{\varsigma}_i, \tilde{\tau}_{i+1}] \right) \bigcup \tilde{r}(\tilde{\varsigma}_k, T].
$$

Now apply theorem 2.2.1 and [21, theorem 1.3], and then

$$\begin{aligned}
& H(\tilde{r}[0, T], r[0, T]) \\
&= H\Bigg(\tilde{r}[0, \tilde{\tau}_1] \bigcup \Bigg(\bigcup_{i=1,2,\ldots,k} \tilde{r}(\tilde{\tau}_i, \tilde{\varsigma}_i]\Bigg) \bigcup \Bigg(\bigcup_{i=1,2,\ldots,k-1} \tilde{r}(\tilde{\varsigma}_i, \tilde{\tau}_{i+1}]\Bigg) \bigcup \tilde{r}(\tilde{\varsigma}_k, T], \\
& \quad r[0, \tau_1] \bigcup \Bigg(\bigcup_{i=1,2,\ldots,k} r(\tau_i, \varsigma_i]\Bigg) \bigcup \Bigg(\bigcup_{i=1,2,\ldots,k-1} r(\varsigma_i, \tau_{i+1}]\Bigg) \bigcup r(\varsigma_k, T]\Bigg) \\
&\leqslant H\Bigg(\tilde{r}[0, \tilde{\tau}_1] \bigcup \Bigg(\bigcup_{i=1,2,\ldots,k-1} \tilde{r}(\tilde{\varsigma}_i, \tilde{\tau}_{i+1}]\Bigg) \bigcup \tilde{r}(\tilde{\varsigma}_k, T], \\
& \quad r[0, \tau_1] \bigcup \Bigg(\bigcup_{i=1,2,\ldots,k-1} r(\varsigma_i, \tau_{i+1}]\Bigg) \bigcup r(\varsigma_k, T]\Bigg) \\
&\leqslant \max\big\{H(\tilde{r}[0, \tilde{\tau}_1], r[0, \tau_1]), H(\tilde{r}(\tilde{\varsigma}_i, \tilde{\tau}_{i+1}], r(\varsigma_i, \tau_{i+1}]), \ i = 1, 2, \ldots, k-1, \\
& \quad H(\tilde{r}(\tilde{\varsigma}_k, T], r(\varsigma_k, T])\big\} < \varepsilon.
\end{aligned}$$

The proof is completed. □

Next, we present the orbital Hausdorff dependence on the initial condition and the difference between the impulsive points τ_i, $i = 1, 2, \ldots$ and the junction points ς_i, $i = 1, 2, \ldots$ of solutions to the fractional order impulsive differential equation (2.69).

Theorem 2.2.8. *Assume conditions $[H_1] - [H_4]$ are fulfilled. Then, the solution of problem (2.69) is orbital Hausdorff dependent on the initial condition and the difference between the impulsive points τ_i, $i = 1, 2, \ldots$ and the junction points ς_i, $i = 1, 2, \ldots$.*

Proof. Consider the possible location of the distribution of the impulsive points τ_i, $\tilde{\tau}_i$ and the junction points ς_i, $\tilde{\varsigma}_i$, and we divide our proofs into several cases.

Case 1. Let $\tau_i^{\min} = \tau_i$, $\tau_i^{\max} = \tilde{\tau}_i$, $\varsigma_i^{\min} = \varsigma_i$, $\varsigma_i^{\max} = \tilde{\varsigma}_i$, $i = 1, 2, \ldots$; (the case $\tau_i^{\min} = \tilde{\tau}_i$, $\tau_i^{\max} = \tau_i$, $\varsigma_i^{\min} = \tilde{\varsigma}_i$, $\varsigma_i^{\max} = \varsigma_i$, $i = 1, 2, \ldots$ can be considered similarly). In this case, we have the same conclusion with δ_T in case 1 of theorem 2.2.7.

Next, we consider the Hausdorff distance between the trajectories on the corresponding subintervals.

For $\tilde{r}[0, \tilde{\tau}_1]$ and $r[0, \tau_1]$, the inequality (2.76) is satisfied and further $H(\tilde{\chi}(\tilde{\tau}_1; 0, \tilde{\chi}_0), r(\tilde{\tau}_1, \tau_1]) = 0$. We now estimate the other two terms.

Set $0 < \eta_{01} < \varepsilon$, we have that $\exists\ \delta_0 > 0$, $\delta_0 < \min\{\delta_T, (\frac{\Gamma(\alpha+1)\eta_{01}}{6M})^{\frac{1}{\alpha}}\}$, $\forall\ \tilde{\chi}_0 \in \mathbb{D}$, $\|\tilde{\chi}_0 - \chi_0\| < \delta_0$, $\forall\ \tilde{d}_{\tau_1} > 0$, $|\tilde{d}_{\tau_1} - d| < \delta_0$, then $\|\tilde{\chi}(\tau; 0, \tilde{\chi}_0) - \chi(\tau; 0, \chi_0)\| < \frac{1}{2}\eta_{01}$, $0 < \tau \leqslant \tau_1^{\min}$, that is $R(\tilde{r}(0, \tau_1^{\min}], r(0, \tau_1^{\min}]) < \frac{1}{2}\eta_{01} < \varepsilon$.

Now $|\tilde{\tau}_1 - \tau_1| = |\tilde{d}_{\tau_1} - d| < \delta_0 < (\frac{\Gamma(\alpha+1)\eta_{01}}{6M})^{\frac{1}{\alpha}}$. For $\tau_1 < \tau \leqslant \tilde{\tau}_1$,

$$
\begin{aligned}
&\left\|\chi(\tau_1; 0, \chi_0) - \tilde{\chi}(\tau; 0, \tilde{\chi}_0)\right\| \\
\leqslant\ &\left\|\chi(\tau_1; 0, \chi_0) - \tilde{\chi}(\tau_1; 0, \tilde{\chi}_0)\right\| + \left\|\tilde{\chi}(\tau; 0, \tilde{\chi}_0) - \tilde{\chi}(\tau_1; 0, \tilde{\chi}_0)\right\| \\
<\ &\frac{1}{2}\eta_{01} + \left\|(\mathbf{I}^{\alpha}_{0,\tau}f)(\tau, \tilde{\chi}) - (\mathbf{I}^{\alpha}_{0,\tau_1}f)(\tau_1, \tilde{\chi})\right\| \\
\leqslant\ &\frac{1}{2}\eta_{01} + \frac{1}{\Gamma(\alpha)}\int_0^{\tau_1} |(\tau - \varsigma)^{\alpha-1} - (\tau_1 - \varsigma)^{\alpha-1}|\|f(\varsigma, \tilde{\chi}(\varsigma))\|ds + \|(\mathbf{I}^{\alpha}_{\tau_1,\tau}f)(\tau, \tilde{\chi})\| \\
\leqslant\ &\frac{1}{2}\eta_{01} + \frac{3M}{\Gamma(\alpha+1)}(\tilde{\tau}_1 - \tau_1)^{\alpha} \\
<\ &\frac{1}{2}\eta_{01} + \frac{3M}{\Gamma(\alpha+1)}\delta_0^{\alpha} < \eta_{01} < \varepsilon,
\end{aligned}
$$

i.e. $H(\chi(\tau_1; 0, \chi_0), \tilde{r}(\tau_1, \tilde{\tau}_1]) < \varepsilon$.

Let $\delta_{\tau_1} > 0$ be an arbitrary constant, assume that $\eta_{01} < \frac{6M}{\Gamma(\alpha+1)}\delta^{\alpha}_{\tau_1}$, and then $|\tilde{\tau}_1 - \tau_1| < \delta_{\tau_1}$.

Therefore, $\forall\ \delta_{\tau_1} > 0$, $\exists\ \delta_0 > 0$, $\forall\ \tilde{\chi}_0 \in \mathbb{D}$, $\|\tilde{\chi}_0 - \chi_0\| < \delta_0$, $\forall\ \tilde{d}_{\tau_1} > 0$, $|\tilde{d}_{\tau_1} - d| < \delta_0$, then $H(\tilde{r}[0, \tilde{\tau}_1], r[0, \tau_1]) < \varepsilon$, $|\tilde{\tau}_1 - \tau_1| < \delta_{\tau_1}$.

For the trajectories $\tilde{r}(\tilde{\tau}_i, \tilde{\varsigma}_i]$ and $r(\tau_i, \varsigma_i]$, $i = 1, 2, \ldots, k$,

$$
\begin{aligned}
&H(\tilde{r}(\tilde{\tau}_i, \tilde{\varsigma}_i], r(\tau_i, \varsigma_i]) \\
&\leqslant \max\Big\{R\Big(\tilde{r}\Big(\tau_i^{\max}, \varsigma_i^{\min}\Big], r\Big(\tau_i^{\max}, \varsigma_i^{\min}\Big]\Big), H\big(\chi(\tau_i + 0; 0, \chi_0), \tilde{r}(\tilde{\tau}_i, \tau_i]\big), \\
&H\big(\tilde{\chi}(\tilde{\tau}_i + 0; 0, \tilde{\chi}_0), r(\tau_i, \tilde{\tau}_i]\big), H\big(\chi(\varsigma_i; 0, \chi_0), \tilde{r}(\varsigma_i, \tilde{\varsigma}_i]\big), H\big(\tilde{\chi}(\tilde{\varsigma}_i; 0, \tilde{\chi}_0), r(\tilde{\varsigma}_i, \varsigma_i]\big)\Big\}.
\end{aligned}
$$

Since $(\tilde{\tau}_i, \tau_i] = \varnothing$ and $(\tilde{\varsigma}_i, \varsigma_i] = \varnothing$, $H(\chi(\tau_i + 0; 0, \chi_0), \tilde{r}(\tilde{\tau}_i, \tau_i]) = 0$, $H(\tilde{\chi}(\tilde{\varsigma}_i; 0, \tilde{\chi}_0), r(\tilde{\varsigma}_i, \varsigma_i]) = 0$.

Let $0 < \eta_{ii} < \varepsilon$, we have that $\exists\ \delta_{\tau_i} > 0$, $\delta_{\tau_i} < \frac{\eta_{ii}}{4L_{g_i}}$, $\forall\ \tilde{d}_{\varsigma_i} > 0$, $|\tilde{d}_{\varsigma_i} - d| < \delta_{\tau_i}$, then $\|\tilde{\chi}(\tau; 0, \tilde{\chi}_0) - \chi(\tau; 0, \chi_0)\| < \frac{1}{2}\eta_{ii}$, $\tau_i^{\max} < \tau \leqslant \varsigma_i^{\min}$, that is $R(\tilde{r}(\tau_i^{\max}, \varsigma_i^{\min}]$, $r(\tau_i^{\max}, \varsigma_i^{\min}]) < \frac{1}{2}\eta_{ii} < \varepsilon$.

For $\tau_i < \tau \leqslant \tilde{\tau}_i$,

$$
\begin{aligned}
&\left\|\tilde{\chi}(\tilde{\tau}_i + 0; 0, \tilde{\chi}_0) - \chi(\tau; 0, \chi_0)\right\| \\
\leqslant\ &\left\|\tilde{\chi}(\tilde{\tau}_i; 0, \tilde{\chi}_0) - \chi(\tilde{\tau}_i; 0, \chi_0)\right\| + \left\|\chi(\tilde{\tau}_i; 0, \chi_0) - \chi(\tau; 0, \chi_0)\right\| \\
<\ &\frac{1}{2}\eta_{ii} + \left\|g_i(\tilde{\tau}_i, \chi(\tau_i^-)) - g_i(\tau, \chi(\tau_i^-))\right\| \\
\leqslant\ &\frac{1}{2}\eta_{ii} + L_{g_i}|\tilde{\tau}_i - \tau_i| \\
<\ &\frac{1}{2}\eta_{ii} + L_{g_i}\delta_{\tau_i} < \eta_{ii} < \varepsilon,
\end{aligned}
$$

i.e. $H(\tilde{\chi}(\tilde{\tau}_i + 0; 0, \tilde{\chi}_0), r(\tau_i, \tilde{\tau}_i]) < \varepsilon$.

Since $|\tilde{\varsigma}_i - \varsigma_i| \leqslant |\tilde{\tau}_i - \tau_i| + |\tilde{d}_{\varsigma_i} - d| < 2\delta_{\tau_i}$, for $\varsigma_i < \tau \leqslant \tilde{\varsigma}_i$,

$$\begin{aligned}
&\left\|\chi(\varsigma_i; 0, \chi_0) - \tilde{\chi}(\tau; 0, \tilde{\chi}_0)\right\| \\
\leqslant\ & \left\|\chi(\varsigma_i; 0, \chi_0) - \tilde{\chi}(\varsigma_i; 0, \tilde{\chi}_0)\right\| + \left\|\tilde{\chi}(\tau; 0, \tilde{\chi}_0) - \tilde{\chi}(\varsigma_i; 0, \tilde{\chi}_0)\right\| \\
<\ & \frac{1}{2}\eta_{ii} + \left\|g_i(\tau, \tilde{\chi}(\tilde{\tau}_i^-)) - g_i(\varsigma_i, \tilde{\chi}(\tilde{\tau}_i^-))\right\| \\
\leqslant\ & \frac{1}{2}\eta_{ii} + L_{g_i}|\tilde{\varsigma}_i - \varsigma_i| \\
<\ & \frac{1}{2}\eta_{ii} + 2L_{g_i}\delta_{\tau_i} < \eta_{ii} < \varepsilon,
\end{aligned}$$

i.e. $H(\chi(\varsigma_i; 0, \chi_0), \tilde{r}(\varsigma_i, \tilde{\varsigma}_i]) < \varepsilon$.

Set $\eta_{ii} < 2L_{g_i}\delta_{\varsigma_i}$, and then $|\tilde{\varsigma}_i - \varsigma_i| < \delta_{\varsigma_i}$, where δ_{ς_i} denotes arbitrary positive constants. Therefore, $\forall\ \delta_{\varsigma_i} > 0,\ \exists\ \delta_{\tau_i} > 0,\ \forall\ \tilde{d}_{\varsigma_i} > 0,\ |\tilde{d}_{\varsigma_i} - d| < \delta_{\tau_i}$, then $H(\tilde{r}(\tilde{\tau}_i, \tilde{\varsigma}_i], r(\tau_i, \varsigma_i]) < \varepsilon$, $|\tilde{\varsigma}_i - \varsigma_i| < \delta_{\varsigma_i}$.

For $\tilde{r}(\tilde{\varsigma}_1, \tilde{\tau}_2]$ and $r(\varsigma_1, \tau_2]$, we have the inequality (2.79) and further $H(\chi(\varsigma_1 + 0; 0, \chi_0), \tilde{r}(\tilde{\varsigma}_1, \varsigma_1]) = 0$, $H(\tilde{\chi}(\tilde{\tau}_2; 0, \tilde{\chi}_0), r(\tilde{\tau}_2, \tau_2]) = 0$.

Let $0 < \eta_{12} < \varepsilon$, we have that $\exists\ \delta_{\varsigma_1}$, $0 < \delta_{\varsigma_1} < (\frac{\Gamma(\alpha+1)\eta_{12}}{3M2^{\alpha+1}})^{\frac{1}{\alpha}}$, $\forall\ \tilde{\varsigma}_1 \in \mathbb{R}^+$, $|\tilde{\varsigma}_1 - \varsigma_1| < \delta_{\varsigma_1}$, $\forall\ \tilde{d}_{\tau_2} > 0$, $|\tilde{d}_{\tau_2} - d| < \delta_{\varsigma_1}$, then $\|\tilde{\chi}(\tau; 0, \tilde{\chi}_0) - \chi(\tau; 0, \chi_0)\| < \frac{1}{2}\eta_{12}$, $\varsigma_1^{\max} < \tau \leqslant \tau_2^{\min}$, that is $R(\tilde{r}(\varsigma_1^{\max}, \tau_2^{\min}], r(\varsigma_1^{\max}, \tau_2^{\min}]) < \frac{\eta_{12}}{2} < \varepsilon$.

For $\varsigma_1 < \tau \leqslant \tilde{\varsigma}_1$,

$$\begin{aligned}
&\left\|\tilde{\chi}(\tilde{\varsigma}_1 + 0; 0, \tilde{\chi}_0) - \chi(\tau; 0, \chi_0)\right\| \\
\leqslant\ & \left\|\tilde{\chi}(\tilde{\varsigma}_1; 0, \tilde{\chi}_0) - \chi(\tilde{\varsigma}_1; 0, \chi_0)\right\| + \left\|\chi(\tilde{\varsigma}_1; 0, \chi_0) - \chi(\tau; 0, \chi_0)\right\| \\
<\ & \frac{1}{2}\eta_{12} + \left\|(\mathbf{I}^{\alpha}_{\varsigma_1, \tilde{\varsigma}_1}f)(\tilde{\varsigma}_1, \chi) - (\mathbf{I}^{\alpha}_{\varsigma_1, \tau}f)(\tau, \chi)\right\| \\
\leqslant\ & \frac{1}{2}\eta_{12} + \frac{1}{\Gamma(\alpha)}\int_{\varsigma_1}^{\tau} |(\tilde{\varsigma}_1 - \varsigma)^{\alpha-1} - (\tau - \varsigma)^{\alpha-1}|\left\|f(\varsigma, \chi(\varsigma))\right\|ds + \left\|(\mathbf{I}^{\alpha}_{\tau, \tilde{\varsigma}_1}f)(\tilde{\varsigma}_1, \chi)\right\| \\
\leqslant\ & \frac{1}{2}\eta_{12} + \frac{3M}{\Gamma(\alpha+1)}(\tilde{\varsigma}_1 - \varsigma_1)^{\alpha} \\
<\ & \frac{1}{2}\eta_{12} + \frac{3M}{\Gamma(\alpha+1)}\delta_{\varsigma_1}^{\alpha} < \eta_{12} < \varepsilon,
\end{aligned}$$

i.e. $H(\tilde{\chi}(\tilde{\varsigma}_1 + 0; 0, \tilde{\chi}_0), r(\varsigma_1, \tilde{\varsigma}_1]) < \varepsilon$.

Consider $|\tilde{\tau}_2 - \tau_2| \leqslant |\tilde{\varsigma}_1 - \varsigma_1| + |\tilde{d}_{\tau_2} - d| < 2\delta_{\varsigma_1}$. For $\tau_2 < \tau \leqslant \tilde{\tau}_2$,

$$\begin{aligned}
&\left\|\chi(\tau_2; 0, \chi_0) - \tilde{\chi}(\tau; 0, \tilde{\chi}_0)\right\| \\
\leqslant\ &\left\|\chi(\tau_2; 0, \chi_0) - \tilde{\chi}(\tau_2; 0, \tilde{\chi}_0)\right\| + \left\|\tilde{\chi}(\tau; 0, \tilde{\chi}_0) - \tilde{\chi}(\tau_2; 0, \tilde{\chi}_0)\right\| \\
<\ &\frac{1}{2}\eta_{12} + \left\|(\mathbf{I}^{\alpha}_{\tilde{\varsigma}_1,\tau}f)(\tau, \tilde{\chi}) - (\mathbf{I}^{\alpha}_{\tilde{\varsigma}_1,\tau_2}f)(\tau_2, \tilde{\chi})\right\| \\
\leqslant\ &\frac{1}{2}\eta_{12} + \frac{3M}{\Gamma(\alpha+1)}(\tilde{\tau}_2 - \tau_2)^{\alpha} \\
<\ &\frac{1}{2}\eta_{12} + \frac{3M}{\Gamma(\alpha+1)}2^{\alpha}\delta^{\alpha}_{\varsigma_1} < \eta_{12} < \varepsilon,
\end{aligned}$$

i.e. $H(\chi(\tau_2; 0, \chi_0), \tilde{r}(\tau_2, \tilde{\tau}_2]) < \varepsilon$.

Let $\delta_{\tau_2} > 0$ denote an arbitrary constant, set $\eta_{12} < \frac{6M}{\Gamma(\alpha+1)}\delta^{\alpha}_{\tau_2}$, and then $|\tilde{\tau}_2 - \tau_2| < \delta_{\tau_2}$.

Therefore, $\forall\ \delta_{\tau_2} > 0$, $\exists\ \delta_{\varsigma_1} > 0, \forall\ \tilde{\varsigma}_1 \in \mathbb{R}^+, |\tilde{\varsigma}_1 - \varsigma_1| < \delta_{\varsigma_1}, \forall\ \tilde{d}_{\tau_2} > 0, |\tilde{d}_{\tau_2} - d| < \delta_{\varsigma_1}$, then $H(\tilde{r}(\tilde{\varsigma}_1, \tilde{\tau}_2], r(\varsigma_1, \tau_2]) < \varepsilon$, $|\tilde{\tau}_2 - \tau_2| < \delta_{\tau_2}$.

For $\tilde{r}(\tilde{\varsigma}_2, \tilde{\tau}_3]$ and $r(\varsigma_2, \tau_3]$, inequality (2.82) is satisfied and we have $H(\chi(\varsigma_2 + 0; 0, \chi_0), \tilde{r}(\tilde{\varsigma}_2, \varsigma_2]) = 0$ and $H(\tilde{\chi}(\tilde{\tau}_3; 0, \tilde{\chi}_0), r(\tilde{\tau}_3, \tau_3]) = 0$.

Let $0 < \eta_{23} < \varepsilon$, we have that $\exists\ \delta_{\varsigma_2}$, $0 < \delta_{\varsigma_2} < (\frac{\Gamma(\alpha+1)\eta_{23}}{3M2^{\alpha+1}})^{\frac{1}{\alpha}}$, $\forall\ \tilde{\varsigma}_2 \in \mathbb{R}^+$, $|\tilde{\varsigma}_2 - \varsigma_2| < \delta_{\varsigma_2}$, $\forall\ \tilde{d}_{\tau_3} > 0$, $|\tilde{d}_{\tau_3} - d| < \delta_{\varsigma_2}$, then $\|\tilde{\chi}(\tau; 0, \tilde{\chi}_0) - \chi(\tau; 0, \chi_0)\| < \frac{1}{2}\eta_{23}$, $\varsigma_2^{\max} < \tau \leqslant \tau_3^{\min}$, that is $R(\tilde{r}(\varsigma_2^{\max}, \tau_3^{\min}], r(\varsigma_2^{\max}, \tau_3^{\min}]) < \frac{\eta_{23}}{2} < \varepsilon$.

For $\varsigma_2 < \tau \leqslant \tilde{\varsigma}_2$,

$$\begin{aligned}
&\left\|\tilde{\chi}(\tilde{\varsigma}_2 + 0; 0, \tilde{\chi}_0) - \chi(\tau; 0, \chi_0)\right\| \\
\leqslant\ &\left\|\tilde{\chi}(\tilde{\varsigma}_2; 0, \tilde{\chi}_0) - \chi(\tilde{\varsigma}_2; 0, \chi_0)\right\| + \left\|\chi(\tilde{\varsigma}_2; 0, \chi_0) - \chi(\tau; 0, \chi_0)\right\| \\
<\ &\frac{1}{2}\eta_{23} + \left\|(\mathbf{I}^{\alpha}_{\varsigma_2,\tilde{\varsigma}_2}f)(\tilde{\varsigma}_2, \chi) - (\mathbf{I}^{\alpha}_{\varsigma_1,\tau}f)(\tau, \chi)\right\| \\
\leqslant\ &\frac{1}{2}\eta_{23} + \frac{1}{\Gamma(\alpha)}\int_{\varsigma_2}^{\tau} |(\tilde{\varsigma}_2 - \varsigma)^{\alpha-1} - (\tau - \varsigma)^{\alpha-1}|\left\|f(\varsigma, \chi(\varsigma))\right\|ds + \left\|(\mathbf{I}^{\alpha}_{\tau,\tilde{\varsigma}_2}f)(\tilde{\varsigma}_2, \chi)\right\| \\
\leqslant\ &\frac{1}{2}\eta_{23} + \frac{3M}{\Gamma(\alpha+1)}(\tilde{\varsigma}_2 - \varsigma_2)^{\alpha} \\
<\ &\frac{1}{2}\eta_{23} + \frac{3M}{\Gamma(\alpha+1)}\delta^{\alpha}_{\varsigma_2} < \eta_{23} < \varepsilon,
\end{aligned}$$

i.e. $H(\tilde{\chi}(\tilde{\varsigma}_2 + 0; 0, \tilde{\chi}_0), r(\varsigma_2, \tilde{\varsigma}_2]) < \varepsilon$.

Consider $|\tilde{\tau}_3 - \tau_3| \leqslant |\tilde{\varsigma}_2 - \varsigma_2| + |\tilde{d}_{\tau_3} - d| < 2\delta_{\varsigma_2}$, and for $\tau_3 < \tau \leqslant \tilde{\tau}_3$,

$$\begin{aligned}
&\left\|\chi(\tau_3; 0, \chi_0) - \tilde{\chi}(\tau; 0, \tilde{\chi}_0)\right\| \\
\leqslant\ &\left\|\chi(\tau_3; 0, \chi_0) - \tilde{\chi}(\tau_3; 0, \tilde{\chi}_0)\right\| + \left\|\tilde{\chi}(\tau; 0, \tilde{\chi}_0) - \tilde{\chi}(\tau_3; 0, \tilde{\chi}_0)\right\| \\
<\ &\frac{1}{2}\eta_{23} + \left\|(\mathbf{I}^{\alpha}_{\tilde{\varsigma}_2,\tau}f)(\tau, \tilde{\chi}) - (\mathbf{I}^{\alpha}_{\tilde{\varsigma}_2,\tau_3}f)(\tau_3, \tilde{\chi})\right\| \\
\leqslant\ &\frac{1}{2}\eta_{23} + \frac{3M}{\Gamma(\alpha + 1)}(\tilde{\tau}_3 - \tau_3)^{\alpha} \\
<\ &\frac{1}{2}\eta_{23} + \frac{3M}{\Gamma(\alpha + 1)}2^{\alpha}\delta^{\alpha}_{\varsigma_2} < \eta_{23} < \varepsilon,
\end{aligned}$$

i.e. $H(\chi(\tau_3; 0, \chi_0), \tilde{r}(\tau_3, \tilde{\tau}_3]) < \varepsilon$.

Set $\eta_{23} < \frac{6M}{\Gamma(\alpha+1)}\delta^{\alpha}_{\tau_3}$, and then $|\tilde{\tau}_3 - \tau_3| < \delta_{\tau_3}$; here δ_{τ_3} denotes an arbitrary positive constant.

Therefore, $\forall\ \delta_{\tau_3} > 0$, $\exists\ \delta_{\varsigma_2} > 0$, $\forall\ \tilde{\varsigma}_2 \in \mathbb{R}^+$, $|\tilde{\varsigma}_2 - \varsigma_2| < \delta_{\varsigma_2}$, $\forall\ \tilde{d}_{\tau_3} > 0$, $|\tilde{d}_{\tau_3} - d| < \delta_{\varsigma_2}$, then $H(\tilde{r}(\tilde{\varsigma}_2, \tilde{\tau}_3], r(\varsigma_2, \tau_3]) < \varepsilon$, $|\tilde{\tau}_3 - \tau_3| < \delta_{\tau_3}$.

Repeat the above procedure, and we obtain the same conclusion in (2.83) and (2.84).

Consider the trajectories $\tilde{r}(\tilde{\varsigma}_k, T]$ and $r(\varsigma_k, T]$, and we have the inequality (2.85) and further $H(\chi(\varsigma_k + 0; 0, \chi_0), \tilde{r}(\tilde{\varsigma}_k, \varsigma_k]) = 0$.

One can deduce that $\forall\ \varepsilon > 0$, $\exists\ \delta_{\varsigma_k}$, $0 < \delta_{\varsigma_k} < (\frac{\Gamma(\alpha+1)\varepsilon}{6M})^{\frac{1}{\alpha}}$, $\forall\ \tilde{\varsigma}_k \in R^+$, $|\tilde{\varsigma}_k - \varsigma_k| < \delta_{\varsigma_k}$, $\forall \tilde{d}_{\tau_{k+1}} > 0$, $|\tilde{d}_{\tau_{k+1}} - d| < \delta_{\varsigma_k}$, then $\|\tilde{\chi}(\tau; 0, \tilde{\chi}_0) - \chi(\tau; 0, \chi_0)\| < \frac{\varepsilon}{2}$, $\varsigma_k^{\max} < \tau \leqslant T$, that is $R(\tilde{r}(\varsigma_k^{\max}, T], r(\varsigma_k^{\max}, T]) < \varepsilon$.

For $\varsigma_k < \tau \leqslant \tilde{\varsigma}_k$,

$$\begin{aligned}
&\left\|\tilde{\chi}(\tilde{\varsigma}_k + 0; 0, \tilde{\chi}_0) - \chi(\tau; 0, \chi_0)\right\| \\
\leqslant\ &\left\|\tilde{\chi}(\tilde{\varsigma}_k; 0, \tilde{\chi}_0) - \chi(\tilde{\varsigma}_k; 0, \chi_0)\right\| + \left\|\chi(\tilde{\varsigma}_k; 0, \chi_0) - \chi(\tau; 0, \chi_0)\right\| \\
<\ &\frac{\varepsilon}{2} + \left\|(\mathbf{I}^{\alpha}_{\varsigma_k,\tilde{\varsigma}_k}f)(\tilde{\varsigma}_k, \chi) - (\mathbf{I}^{\alpha}_{\varsigma_k,\tau}f)(\tau, \chi)\right\| \\
\leqslant\ &\frac{\varepsilon}{2} + \frac{3M}{\Gamma(\alpha + 1)}(\tilde{\varsigma}_k - \varsigma_k)^{\alpha} \\
<\ &\frac{\varepsilon}{2} + \frac{3M}{\Gamma(\alpha + 1)}\delta^{\alpha}_{\varsigma_k} < \varepsilon,
\end{aligned}$$

i.e. $H(\tilde{\chi}(\tilde{\varsigma}_k + 0; 0, \tilde{\chi}_0), r(\varsigma_k, \tilde{\varsigma}_k]) < \varepsilon$.

Therefore, $\forall\ \varepsilon > 0$, $\exists\ \delta_{\varsigma_k} > 0$, $\forall\ \tilde{\varsigma}_k \in \mathbb{R}^+$, $|\tilde{\varsigma}_k - \varsigma_k| < \delta_{\varsigma_k}$, $\forall \tilde{d}_{\tau_k+1} > 0$, $|\tilde{d}_{\tau_k+1} - d| < \delta_{\varsigma_k}$, then $H(\tilde{r}(\tilde{\varsigma}_k, T], r(\varsigma_k, T]) < \varepsilon$.

Similar to case 1 of theorem 2.2.7, we have $\forall\ \varepsilon > 0$, $\exists\ \delta_0 > 0$, $\forall\ \tilde{\chi}_0 \in \mathbb{D}$, $\left\|\tilde{\chi}_0 - \chi_0\right\| < \delta_0$, $\forall\ \tilde{d}_{\tau_i} > 0$, $\forall\ \tilde{d}_{\varsigma_i} > 0$, $|\tilde{d}_{\tau_i} - d| < \delta_0$, $|\tilde{d}_{\varsigma_i} - d| < \delta_0$,

$i = 1, 2, \ldots, k$, then $H(\tilde{r}[0, \tilde{\tau}_1], r[0, \tau_1]) < \varepsilon$, $H(\tilde{r}(\tilde{\tau}_i, \tilde{\varsigma}_i], r(\tau_i, \varsigma_i]) < \varepsilon$, $i = 1, \ldots, k$, $H(\tilde{r}(\tilde{\varsigma}_i, \tilde{\tau}_{i+1}], r(\varsigma_i, \tau_{i+1}]) < \varepsilon$, $i = 1, \ldots, k-1$, $H(\tilde{r}(\tilde{\varsigma}_k, T], r(\varsigma_k, T]) < \varepsilon$.

Furthermore,

$$\begin{aligned} &H(\tilde{r}[0, T], r[0, T]) \\ &\leqslant \max\{H(\tilde{r}[0, \tilde{\tau}_1], r[0, \tau_1]),\ H(\tilde{r}(\tilde{\tau}_i, \tilde{\varsigma}_i], r(\tau_i, \varsigma_i]),\ i = 1, 2, \ldots, k, \\ &H(\tilde{r}(\tilde{\varsigma}_i, \tilde{\tau}_{i+1}], r(\varsigma_i, \tau_{i+1}]),\ i = 1, 2, \ldots, k-1,\ H(\tilde{r}(\tilde{\varsigma}_k, T], r(\varsigma_k, T])\} < \varepsilon. \end{aligned}$$

Case 2. Let $\tau_i^{\min} = \tau_i$, $\tau_i^{\max} = \tilde{\tau}_i$, $\varsigma_i^{\min} = \tilde{\varsigma}_i$, $\varsigma_i^{\max} = \varsigma_i$, $i = 1, 2, \ldots$; (the case $\tau_i^{\min} = \tilde{\tau}_i, \tau_i^{\max} = \tau_i, \varsigma_i^{\min} = \varsigma_i, \varsigma_i^{\max} = \tilde{\varsigma}_i, i = 1, 2, \ldots$ can be considered analogously).

In the case, we still have formula (2.76), and follow the proof in case 1 of theorem 2.2.8, for the trajectories $\tilde{r}[0, \tilde{\tau}_1]$ and $r[0, \tau_1]$, and we obtain the same conclusion that $\forall\ \delta_{\tau_1} > 0$, $\exists\ \delta_0 > 0$, $\forall\ \tilde{\chi}_0 \in \mathbb{D}$, $\|\tilde{\chi}_0 - \chi_0\| < \delta_0$, $\forall\ \tilde{d}_{\tau_1} > 0$, $|\tilde{d}_{\tau_1} - d| < \delta_0$, then $H(\tilde{r}[0, \tilde{\tau}_1], r[0, \tau_1]) < \varepsilon$, $|\tilde{\tau}_1 - \tau_1| < \delta_{\tau_1}$.

Consider the trajectories $\tilde{r}(\tilde{\tau}_1, \tilde{\varsigma}_1]$ and $r(\tau_1, \varsigma_1]$, and we also have the inequality (2.77). Since $(\tilde{\tau}_1, \tau_1] = \varnothing$ and $(\varsigma_1, \tilde{\varsigma}_1] = \varnothing$, then $H(\chi(\tau_1 + 0; 0, \chi_0), \tilde{r}(\tilde{\tau}_1, \tau_1]) = 0$, $H(\chi(\varsigma_1; 0, \chi_0), \tilde{r}(\varsigma_1, \tilde{\varsigma}_1]) =0$.

Let $0 < \eta_{11} < \varepsilon$, we have that $\exists\ \delta_{\tau_1} > 0$, $\delta_{\tau_1} < \frac{\eta_{11}}{4L_{g_1}}$, $\forall\ \tilde{\chi}_0 \in \mathbb{D}$, $\|\tilde{\chi}_0 - \chi_0\| < \delta_0$, $\forall\ \tilde{d}_{\varsigma_1} > 0$, $|d - \tilde{d}_{\varsigma_1}| < \delta_{\tau_1}$, then $\|\tilde{\chi}(\tau; 0, \tilde{\chi}_0) - \chi(\tau; 0, \chi_0)\| < \frac{1}{2}\eta_{11}$, $\tau_1^{\max} < \tau \leqslant \varsigma_1^{\min}$, that is $R(\tilde{r}(\tau_1^{\max}, \varsigma_1^{\min}], r(\tau_1^{\max}, \varsigma_1^{\min}]) < \frac{1}{2}\eta_{11} < \varepsilon$.

For $\tau_1 < \tau \leqslant \tilde{\tau}_1$, similar to (2.78), we get $H(\tilde{\chi}(\tilde{\tau}_1 + 0; 0, \tilde{\chi}_0), r(\tau_1, \tilde{\tau}_1]) < \varepsilon$.

Now $|\varsigma_1 - \tilde{\varsigma}_1| \leqslant |\tau_1 - \tilde{\tau}_1| + |d - \tilde{d}_{\varsigma_1}| < 2\delta_{\tau_1}$. For $\tilde{\varsigma}_1 < \tau \leqslant \varsigma_1$, similar to (2.87) we have that $H(\tilde{\chi}(\tilde{\varsigma}_1; 0, \tilde{\chi}_0), r(\tilde{\varsigma}_1, \varsigma_1]) < \varepsilon$.

Therefore, $\forall\ \delta_{\varsigma_1} > 0$, $\exists\ \delta_{\tau_1} > 0$, $\forall\ \tilde{\chi}_0 \in \mathbb{D}$, $\|\tilde{\chi}_0 - \chi_0\| < \delta_0$, $\forall\ \tilde{d}_{\varsigma_1} > 0$, $|d - \tilde{d}_{\varsigma_1}| < \delta_{\tau_1}$, then $H(\tilde{r}(\tilde{\tau}_1, \tilde{\varsigma}_1], r(\tau_1, \varsigma_1]) < \varepsilon$, $|\varsigma_1 - \tilde{\varsigma}_1| < \delta_{\varsigma_1}$, where δ_{ς_1} denotes an arbitrary positive constant.

For the Hausdorff distance about the trajectories $\tilde{r}(\tilde{\varsigma}_1, \tilde{\tau}_2]$ and $r(\varsigma_1, \tau_2]$, the inequality (2.79) holds. Since $(\varsigma_1, \tilde{\varsigma}_1] = \varnothing$ and $(\tilde{\tau}_2, \tau_2] = \varnothing$, then $H(\tilde{\chi}(\tilde{\varsigma}_1 + 0; 0, \tilde{\chi}_0), r(\varsigma_1, \tilde{\varsigma}_1]) = 0$, $H(\tilde{\chi}(\tilde{\tau}_2; 0, \tilde{\chi}_0), r(\tilde{\tau}_2, \tau_2]) = 0$.

Let $0 < \eta_{12} < \varepsilon$, we have that $\exists\ \delta_{\varsigma_1} > 0$, $\delta_{\varsigma_1} < (\frac{\Gamma(\alpha+1)\eta_{12}}{3M2^{\alpha+1}})^{\frac{1}{\alpha}}$, $\forall\ \tilde{\varsigma}_1 \in \mathbb{R}^+$, $|\varsigma_1 - \tilde{\varsigma}_1| < \delta_{\varsigma_1}$, $\forall\ \tilde{d}_{\tau_2} > 0$, $|\tilde{d}_{\tau_2} - d| < \delta_{\varsigma_1}$, then $\|\tilde{\chi}(\tau; 0, \tilde{\chi}_0) - \chi(\tau; 0, \chi_0)\| < \frac{1}{2}\eta_{12}$, $\varsigma_1^{\max} < \tau \leqslant \tau_2^{\min}$, that is $R(\tilde{r}(\varsigma_1^{\max}, \tau_2^{\min}], r(\varsigma_1^{\max}, \tau_2^{\min}]) < \frac{\eta_{12}}{2} < \varepsilon$.

For $\tilde{\varsigma}_1 < \tau \leqslant \varsigma_1$,

$$\begin{aligned}
&\left\|\chi(\varsigma_1 + 0; 0, \chi_0) - \tilde{\chi}(\tau; 0, \tilde{\chi}_0)\right\| \\
\leqslant &\left\|\chi(\varsigma_1; 0, \chi_0) - \tilde{\chi}(\varsigma_1; 0, \tilde{\chi}_0)\right\| + \left\|\tilde{\chi}(\varsigma_1; 0, \tilde{\chi}_0) - \tilde{\chi}(\tau; 0, \tilde{\chi}_0)\right\| \\
< &\frac{1}{2}\eta_{12} + \left\|(\mathbf{I}^{\alpha}_{\tilde{\varsigma}_1, \varsigma_1} f)(\varsigma_1, \tilde{\chi}) - (\mathbf{I}^{\alpha}_{\tilde{\varsigma}_1, \tau} f)(\tau, \tilde{\chi})\right\| \\
\leqslant &\frac{1}{2}\eta_{12} + \frac{3M}{\Gamma(\alpha + 1)}(\varsigma_1 - \tilde{\varsigma}_1)^{\alpha} \\
< &\frac{1}{2}\eta_{12} + \frac{3M}{\Gamma(\alpha + 1)}\delta^{\alpha}_{\varsigma_1} < \eta_{12} < \varepsilon,
\end{aligned}$$

i.e. $H(\chi(\varsigma_1 + 0; 0, \chi_0), \tilde{r}(\tilde{\varsigma}_1, \varsigma_1]) < \varepsilon$.

For $\tau_2 < \tau \leqslant \tilde{\tau}_2$, similar to (2.80), we have that $H(\chi(\tau_2; 0, \chi_0), \tilde{r}(\tau_2, \tilde{\tau}_2]) < \varepsilon$.

Let δ_{τ_2} be an arbitrary positive constant, assume that $\eta_{12} < \frac{6M}{\Gamma(\alpha+1)}\delta^{\alpha}_{\tau_2}$, and then $|\tilde{\tau}_2 - \tau_2| < \delta_{\tau_2}$.

Therefore, $\forall\ \delta_{\tau_2} > 0,\ \exists\ \delta_{\varsigma_1} > 0,\quad \forall\ \tilde{\varsigma}_1 \in \mathbb{R}^+,\ |\varsigma_1 - \tilde{\varsigma}_1| < \delta_{\varsigma_1},\quad \forall\ \tilde{d}_{\tau_2} > 0$, $|\tilde{d}_{\tau_2} - d| < \delta_{\varsigma_1}$, then $H(\tilde{r}(\tilde{\varsigma}_1, \tilde{\tau}_2], r(\varsigma_1, \tau_2]) < \varepsilon,\ |\tilde{\tau}_2 - \tau_2| < \delta_{\tau_2}$.

Repeat the procedure, and we obtain the same conclusion in (2.83) and (2.84).

Considering the trajectories $\tilde{r}(\tilde{\varsigma}_k, T]$ and $r(\varsigma_k, T]$, we still have the expression (2.85) and $H(\tilde{\chi}(\tilde{\varsigma}_k + 0; 0, \tilde{\chi}_0), r(\varsigma_k, \tilde{\varsigma}_k]) = 0$.

We have that $\forall\ \varepsilon > 0,\ \exists\ \delta_{\varsigma_k},\ 0 < \delta_{\varsigma_k} < (\frac{\Gamma(\alpha+1)\varepsilon}{6M})^{\frac{1}{\alpha}},\ \forall\ \tilde{\varsigma}_k \in R^+,\ |\varsigma_k - \tilde{\varsigma}_k| < \delta_{\varsigma_k}$, $\forall\, \tilde{d}_{\tau_{k+1}} > 0,\ |\tilde{d}_{\tau_{k+1}} - d| < \delta_{\varsigma_k}$, then $\|\tilde{\chi}(\tau; 0, \tilde{\chi}_0) - \chi(\tau; 0, \chi_0)\| < \frac{\varepsilon}{2},\ \varsigma_k^{\max} < \tau \leqslant T$, that is $R(\tilde{r}(\varsigma_k^{\max}, T], r(\varsigma_k^{\max}, T]) < \varepsilon$.

For $\tilde{\varsigma}_k < \tau \leqslant \varsigma_k$,

$$\begin{aligned}
&\left\|\chi(\varsigma_k + 0; 0, \chi_0) - \tilde{\chi}(\tau; 0, \tilde{\chi}_0)\right\| \\
\leqslant &\left\|\chi(\varsigma_k; 0, \chi_0) - \tilde{\chi}(\varsigma_k; 0, \tilde{\chi}_0)\right\| + \left\|\tilde{\chi}(\varsigma_k; 0, \tilde{\chi}_0) - \tilde{\chi}(\tau; 0, \tilde{\chi}_0)\right\| \\
< &\frac{\varepsilon}{2} + \left\|(\mathbf{I}^{\alpha}_{\tilde{\varsigma}_k, \varsigma_k} f)(\varsigma_k, \tilde{\chi}) - (\mathbf{I}^{\alpha}_{\tilde{\varsigma}_k, \tau} f)(\tau, \tilde{\chi})\right\| \\
\leqslant &\frac{\varepsilon}{2} + \frac{3M}{\Gamma(\alpha + 1)}(\varsigma_k - \tilde{\varsigma}_k)^{\alpha} \\
< &\frac{\varepsilon}{2} + \frac{3M}{\Gamma(\alpha + 1)}\delta^{\alpha}_{\varsigma_k} < \varepsilon,
\end{aligned}$$

i.e. $H(\chi(\varsigma_k + 0; 0, \chi_0), \tilde{r}(\tilde{\varsigma}_k, \varsigma_k]) < \varepsilon$.

Therefore, $\forall\ \varepsilon > 0,\ \exists\ \delta_{\varsigma_k} > 0,\quad \forall\ \tilde{\varsigma}_k \in \mathbb{R}^+,\quad |\varsigma_k - \tilde{\varsigma}_k| < \delta_{\varsigma_k},\quad \forall\, \tilde{d}_{\tau_k+1} > 0$, $|\tilde{d}_{\tau_k+1} - d| < \delta_{\varsigma_k}$, then $H(\tilde{r}(\tilde{\varsigma}_k, T], r(\varsigma_k, T]) < \varepsilon$.

Now $\delta_{\varsigma_k} = \delta_{\varsigma_k}(\varepsilon),\ \ \delta_{\tau_k} = \delta_{\tau_k}(\delta_{\varsigma_k}, \varepsilon),\ \ \delta_{\varsigma_{k-1}} = \delta_{\varsigma_{k-1}}(\delta_{\tau_k}, \varepsilon),\ \ \cdots,\ \ \delta_{\tau_1} = \delta_{\tau_1}(\delta_{\varsigma_1}, \varepsilon)$ and $\delta_0 = \delta_0(\delta_{\tau_1}, \varepsilon)$.

Consequently one has the conclusion: $\forall\, \varepsilon > 0$, $\exists\, \delta_0 > 0$, $\forall\, \tilde{\chi}_0 \in \mathbb{D}$, $\|\tilde{\chi}_0 - \chi_0\| < \delta_0$, $\forall\, \tilde{d}_{\tau_i} > 0$, $\forall\, \tilde{d}_{\varsigma_i} > 0$, $|\tilde{d}_{\tau_i} - d| < \delta_0$, $|d - \tilde{d}_{\varsigma_i}| < \delta_0$, $i = 1, 2, \ldots, k$, then $H(\tilde{r}[0, \tilde{\tau}_1], r[0, \tau_1]) < \varepsilon$, $H(\tilde{r}(\tilde{\tau}_i, \tilde{\varsigma}_i], r(\tau_i, \varsigma_i]) < \varepsilon$, $i = 1, 2, \ldots, k$, $H(\tilde{r}(\tilde{\varsigma}_i, \tilde{\tau}_{i+1}], r(\varsigma_i, \tau_{i+1}]) < \varepsilon$, $i = 1, 2, \ldots, k-1$, $H(\tilde{r}(\tilde{\varsigma}_k, T], r(\varsigma_k, T]) < \varepsilon$.

Now apply theorem 2.2.1 and [21, theorem 1.3], and we have the result (2.86), that is $H(\tilde{r}[0, T], r[0, T]) < \varepsilon$.

Case 3. Let $\tau_i^{\min} = \tau_i$, $\varsigma_i^{\min} = \varsigma_i$, $\tau_i^{\max} = \tilde{\tau}_i$, $\varsigma_i^{\max} = \tilde{\varsigma}_i$, $i = 1, 2, \ldots$; (the case $\tau_i^{\min} = \tilde{\tau}_i$, $\varsigma_i^{\min} = \tilde{\varsigma}_i$, $\tau_i^{\max} = \tau_i$, $\varsigma_i^{\max} = \varsigma_i$, $i = 1, 2, \ldots$ can be considered similarly). In the case, we have the same conclusion with δ_T in case 3 of theorem 2.2.7.

For $\tilde{r}[0, \tilde{\tau}_1]$ and $r[0, \tau_1]$, we still have the inequality (2.88) and further $H(\tilde{r}(\varsigma_1, \tilde{\tau}_1], r(\tilde{\tau}_1, \varsigma_1]) = 0$, $H(\tilde{\chi}(\tilde{\tau}_1; 0, \tilde{\chi}_0), r(\varsigma_1, \tau_1]) = 0$.

Set $0 < \eta_{01} < \varepsilon$, we have that $\exists\, \delta_0$, $0 < \delta_0 < \min\{\delta_T, (\frac{\Gamma(\alpha+1)\eta_{01}}{6M})^{\frac{1}{\alpha}}\}$, $\forall\, \tilde{\chi}_0 \in \mathbb{D}$, $\|\tilde{\chi}_0 - \chi_0\| < \delta_0$, $\forall\, \tilde{d}_{\tau_1} > 0$, $|\tilde{d}_{\tau_1} - d| < \delta_0$, then $\|\tilde{\chi}(\tau; 0, \tilde{\chi}_0) - \chi(\tau; 0, \chi_0)\| < \frac{1}{2}\eta_{01}$, $0 < \tau \leqslant \tau_1^{\min}$, that is $R(\tilde{r}(0, \tau_1^{\min}], r(0, \tau_1^{\min}]) < \frac{1}{2}\eta_{01} < \varepsilon$.

Now $|\tilde{\tau}_1 - \tau_1| = |\tilde{d}_{\tau_1} - d| < \delta_0 < (\frac{\Gamma(\alpha+1)\eta_{01}}{6M})^{\frac{1}{\alpha}}$. For $\tau_1 < \tau \leqslant \varsigma_1$,

$$\begin{aligned}
&\left\|\chi(\tau_1; 0, \chi_0) - \tilde{\chi}(\tau; 0, \tilde{\chi}_0)\right\| \\
\leqslant\; &\left\|\chi(\tau_1; 0, \chi_0) - \tilde{\chi}(\tau_1; 0, \tilde{\chi}_0)\right\| + \left\|\tilde{\chi}(\tau; 0, \tilde{\chi}_0) - \tilde{\chi}(\tau_1; 0, \tilde{\chi}_0)\right\| \\
<\; &\frac{1}{2}\eta_{01} + \left\|(\mathbf{I}^{\alpha}_{0,\tau} f)(\tau, \tilde{\chi}) - (\mathbf{I}^{\alpha}_{0,\tau_1} f)(\tau_1, \tilde{\chi})\right\| \\
\leqslant\; &\frac{1}{2}\eta_{01} + \frac{3M}{\Gamma(\alpha+1)}(\varsigma_1 - \tau_1)^{\alpha} < \frac{1}{2}\eta_{01} + \frac{3M}{\Gamma(\alpha+1)}(\tilde{\tau}_1 - \tau_1)^{\alpha} \\
<\; &\frac{1}{2}\eta_{01} + \frac{3M}{\Gamma(\alpha+1)}\delta_0^{\alpha} < \eta_{01} < \varepsilon,
\end{aligned}$$

i.e. $H(\chi(\tau_1; 0, \chi_0), \tilde{r}(\tau_1, \varsigma_1]) < \varepsilon$.

Assume that $\eta_{01} < \frac{6M}{\Gamma(\alpha+1)}\delta_{\tau_1}^{\alpha}$, and then $|\tilde{\tau}_1 - \tau_1| < \delta_{\tau_1}$, where δ_{τ_1} is an arbitrary positive constant. Therefore, $\forall\, \delta_{\tau_1} > 0$, $\exists\, \delta_0 > 0$, $\forall\, \tilde{\chi}_0 \in \mathbb{D}$, $\|\tilde{\chi}_0 - \chi_0\| < \delta_0$, $\forall\, \tilde{d}_{\tau_1} > 0$, $|\tilde{d}_{\tau_1} - d| < \delta_0$, then $H(\tilde{r}[0, \tilde{\tau}_1], r[0, \tau_1]) < \varepsilon$, $|\tilde{\tau}_1 - \tau_1| < \delta_{\tau_1}$.

For $\tilde{r}(\tilde{\varsigma}_1, \tilde{\tau}_2]$ and $r(\varsigma_1, \tau_2]$, the inequality (2.89) is satisfied, and further $H(\chi(\varsigma_1 + 0; 0, \chi_0), \tilde{r}(\tilde{\varsigma}_1, \tilde{\tau}_1]) = 0$, $H(r(\varsigma_1, \tilde{\tau}_1], \tilde{r}(\tilde{\tau}_1, \varsigma_1]) = 0$, $H(\tilde{r}(\varsigma_2, \tilde{\tau}_2], r(\tilde{\tau}_2, \varsigma_2]) = 0$ and $H(\tilde{\chi}(\tilde{\tau}_2; 0, \tilde{\chi}_0), r(\varsigma_2, \tau_2]) = 0$.

Set $|\tilde{d}_{\varsigma_1} - d| < \delta_{\tau_1}$, and then $|\tilde{\varsigma}_1 - \varsigma_1| \leqslant |\tilde{\tau}_1 - \tau_1| + |\tilde{d}_{\varsigma_1} - d| < 2\delta_{\tau_1}$, so if δ_{τ_1} is sufficiently small, then $|\tilde{\varsigma}_1 - \varsigma_1| < \delta_{\varsigma_1}$, where δ_{ς_1} denotes an arbitrary positive number.

Let $0 < \eta_{12} < \varepsilon$, we have that $\exists\, \delta_{\varsigma_1} > 0$, $\delta_{\varsigma_1} < (\frac{\Gamma(\alpha+1)\eta_{12}}{3M2^{\alpha+1}})^{\frac{1}{\alpha}}$, $\forall\, \tilde{\varsigma}_1 \in \mathbb{R}^+$, $|\tilde{\varsigma}_1 - \varsigma_1| < \delta_{\varsigma_1}$, $\forall\, \tilde{d}_{\tau_2} > 0$, $|\tilde{d}_{\tau_2} - d| < \delta_{\varsigma_1}$, then $\|\tilde{\chi}(\tau; 0, \tilde{\chi}_0) - \chi(\tau; 0, \chi_0)\| < \frac{1}{2}\eta_{12}$, $\varsigma_1^{\max} < \tau \leqslant \tau_2^{\min}$, that is $R(\tilde{r}(\varsigma_1^{\max}, \tau_2^{\min}], r(\varsigma_1^{\max}, \tau_2^{\min}]) < \frac{\eta_{12}}{2} < \varepsilon$.

For $\tilde{\tau}_1 < \tau \leqslant \tilde{\varsigma}_1$,

$$\begin{aligned}
&\left\|\tilde{\chi}(\tilde{\varsigma}_1 + 0; 0, \tilde{\chi}_0) - \chi(\tau; 0, \chi_0)\right\| \\
\leqslant\ &\left\|\tilde{\chi}(\tilde{\varsigma}_1; 0, \tilde{\chi}_0) - \chi(\tilde{\varsigma}_1; 0, \chi_0)\right\| + \left\|\chi(\tilde{\varsigma}_1; 0, \chi_0) - \chi(\tau; 0, \chi_0)\right\| \\
<\ &\frac{1}{2}\eta_{12} + \left\|(\mathbf{I}^{\alpha}_{\varsigma_1, \tilde{\varsigma}_1} f)(\tilde{\varsigma}_1, \chi) - (\mathbf{I}^{\alpha}_{\varsigma_1, \tau} f)(\tau, \chi)\right\| \\
\leqslant\ &\frac{1}{2}\eta_{12} + \frac{3M}{\Gamma(\alpha + 1)}(\tilde{\varsigma}_1 - \tilde{\tau}_1)^{\alpha} < \frac{1}{2}\eta_{12} + \frac{3M}{\Gamma(\alpha + 1)}(\tilde{\varsigma}_1 - \varsigma_1)^{\alpha} \\
<\ &\frac{1}{2}\eta_{12} + \frac{3M}{\Gamma(\alpha + 1)}\delta^{\alpha}_{\varsigma_1} < \eta_{12} < \varepsilon,
\end{aligned}$$

i.e. $H(\tilde{\chi}(\tilde{\varsigma}_1 + 0; 0, \tilde{\chi}_0), r(\tilde{\tau}_1, \tilde{\varsigma}_1]) < \varepsilon$.

Now $|\tilde{\tau}_2 - \tau_2| \leqslant |\tilde{\varsigma}_1 - \varsigma_1| + |\tilde{d}_{\tau_2} - d| < 2\delta_{\varsigma_1}$. For $\tau_2 < \tau \leqslant \varsigma_2$,

$$\begin{aligned}
&\left\|\chi(\tau_2; 0, \chi_0) - \tilde{\chi}(\tau; 0, \tilde{\chi}_0)\right\| \\
\leqslant\ &\left\|\chi(\tau_2; 0, \chi_0) - \tilde{\chi}(\tau_2; 0, \tilde{\chi}_0)\right\| + \left\|\tilde{\chi}(\tau; 0, \tilde{\chi}_0) - \tilde{\chi}(\tau_2; 0, \tilde{\chi}_0)\right\| \\
<\ &\frac{1}{2}\eta_{12} + \left\|(\mathbf{I}^{\alpha}_{\tilde{\varsigma}_1, \tau} f)(\tau, \tilde{\chi}) - (\mathbf{I}^{\alpha}_{\tilde{\varsigma}_1, \tau_2} f)(\tau_2, \tilde{\chi})\right\| \\
\leqslant\ &\frac{1}{2}\eta_{12} + \frac{3M}{\Gamma(\alpha + 1)}(\varsigma_2 - \tau_2)^{\alpha} < \frac{1}{2}\eta_{12} + \frac{3M}{\Gamma(\alpha + 1)}(\tilde{\tau}_2 - \tau_2)^{\alpha} \\
<\ &\frac{1}{2}\eta_{12} + \frac{3M}{\Gamma(\alpha + 1)}2^{\alpha}\delta^{\alpha}_{\varsigma_1} < \eta_{12} < \varepsilon,
\end{aligned}$$

i.e. $H(\chi(\tau_2; 0, \chi_0), \tilde{r}(\tau_2, \varsigma_2]) < \varepsilon$.

Let δ_{τ_2} be an arbitrary positive number, assume that $\eta_{12} < \frac{6M}{\Gamma(\alpha+1)}\delta^{\alpha}_{\tau_2}$, and then $|\tilde{\tau}_2 - \tau_2| < \delta_{\tau_2}$.

Therefore, $\forall\ \delta_{\tau_2} > 0,\ \exists\ \delta_{\varsigma_1} > 0,\ \forall\ \tilde{\varsigma}_1 \in \mathbb{R}^+,\ |\tilde{\varsigma}_1 - \varsigma_1| < \delta_{\varsigma_1},\ \forall\ \tilde{d}_{\tau_2} > 0$, $|\tilde{d}_{\tau_2} - d| < \delta_{\varsigma_1}$, then $H(\tilde{r}(\tilde{\varsigma}_1, \tilde{\tau}_2], r(\varsigma_1, \tau_2]) < \varepsilon$, $|\tilde{\tau}_2 - \tau_2| < \delta_{\tau_2}$.

Consequently, the same conclusion in (2.90) follows.

For the trajectories $\tilde{r}(\tilde{\varsigma}_k, T]$ and $r(\varsigma_k, T]$, the inequality (2.91) holds, and furthermore we have $H(\chi(\varsigma_k + 0; 0, \chi_0), \tilde{r}(\tilde{\varsigma}_k, \tilde{\tau}_k] = 0$ and $H(r(\varsigma_k, \tilde{\tau}_k], \tilde{r}(\tilde{\tau}_k, \varsigma_k] = 0$.

Set $|\tilde{d}_{\varsigma_k} - d| < \delta_{\tau_k}$, and then $|\tilde{\varsigma}_k - \varsigma_k| \leqslant |\tilde{\tau}_k - \tau_k| + |\tilde{d}_{\varsigma_k} - d| < 2\delta_{\tau_k}$, and let δ_{ς_k} denote an arbitrary positive number, so if δ_{τ_k} is sufficiently small, then $|\tilde{\varsigma}_k - \varsigma_k| < \delta_{\varsigma_k}$.

We deduce that $\forall\ \varepsilon > 0,\ \exists\ \delta_{\varsigma_k},\ 0 < \delta_{\varsigma_k} < (\frac{\Gamma(\alpha+1)\varepsilon}{6M})^{\frac{1}{\alpha}},\ \forall\ \tilde{\varsigma}_k \in R^+,\ |\tilde{\varsigma}_k - \varsigma_k| < \delta_{\varsigma_k}$, $\forall\ \tilde{d}_{\tau_{k+1}} > 0,\ |\tilde{d}_{\tau_{k+1}} - d| < \delta_{\varsigma_k}$, then $\left\|\tilde{\chi}(\tau; 0, \tilde{\chi}_0) - \chi(\tau; 0, \chi_0)\right\| < \frac{\varepsilon}{2}$, $\varsigma_k^{\max} < \tau \leqslant T$, that is $R(\tilde{r}(\varsigma_k^{\max}, T], r(\varsigma_k^{\max}, T]) < \varepsilon$.

For $\tilde{\tau}_k < \tau \leqslant \tilde{\varsigma}_k$,

$$
\begin{aligned}
&\left\|\tilde{\chi}(\tilde{\varsigma}_k + 0; 0, \tilde{\chi}_0) - \chi(\tau; 0, \chi_0)\right\| \\
\leqslant\ &\left\|\tilde{\chi}(\tilde{\varsigma}_k; 0, \tilde{\chi}_0) - \chi(\tilde{\varsigma}_k; 0, \chi_0)\right\| + \left\|\chi(\tilde{\varsigma}_k; 0, \chi_0) - \chi(\tau; 0, \chi_0)\right\| \\
<\ &\frac{\varepsilon}{2} + \left\|(\mathbf{I}^{\alpha}_{\varsigma_k, \tilde{\varsigma}_k} f)(\tilde{\varsigma}_k, \chi) - (\mathbf{I}^{\alpha}_{\varsigma_k, \tau} f)(\tau, \chi)\right\| \\
<\ &\frac{\varepsilon}{2} + \frac{3M}{\Gamma(\alpha + 1)}(\tilde{\varsigma}_k - \varsigma_k)^{\alpha} \\
<\ &\frac{\varepsilon}{2} + \frac{3M}{\Gamma(\alpha + 1)}\delta^{\alpha}_{\varsigma_k} < \varepsilon,
\end{aligned}
$$

i.e. $H(\tilde{\chi}(\tilde{\varsigma}_k + 0; 0, \tilde{\chi}_0), r(\tilde{\tau}_k, \tilde{\varsigma}_k]) < \varepsilon$.

Therefore, $\forall\ \varepsilon > 0,\ \exists\ \delta_{\varsigma_k} > 0,\ \ \forall\ \tilde{\varsigma}_k \in \mathbb{R}^+,\ |\tilde{\varsigma}_k - \varsigma_k| < \delta_{\varsigma_k},\ \ \forall\, \tilde{d}_{\tau_k+1} > 0$, $|\tilde{d}_{\tau_k+1} - d| < \delta_{\varsigma_k}$, then $H(\tilde{r}(\tilde{\varsigma}_k, T], r(\varsigma_k, T]) < \varepsilon$.

From an argument similar to that in case 3 of theorem 2.2.7, we obtain $\forall\ \varepsilon > 0,\ \exists\ \delta_0 > 0,\ \forall\ \tilde{\chi}_0 \in \mathbb{D},\ \left\|\tilde{\chi}_0 - \chi_0\right\| < \delta_0,\ \forall\ \tilde{d}_{\tau_i} > 0,\ \forall\ \tilde{d}_{\varsigma_i} > 0,\ |\tilde{d}_{\tau_i} - d| < \delta_0$, $|\tilde{d}_{\varsigma_i} - d| < \delta_0$, $i = 1, 2, \ldots, k$, then $H(\tilde{r}[0, \tilde{\tau}_1], r[0, \tau_1]) < \varepsilon$, $H(\tilde{r}(\tilde{\varsigma}_i, \tilde{\tau}_{i+1}], r(\varsigma_i, \tau_{i+1}]) < \varepsilon$, $i = 1, 2, \ldots, k - 1$, $H(\tilde{r}(\tilde{\varsigma}_k, T], r(\varsigma_k, T]) < \varepsilon$.

Apply theorem 2.2.1 and [21, theorem 1.3], and then

$$
\begin{aligned}
&H(\tilde{r}[0, T], r[0, T]) \\
\leqslant\ &H\left(\tilde{r}[0, \tilde{\tau}_1] \cup \left(\bigcup_{i=1,2,\ldots,k-1} \tilde{r}(\tilde{\varsigma}_i, \tilde{\tau}_{i+1}]\right) \cup \tilde{r}(\tilde{\varsigma}_k, T],\right. \\
&\left. r[0, \tau_1] \cup \left(\bigcup_{i=1,2,\ldots,k-1} r(\varsigma_i, \tau_{i+1}]\right) \cup_r (\varsigma_k, T]\right) \\
\leqslant\ &\max\left\{H(\tilde{r}[0, \tilde{\tau}_1], r[0, \tau_1]),\ H(\tilde{r}(\tilde{\varsigma}_i, \tilde{\tau}_{i+1}], r(\varsigma_i, \tau_{i+1}]),\ i = 1, 2, \ldots, k - 1,\right. \\
&\left. H(\tilde{r}(\tilde{\varsigma}_k, T], r(\varsigma_k, T])\right\} < \varepsilon.
\end{aligned}
$$

The proof is completed. □

2.2.4 Examples

Set $\tau_0 = \varsigma_0 = 0$, $\tau_i = 2i - 1$ and $\varsigma_i = 2i$, $i \in \{0\} \cup \mathbb{N}$. Clearly, $\varsigma_i < \tau_{i+1} \to \infty\ (i \to \infty)$.

Example 2.2.9. *Consider the following NIDEs of integer order*

$$\begin{cases} \chi'(\tau) = \arctan \chi(\tau), \ \ \tau \in (2i, 2i+1], \ \ i \in \{0\} \bigcup \mathbb{N}, \\ \chi((2i-1)^{+}) = \dfrac{2i - 1 + |\chi((2i-1)^{-})|}{2i + |\chi((2i-1)^{-})|}, \ \ i \in \mathbb{N}, \\ \chi(\tau) = \dfrac{\tau + |\chi((2i-1)^{-})|}{1 + \tau + |\chi((2i-1)^{-})|}, \ \ \tau \in (2i-1, 2i], \ \ i \in \mathbb{N}, \\ \chi(0) = \chi_0. \end{cases} \tag{2.92}$$

One can derive the solution of (2.92), namely

$$\chi(\tau) = \begin{cases} \chi_0 + \displaystyle\int_0^{\tau} \arctan \chi(s) ds, \ \ \text{for} \tau \in (0, 1], \\ \dfrac{\tau + |\chi(1^{-})|}{1 + \tau + |\chi(1^{-})|}, \ \ \text{for} \tau \in (1, 2], \\ \dfrac{2 + |\chi(1^{-})|}{3 + |\chi(1^{-})|} + \displaystyle\int_2^{\tau} \arctan \chi(s) ds, \ \ \text{for} \tau \in (2, 3], \\ \vdots \\ \dfrac{\tau + |\chi((2\gamma-1)^{-})|}{1 + \tau + |\chi((2\gamma-1)^{-})|}, \ \ \text{for} \tau \in (2\gamma - 1, 2\gamma], \\ \dfrac{2\gamma + |\chi((2\gamma-1)^{-})|}{1 + 2\gamma + |\chi((2\gamma-1)^{-})|} + \displaystyle\int_{2\gamma}^{\tau} \arctan \chi(s) ds, \ \ \text{for} \tau \in (2\gamma, 2\gamma + 1], \\ \vdots \end{cases} \tag{2.93}$$

Set $f(\tau, \chi) = \arctan \chi$, $g_i(\tau, \chi) = \frac{\tau + |\chi|}{1 + \tau + |\chi|}$. *Note* $g_i \in C([2i-1, 2i] \times \mathbb{D}, \mathbb{R}^n)$, $i = 1, 2, \ldots$.

Let $\tau \in (2i, 2i+1]$. *Clearly,* $\|f(\tau, \chi) - f(\tau, \psi)\| \leqslant \|\chi - \psi\|$ *and* $\left\|f(\tau, \chi)\right\| \leqslant M := \frac{\pi}{2}$, $\forall\, \chi \in \mathbb{R}^n$. *In addition,* $\left\|g_i(\tau_1, \chi) - g_i(\tau_2, \psi)\right\| \leqslant \left\|\tau_1 - \tau_2\right\| + \|\chi - \psi\|$, $\forall\, \chi, \psi \in \mathbb{R}^n$ *so choose* $L_{g_i} = 1$. *Thus,* $[H_1] - [H_4]$ *hold. Therefore Theorem 2.2.7 can be applied to (2.92).*

Example 2.2.10. *Consider the following NIDEs of fractional order*

$$\begin{cases} {}^{c}\mathbf{D}_{2i,\tau}^{\frac{1}{2}} \chi(\tau) = \arctan \chi(\tau), \ \ \tau \in (2i, 2i+1], \ \ i \in \{0\} \bigcup \mathbb{N}, \ \ \alpha = \dfrac{1}{2}, \\ \chi((2i-1)^{+}) = \dfrac{2i - 1 + |\chi((2i-1)^{-})|}{2i + |\chi((2i-1)^{-})|}, \ \ i \in \mathbb{N}, \\ \chi(\tau) = \dfrac{\tau + |\chi((2i-1)^{-})|}{1 + \tau + |\chi((2i-1)^{-})|}, \ \ \tau \in (2i-1, 2i], \ \ i \in \mathbb{N}, \\ \chi(0) = \chi_0. \end{cases} \tag{2.94}$$

Clearly, $f(\tau, \chi) = \arctan \chi$, $g_i(\tau, \chi) = \frac{\tau + |\chi|}{1+\tau+|\chi|}$, *which are the same as in example 2.2.9. One can derive the solution of (2.94), namely*

$$\chi(\tau) = \begin{cases} \chi_0 + \dfrac{1}{\sqrt{\pi}} \displaystyle\int_0^\tau (\tau - s)^{-\frac{1}{2}} \arctan \chi(s) ds, \quad \tau \in (0, 1], \\ \dfrac{\tau + |\chi(1^-)|}{1 + \tau + |\chi(1^-)|}, \quad \text{for} \tau \in (1, 2], \\ \dfrac{2 + |\chi(1^-)|}{3 + |\chi(1^-)|} + \dfrac{1}{\sqrt{\pi}} \displaystyle\int_2^\tau (\tau - s)^{-\frac{1}{2}} \arctan \chi(s) ds, \quad \text{for} \tau \in (2, 3], \\ \vdots \\ \dfrac{\tau + |\chi((2\gamma - 1)^-)|}{1 + \tau + |\chi((2\gamma - 1)^-)|}, \quad \text{for} \tau \in (2\gamma - 1, 2\gamma], \\ \dfrac{2\gamma + |\chi((2\gamma - 1)^-)|}{1 + 2\gamma + |\chi((2\gamma - 1)^-)|} + \dfrac{1}{\sqrt{\pi}} \displaystyle\int_{2\gamma}^\tau (\tau - s)^{-\frac{1}{2}} \arctan \chi(s) ds, \\ \quad \text{for} \tau \in (2\gamma, 2\gamma + 1], \\ \vdots \end{cases} \tag{2.95}$$

Note $[H_1] - [H_4]$ *hold. Therefore Theorem 2.2.8 can be applied to (2.94).*

2.2.5 Notes and remarks

This section transfer the ideas in [21] to investigate orbital Hausdorff continuous dependence of the solutions to integer order and fractional nonlinear non-instantaneous differential equations. The concept of orbital Hausdorff continuous dependence is used to characterize the relations of solutions corresponding to the impulsive points and junction points in the sense of the Hausdorff distance. Then, we establish sufficient conditions to guarantee this specific continuous dependence on their respective trajectories.

The results in section 2.2 are motivated by [26].

2.3 Differentiability of solutions

2.3.1 Introduction

This section is devoted to study asymptotic and smooth properties of solutions of integer order nonlinear non-instantaneous impulsive differential equations involving parameters

$$\begin{cases} x'(t) = f(t, x(t), \lambda_{s_i}), \quad t \in (s_i, t_{i+1}], \ i \in \{0\} \cup \mathbb{N}, \\ x(t_i^+) = g_i(t_i, x(t_i^-), \lambda_{t_i}), \quad i \in \mathbb{N}, \\ x(t) = g_i(t, x(t_i^-), \lambda_{t_i}) + h(t, \lambda_{s_i}), \quad t \in (t_i, s_i], \ i \in \mathbb{N}, \\ x(0) = x_0(\lambda_0); \end{cases} \tag{2.96}$$

and fractional order nonlinear non-instantaneous impulsive differential equations involving parameters:

$$\begin{cases} {}^{c}\mathbf{D}^{\alpha}_{s_i,t}x(t) = f(t, x(t), \lambda_{s_i}), & t \in (s_i, t_{i+1}], \ i \in \{0\} \cup \mathbb{N}, \ \alpha \in (0, 1), \\ x(t_i^+) = g_i(t_i, x(t_i^-), \lambda_{t_i}), & i \in \mathbb{N}, \\ x(t) = g_i(t, x(t_i^-), \lambda_{t_i}) + h(t, \lambda_{s_i}), & t \in (t_i, s_i], \ i \in \mathbb{N}, \\ x(0) = x_0(\lambda_0), \end{cases} \tag{2.97}$$

where λ_0, λ_{t_i}, λ_{s_i} are real parameters, $i \in N$ and ${}^{c}\mathbf{D}^{\alpha}_{s_i,t}$ denotes the classical Caputo fractional derivative of order α by changing the lower limit s_i, and t_i acts as an impulsive point [1] and s_i acts as a junction point satisfying $s_i < t_{i+1} \to \infty$ with $t_0 = s_0 = 0$, and λ_{t_i}, λ_{s_i} are denoted as impulsive parameters and junction parameters, respectively. Set $h(t_i, \lambda_{s_i}) = 0$. Note $x(t_i^+) = \lim_{\varepsilon \to 0^+} x(t_i + \varepsilon)$ and $x(t_i^-) = \lim_{\varepsilon \to 0^+} x(t_i - \varepsilon) := x(t_i)$. Let f: $[0, \infty) \times \mathbb{D} \times \mathbb{R} \to \mathbb{R}^n$, g_i: $[t_i, s_i] \times \mathbb{D} \times \mathbb{R} \to \mathbb{R}^n$ and h: $[t_i, s_i] \times \mathbb{R} \to \mathbb{R}^n$, $i \in N$ where the domain $\mathbb{D} \subset \mathbb{R}^n$.

The main objective is to extend the work in [25] to non-instantaneous impulses and fractional order and present sufficient conditions to guarantee continuous dependence and differentiability of solutions with respect to initial condition, impulsive perturbation and junction perturbation.

The rest of this part is organized as follows. In section 2.3.2, we introduce the definitions of continuous dependence and differentiability of the solutions for (2.96) and (2.97). In section 2.3.3, we present the results for the continuous dependence and differentiability of solutions with respect to initial condition, impulsive parameters and junction parameters. Two examples are given in section 2.3.4 to illustrate our results.

2.3.2 Preliminaries

Let $J = [0, \infty)$ and $C(J, \mathbb{R}^n)$ be the space of all continuous functions from J into $\mathbb{R}^n$. We recall the piecewise continuous functions space $PC(J, \mathbb{R}^n) := \{x: J \to \mathbb{R}^n: x \in C((t_k, t_{k+1}], \mathbb{R}^n), \ k = 0, 1, \ldots$ and there exist $x(t_k^+)$ and $x(t_k^-)$, $k = 1, 2, \ldots$ with $x(t_k^-) = x(t_k)\}$ with the norm $\|x\|_{PC} := \sup\{\|x(t)\|: t \in J\}$, where $\|x(t)\|$ denotes the norm for $x(t) \in \mathbb{R}^n$.

From [20, definition 2.1], one can see the representation of the piecewise continuous solution $x \in PC(J, \mathbb{R}^n)$ of problem (2.96) is given by

$$\begin{aligned} x(t; \lambda_0) &= x_0(\lambda_0) + \int_0^t f(s, x(s; \lambda_0), \lambda_0)ds, \quad t \in [0, t_1], \\ x(t; \lambda_0, \lambda_{t_1}, \ldots, \lambda_{s_{i-1}}, \lambda_{t_i}) &= g_i(t, x(t_i^-; \lambda_0, \lambda_{t_1}, \ldots, \lambda_{t_{i-1}}, \lambda_{s_{i-1}}), \lambda_{t_i}) \\ &\quad + h(t; \lambda_{s_i}), \quad t \in (t_i, s_i], \ i \in \mathbb{N}, \\ x(t; \lambda_0, \lambda_{t_1}, \ldots, \lambda_{s_{i-1}}, \lambda_{t_i}, \lambda_{s_i}) &= g_i(s_i, x(t_i^-; \lambda_0, \lambda_{t_1}, \ldots, \lambda_{t_{i-1}}, \lambda_{s_{i-1}}), \lambda_{t_i}) + h(s_i; \lambda_{s_i}) \\ &\quad + \int_{s_i}^t f(s, x(s; \lambda_0, \lambda_{t_1}, \ldots, \lambda_{t_i}, \lambda_{s_i}), \lambda_{s_i})ds, \\ &\quad t \in (s_i, t_{i+1}], \ i \in \mathbb{N}, \end{aligned}$$

and by [22, 24, section 3] and [19, section 8] one can see the representation of the solution $x \in PC(J, \mathbb{R}^n)$ of problem (2.97) is given by

$$\begin{aligned}
x(t; \lambda_0) &= x_0(\lambda_0) + \frac{1}{\Gamma(\alpha)} \int_0^t (t-s)^{\alpha-1} f(s, x(s; \lambda_0), \lambda_0) ds, \\
&\quad t \in [0, t_1], \\
x(t; \lambda_0, \lambda_{t_1}, \ldots, \lambda_{s_{i-1}}, \lambda_{t_i}) &= g_i(t, x(t_i^-; \lambda_0, \lambda_{t_1}, \ldots, \lambda_{t_{i-1}}, \lambda_{s_{i-1}}), \lambda_{t_i}) \\
&\quad + h(t; \lambda_{s_i}), \ t \in (t_i, s_i], \ i \in \mathbb{N}, \\
x(t; \lambda_0, \lambda_{t_1}, \ldots, \lambda_{s_{i-1}}, \lambda_{t_i}, \lambda_{s_i}) &= g_i(s_i, x(t_i^-; \lambda_0, \lambda_{t_1}, \ldots, \lambda_{t_{i-1}}, \lambda_{s_{i-1}}), \lambda_{t_i}) + h(s_i; \lambda_{s_i}) \\
&\quad + \frac{1}{\Gamma(\alpha)} \int_{s_i}^t (t-s)^{\alpha-1} \\
&\quad f(s, x(s; \lambda_0, \lambda_{t_1}, \ldots, \lambda_{t_i}, \lambda_{s_i}), \lambda_{s_i}) ds, \\
&\quad t \in (s_i, t_{i+1}], \ i \in \mathbb{N}.
\end{aligned}$$

For brevity, we set $x(t; \cdot\,)$ and $x(t; \cdot^*\,)$ be the solutions of the original equations and the perturbation equations respectively, where the perturbation is about the parameters.

Next, inspired by [25], we introduce some definitions of continuous dependence and differentiability of the solutions to (2.96) and (2.97) with respect to the impulsive perturbations, respectively.

Definition 2.3.1. *The solution of problem (2.96) depends continuously on initial parameter λ_0, impulsive parameters λ_{t_i}, $i \in \mathbb{N}$ and junction parameters λ_{s_i}, $i \in \mathbb{N}$, respectively, provided that the following limits*

$$\begin{aligned}
\lim_{\lambda_0^* \to \lambda_0} x(t; \lambda_0^*) &= x(t; \lambda_0), \ t \in [0, t_1], \\
\lim_{\substack{\lambda_{s_{i-1}}^* \to \lambda_{s_{i-1}} \\ \lambda_{t_i}^* \to \lambda_{t_i}}} x(t; \lambda_0, \lambda_{t_1}, \ldots, \lambda_{s_{i-1}}^*, \lambda_{t_i}^*) &= x(t; \lambda_0, \lambda_{t_1}, \ldots, \lambda_{s_{i-1}}, \lambda_{t_i}), \ t \in (t_i, s_i], \ i \in \mathbb{N}, \\
\lim_{\substack{\lambda_{s_{i-1}}^* \to \lambda_{s_{i-1}} \\ \lambda_{t_i}^* \to \lambda_{t_i} \\ \lambda_{s_i}^* \to \lambda_{s_i}}} x(t; \lambda_0, \lambda_{t_1}, \ldots, \lambda_{s_{i-1}}^*, \lambda_{t_i}^*, \lambda_{s_i}^*) &= x(t; \lambda_0, \lambda_{t_1}, \ldots, \lambda_{s_{i-1}}, \lambda_{t_i}, \lambda_{s_i}), \ t \in (s_i, t_{i+1}], \\
&\quad i \in \mathbb{N},
\end{aligned}$$

exist, respectively.

Definition 2.3.2. *The solution of problem (2.97) depends continuously about the order α and initial parameter λ_0, impulsive parameters λ_{t_i}, $i \in \mathbb{N}$ and junction parameters λ_{s_i}, $i \in \mathbb{N}$, respectively, provided that the limits*

$$\lim_{\substack{\beta\to\alpha\\ \lambda_0^*\to\lambda_0}} x(t;\lambda_0^*) = x(t;\lambda_0),\quad t\in[0,t_1],$$

$$\lim_{\substack{\beta\to\alpha\\ \lambda_{s_{i-1}}^*\to\lambda_{s_{i-1}}\\ \lambda_{t_i}^*\to\lambda_{t_i}}} x(t;\lambda_0,\lambda_{t_1},\ldots,\lambda_{s_{i-1}}^*,\lambda_{t_i}^*) = x(t;\lambda_0,\lambda_{t_1},\ldots,\lambda_{s_{i-1}},\lambda_{t_i}),\quad t\in(t_i,s_i],\ i\in\mathbb{N},$$

$$\lim_{\substack{\beta\to\alpha\\ \lambda_{s_{i-1}}^*\to\lambda_{s_{i-1}}\\ \lambda_{t_i}^*\to\lambda_{t_i}\\ \lambda_{s_i}^*\to\lambda_{s_i}}} x(t;\lambda_0,\lambda_{t_1},\ldots,\lambda_{s_{i-1}}^*,\lambda_{t_i}^*,\lambda_{s_i}^*) = x(t;\lambda_0,\lambda_{t_1},\ldots,\lambda_{s_{i-1}},\lambda_{t_i},\lambda_{s_i}),\quad t\in(s_i,t_{i+1}],\ i\in\mathbb{N},$$

exist, respectively.

Definition 2.3.3. *The solution of problems (2.96) and (2.97) is differentiable on initial parameter λ_0, impulsive parameters λ_{t_i}, $i\in\mathbb{N}$, and junction parameters λ_{s_i}, $i\in\mathbb{N}$, respectively, provided that*

$$\lim_{\lambda_0^*\to\lambda_0}\frac{1}{\lambda_0^*-\lambda_0}(x(t;\lambda_0^*)-x(t;\lambda_0)),\quad t\in[0,t_1],$$

$$\lim_{\lambda_{t_i}^*\to\lambda_{t_i}}\frac{1}{\lambda_{t_i}^*-\lambda_{t_i}}\Big(x(t;\lambda_0,\lambda_{t_1},\ldots,\lambda_{s_{i-1}},\lambda_{t_i}^*)-x(t;\lambda_0,\lambda_{t_1},\ldots,\lambda_{s_{i-1}},\lambda_{t_i})\Big),\quad t\in(t_i,s_i],$$

$$\lim_{\lambda_{s_i}^*\to\lambda_{s_i}}\frac{1}{\lambda_{s_i}^*-\lambda_{s_i}}\Big(x(t;\lambda_0,\lambda_{t_1},\ldots,\lambda_{s_{i-1}},\lambda_{t_i},\lambda_{s_i}^*)-x(t;\lambda_0,\lambda_{t_1},\ldots,\lambda_{s_{i-1}},\lambda_{t_i},\lambda_{s_i})\Big),$$

$t\in(s_i,t_{i+1}]$,

i.e. the derivatives

$$\frac{\partial}{\partial\lambda_0}x(t;\lambda_0),\ \frac{\partial}{\partial\lambda_{t_i}}x(t;\lambda_0,\lambda_{t_1},\ldots,\lambda_{s_{i-1}},\lambda_{t_i}),\ \frac{\partial}{\partial\lambda_{s_i}}x(t;\lambda_0,\lambda_{t_1},\ldots,\lambda_{s_{i-1}},\lambda_{t_i},\lambda_{s_i})$$

exist, respectively.

In this section, we present sufficient conditions to guarantee that the solution of (2.96) and (2.97) depends continuously and is differentiable on the initial condition, impulsive perturbation and junction perturbation.

Consider the following assumptions:

[H_1] The function $f\colon J\times\mathbb{D}\times\mathbb{R}\to\mathbb{R}^n$ is continuous, $g_i\in C([t_i,s_i]\times\mathbb{D}\times\mathbb{R},\mathbb{R}^n)$, and $h\in C([t_i,s_i]\times\mathbb{R},\mathbb{R}^n)$, $i\in\mathbb{N}$.

[H_2] There exists a positive constant L_f such that

$$\|f(t,x,\mu)-f(t,y,\nu)\|\leqslant L_f(\|x-y\|+|\mu-\nu|),$$

for each $t\in[s_i,t_{i+1}]$, $i\in\{0\}\cup\mathbb{N}$, for all $x,y\in\mathbb{R}^n$ and $\mu,\nu\in\mathbb{R}$.

$[H_3]$ There exist positive constants L_{g_i}, $i \in \mathbb{N}$ such that

$$\left\| g_i(t, x, \mu) - g_i(t, y, \nu) \right\| \leqslant L_{g_i}(\left\| x - y \right\| + |\mu - \nu|),$$

for each $t \in [t_i, s_i]$, $i \in \mathbb{N}$, for all $x, y \in \mathbb{R}^n$ and $\mu, \nu \in \mathbb{R}$.

$[H_4]$ There exists a positive constant L_h, such that

$$\|h(t, \xi) - h(t, \varsigma)\| \leqslant L_h|\xi - \varsigma|,$$

for each $t \in J$, for all $\xi, \varsigma \in \mathbb{R}$.

2.3.3 Differentiability of solutions

We need the following existence and uniqueness result.

Theorem 2.3.4. *Problems (2.96) and (2.97) have a unique solution $x \in PC(J, \mathbb{R}^n)$ if assumptions $[H_1] - [H_4]$ hold.*

Proof. The proof is a slight modification of the argument in [20, theorem 2.2], so we omit it.

To study differentiability of solutions we will also need the following assumptions:

$[H_5]$ The functions $\frac{\partial f}{\partial x}, \frac{\partial f}{\partial \lambda_{s_i}} \in C(J \times \mathbb{D} \times \mathbb{R}, \mathbb{R}^n)$, $i \in \{0\} \cup \mathbb{N}$.

$[H_6]$ There exists a positive constant M such that

$$\left\| \frac{\partial f(t, x, \lambda)}{\partial x} \right\| \leqslant M,$$

for any $(t, x, \lambda) \in J \times \mathbb{D} \times \mathbb{R}$.

$[H_7]$ The function $\frac{\partial g_i}{\partial \lambda_{t_i}} \in C([t_i, s_i] \times \mathbb{D} \times \mathbb{R}, \mathbb{R}^n)$, $i \in \mathbb{N}$.

$[H_8]$ The initial function $x_0 \in C^1(\mathbb{R}, \mathbb{R}^n)$ and the function $\frac{\partial h}{\partial \lambda_{s_i}} \in C([t_i, s_i] \times \mathbb{R}, \mathbb{R}^n)$, $i \in \mathbb{N}$.

We first study continuous dependence and differentiable of solution to (2.96).

Theorem 2.3.5. *Assume $[H_1] - [H_4]$ hold. Then, the solution of problem (2.96) depends continuously on the initial parameter λ_0, impulsive parameters λ_{t_i} and junction parameters λ_{s_i}, $i \in \mathbb{N}$, respectively.*

Proof. For $t \in [0, t_1]$, and $\lambda_0^* \in \mathbb{R}$, the solution of problem (2.96) with λ_0^* (respectively λ_0) coincides with the solution of the problem without impulses which we denote by $x(t; \lambda_0^*)$ (respectively $x(t; \lambda_0)$) is as follows:

$$x(t; \lambda_0^*) = x_0(\lambda_0^*) + \int_0^t f(s, x(s; \lambda_0^*), \lambda_0^*)ds,$$

and

$$x(t; \lambda_0) = x_0(\lambda_0) + \int_0^t f(s, x(s; \lambda_0), \lambda_0)ds.$$

Clearly,

$$\begin{aligned}
&\|x(t; \lambda_0^*) - x(t; \lambda_0)\| \\
\leqslant\ & \left\|x_0(\lambda_0^*) - x_0(\lambda_0)\right\| + \int_0^t L_f(\left\|x(s; \lambda_0^*) - x(s; \lambda_0)\right\| + |\lambda_0^* - \lambda_0|)ds \\
\leqslant\ & \left\|x_0(\lambda_0^*) - x_0(\lambda_0)\right\| + L_f|\lambda_0^* - \lambda_0|t_1 + \int_0^t L_f\left\|x(s; \lambda_0^*) - x(s; \lambda_0)\right\|ds \\
\leqslant\ & (\left\|x_0(\lambda_0^*) - x_0(\lambda_0)\right\| + L_f|\lambda_0^* - \lambda_0|t_1)e^{L_f t_1},
\end{aligned} \tag{2.98}$$

where we use [17, theorem 1.1]. Thus

$$\lim_{\lambda_0^* \to \lambda_0}\left\|x(t; \lambda_0^*) - x(t; \lambda_0)\right\| = 0, \ \ t \in [0, t_1]. \tag{2.99}$$

For $t \in (t_1, s_1]$, and $\lambda_0^*, \lambda_{t_1}^* \in \mathbb{R}$, the solution of problem (2.96) which we denote respectively by $x(t; \lambda_0^*, \lambda_{t_1}^*)$ and $x(t; \lambda_0, \lambda_{t_1})$ is

$$x(t; \lambda_0^*, \lambda_{t_1}^*) = g_1(t, x(t_1^-; \lambda_0^*), \lambda_{t_1}^*) + h(t; \lambda_{s_1}),$$

and

$$x(t; \lambda_0, \lambda_{t_1}) = g_1(t, x(t_1^-; \lambda_0), \lambda_{t_1}) + h(t; \lambda_{s_1}).$$

Then

$$\left\|x(t; \lambda_0^*, \lambda_{t_1}^*) - x(t; \lambda_0, \lambda_{t_1})\right\| \leqslant L_{g_1}(\left\|x(t_1^-; \lambda_0^*) - x(t_1^-; \lambda_0)\right\| + |\lambda_{t_1}^* - \lambda_{t_1}|). \tag{2.100}$$

From (2.99) and (2.100), we have

$$\lim_{\substack{\lambda_0^* \to \lambda_0 \\ \lambda_{t_1}^* \to \lambda_{t_1}}}\left\|x(t; \lambda_0^*, \lambda_{t_1}^*) - x(t; \lambda_0, \lambda_{t_1})\right\| = 0, \ \ t \in (t_1, s_1]. \tag{2.101}$$

For $t \in (s_1, t_2]$, $\lambda_0^*, \lambda_{t_1}^*, \lambda_{s_1}^* \in \mathbb{R}$, the solution of problem (2.96) which we denote respectively by $x(t; \lambda_0^*, \lambda_{t_1}^*, \lambda_{s_1}^*)$ and $x(t; \lambda_0, \lambda_{t_1}, \lambda_{s_1}^*)$ is

$$\begin{aligned}
x(t; \lambda_0^*, \lambda_{t_1}^*, \lambda_{s_1}^*) \ = \ & g_1(s_1, x(t_1^-; \lambda_0^*), \lambda_{t_1}^*) \\
& + h(s_1; \lambda_{s_1}^*) + \int_{s_1}^t f(s, x(s; \lambda_0^*, \lambda_{t_1}^*, \lambda_{s_1}^*), \lambda_{s_1}^*)ds,
\end{aligned}$$

and

$$x(t;\lambda_0,\lambda_{t_1},\lambda_{s_1}) = g_1(s_1, x(t_1^-;\lambda_0),\lambda_{t_1}) + h(s_1;\lambda_{s_1}) + \int_{s_1}^{t} f(s, x(s;\lambda_0,\lambda_{t_1},\lambda_{s_1}),\lambda_{s_1})ds.$$

Applying [17, theorem 1.1], we obtain that

$$\begin{aligned}
&\left\|x(t;\lambda_0^*,\lambda_{t_1}^*,\lambda_{s_1}^*) - x(t;\lambda_0,\lambda_{t_1},\lambda_{s_1})\right\| \\
\leqslant\ & L_{g_1}(\left\|x(t_1^-;\lambda_0^*) - x(t_1^-;\lambda_0)\right\| + |\lambda_{t_1}^* - \lambda_{t_1}|) + L_h|\lambda_{s_1}^* - \lambda_{s_1}| \\
&+ \int_{s_1}^{t} L_f(\left\|x(s;\lambda_0^*,\lambda_{t_1}^*,\lambda_{s_1}^*) - x(s;\lambda_0,\lambda_{t_1},\lambda_{s_1})\right\| + |\lambda_{s_1}^* - \lambda_{s_1}|) \\
=\ & L_{g_1}(\left\|x(t_1^-;\lambda_0^*) - x(t_1^-;\lambda_0)\right\| + |\lambda_{t_1}^* - \lambda_{t_1}|) + L_h|\lambda_{s_1}^* - \lambda_{s_1}| \\
&+ \int_{s_1}^{t} L_f|\lambda_{s_1}^* - \lambda_{s_1}|ds + \int_{s_1}^{t} L_f\left\|x(s;\lambda_0^*,\lambda_{t_1}^*,\lambda_{s_1}^*) - x(s;\lambda_0,\lambda_{t_1},\lambda_{s_1})\right\|ds \\
\leqslant\ & \Big(L_{g_1}(\left\|x(t_1^-;\lambda_0^*) - x(t_1^-;\lambda_0)\right\| + |\lambda_{t_1}^* - \lambda_{t_1}|) \\
&+ (L_f(t_2 - s_1) + L_h)|\lambda_{s_1}^* - \lambda_{s_1}|\Big)e^{L_f(t_2-s_1)},
\end{aligned}$$

from (2.101), we have

$$\lim_{\substack{\lambda_0^*\to\lambda_0\\ \lambda_{t_1}^*\to\lambda_{t_1}\\ \lambda_{s_1}^*\to\lambda_{s_1}}} \left\|x(t;\lambda_0^*,\lambda_{t_1}^*,\lambda_{s_1}^*) - x(t;\lambda_0,\lambda_{t_1},\lambda_{s_1})\right\| = 0,\ \ t\in(s_1,t_2]. \tag{2.102}$$

For $t \in (t_2, s_2]$, we get

$$\begin{aligned}
&\left\|x(t;\lambda_0,\lambda_{t_1},\lambda_{s_1}^*,\lambda_{t_2}^*) - x(t;\lambda_0,\lambda_{t_1},\lambda_{s_1},\lambda_{t_2})\right\| \\
\leqslant\ & L_{g_2}(\left\|x(t_2^-;\lambda_0,\lambda_{t_1},\lambda_{s_1}^*) - x(t_2^-;\lambda_0,\lambda_{t_1},\lambda_{s_1})\right\| + |\lambda_{t_2}^* - \lambda_{t_2}|),
\end{aligned}$$

and use (2.102) to have

$$\lim_{\substack{\lambda_{s_1}^*\to\lambda_{s_1}\\ \lambda_{t_2}^*\to\lambda_{t_2}}} \left\|x(t;\lambda_0,\lambda_{t_1},\lambda_{s_1}^*,\lambda_{t_2}^*) - x(t;\lambda_0,\lambda_{t_1},\lambda_{s_1},\lambda_{t_2})\right\| = 0,\ \ t\in(t_2,s_2]. \tag{2.103}$$

For $t \in (s_2, t_3]$, and $\lambda_0^*, \lambda_{t_i}^*, \lambda_{s_i}^* \in \mathbb{R}$, $i = 1, 2$, we have

$$x(t;\lambda_0,\lambda_{t_1},\lambda_{s_1}^*,\lambda_{t_2}^*,\lambda_{s_2}^*) = g_2(s_2, x(t_2^-;\lambda_0,\lambda_{t_1},\lambda_{s_1}^*),\lambda_{t_2}^*) + h(s_2;\lambda_{s_2}^*) + \int_{s_2}^{t} f(s, x(s;\lambda_0,\lambda_{t_1},\lambda_{s_1}^*,\lambda_{t_2}^*,\lambda_{s_2}^*),\lambda_{s_2}^*)ds,$$

and

$$x(t;\lambda_0,\lambda_{t_1},\lambda_{s_1},\lambda_{t_2},\lambda_{s_2}) = g_2(s_2,x(t_2^-;\lambda_0,\lambda_{t_1},\lambda_{s_1}),\lambda_{t_2}) + h(s_2;\lambda_{s_2}) + \int_{s_2}^{t} f(s,x(s;\lambda_0,\lambda_{t_1},\lambda_{s_1},\lambda_{t_2},\lambda_{s_2}),\lambda_{s_2})ds.$$

Then

$$\begin{aligned}
&\left\|x(t;\lambda_0,\lambda_{t_1},\lambda_{s_1}^*,\lambda_{t_2}^*,\lambda_{s_2}^*) - x(t;\lambda_0,\lambda_{t_1},\lambda_{s_1},\lambda_{t_2},\lambda_{s_2})\right\| \\
\leqslant\ & L_{g_2}(\left\|x(t_2^-;\lambda_0,\lambda_{t_1},\lambda_{s_1}^*) - x(t_2^-;\lambda_0,\lambda_{t_1},\lambda_{s_1})\right\| + |\lambda_{t_2}^* - \lambda_{t_2}|) + L_h|\lambda_{s_2}^* - \lambda_{s_2}| \\
&+ \int_{s_2}^{t} L_f(\left\|x(s;\lambda_0,\lambda_{t_1},\lambda_{s_1}^*,\lambda_{t_2}^*,\lambda_{s_2}^*) - x(s;\lambda_0,\lambda_{t_1},\lambda_{s_1},\lambda_{t_2},\lambda_{s_2})\right\| + |\lambda_{s_2}^* - \lambda_{s_2}|)ds \\
=\ & L_{g_2}(\left\|x(t_2^-;\lambda_0,\lambda_{t_1},\lambda_{s_1}^*) - x(t_2^-;\lambda_0,\lambda_{t_1},\lambda_{s_1})\right\| + |\lambda_{t_2}^* - \lambda_{t_2}|) + L_h|\lambda_{s_2}^* - \lambda_{s_2}| \\
&+ \int_{s_2}^{t} L_f|\lambda_{s_2}^* - \lambda_{s_2}|ds + \int_{s_2}^{t} L_f\left\|x(s;\lambda_0,\lambda_{t_1},\lambda_{s_1}^*,\lambda_{t_2}^*,\lambda_{s_2}^*)\right. \\
&\left. - x(s;\lambda_0,\lambda_{t_1},\lambda_{s_1},\lambda_{t_2},\lambda_{s_2})\right\|ds,
\end{aligned}$$

and using [17, theorem 1.1], we have

$$\begin{aligned}
&\left\|x(t;\lambda_0,\lambda_{t_1},\lambda_{s_1}^*,\lambda_{t_2}^*,\lambda_{s_2}^*) - x(t;\lambda_0,\lambda_{t_1},\lambda_{s_1},\lambda_{t_2},\lambda_{s_2})\right\| \\
\leqslant\ & \Big(L_{g_2}(\left\|x(t_2^-;\lambda_0,\lambda_{t_1},\lambda_{s_1}^*) - x(t_2^-;\lambda_0,\lambda_{t_1},\lambda_{s_1})\right\| + |\lambda_{t_2}^* - \lambda_{t_2}|) \\
&+(L_f(t_3 - s_2) + L_h)|\lambda_{s_2}^* - \lambda_{s_2}|\Big)e^{L_f(t_3-s_2)}.
\end{aligned}$$

Consequently using (2.102) we have

$$\lim_{\substack{\lambda_{s_1}^*\to\lambda_{s_1}\\ \lambda_{t_2}^*\to\lambda_{t_2}\\ \lambda_{s_2}^*\to\lambda_{s_2}}} \left\|x(t;\lambda_0,\lambda_{t_1},\lambda_{s_1}^*,\lambda_{t_2}^*,\lambda_{s_2}^*) - x(t;\lambda_0,\lambda_{t_1},\lambda_{s_1},\lambda_{t_2},\lambda_{s_2})\right\| = 0. \tag{2.104}$$

Repeating the above procedure we have

$$\lim_{\substack{\lambda_{s_{i-1}}^*\to\lambda_{s_{i-1}}\\ \lambda_{t_i}^*\to\lambda_{t_i}}} \left\|x(t;\lambda_0,\lambda_{t_1},\ldots,\lambda_{s_{i-1}}^*,\lambda_{t_i}^*) - x(t;\lambda_0,\lambda_{t_1},\ldots,\lambda_{s_{i-1}},\lambda_{t_i})\right\| = 0,$$

$$t \in (t_i, s_i],\ i \in \mathbb{N},$$

and

$$\lim_{\substack{\lambda_{s_{i-1}}^*\to\lambda_{s_{i-1}}\\ \lambda_{t_i}^*\to\lambda_{t_i}\\ \lambda_{s_i}^*\to\lambda_{s_i}}} \left\|x(t;\lambda_0,\lambda_{t_1},\ldots,\lambda_{s_{i-1}}^*,\lambda_{t_i}^*,\lambda_{s_i}^*) - x(t;\lambda_0,\lambda_{t_1},\ldots,\lambda_{s_{i-1}},\lambda_{t_i},\lambda_{s_i})\right\| = 0,$$

$$t \in (s_i, t_{i+1}],\ i \in \mathbb{N}.$$

The proof is completed. □

Now we study continuous dependence and differentiable of solutions to (2.96).

Theorem 2.3.6. *Assume* $[H_1] - [H_8]$ *hold. Then, the solution of problem (2.96) is continuously differentiable with respect to the parameters* λ_0, λ_{t_i} *and* λ_{s_i}, $i \in \mathbb{N}$, *respectively.*

Proof. For $t \in [0, t_1]$ we have

$$\begin{aligned}
& x(t; \lambda_0^*) - x(t; \lambda_0) \\
= \; & x_0(\lambda_0^*) - x_0(\lambda_0) + \int_0^t \big(f(s, x(s; \lambda_0^*), \lambda_0^*) - f(s, x(s; \lambda_0), \lambda_0)\big)ds \\
= \; & \frac{\partial}{\partial \lambda_0} x_0(\lambda_0 + \theta_0(\lambda_0^* - \lambda_0))(\lambda_0^* - \lambda_0) \\
& + \int_0^t \Big(\frac{\partial}{\partial x} f(s, x(s; \lambda_0) + \theta_1(x(s; \lambda_0^*) - x(s; \lambda_0)), \lambda_0^*)(x(s; \lambda_0^*) - x(s; \lambda_0)) \\
& + \frac{\partial}{\partial \lambda_0} f(s, x(s; \lambda_0), \lambda_0 + \theta_2(\lambda_0^* - \lambda_0))(\lambda_0^* - \lambda_0)\Big)ds,
\end{aligned}$$

where $\theta_0, \theta_1, \theta_2 \in (0, 1)$.

Note the continuity of $\frac{\partial x_0}{\partial \lambda_0}, \frac{\partial f}{\partial x}, \frac{\partial f}{\partial \lambda_0}$ and

$$\begin{aligned}
& \frac{x(t; \lambda_0^*) - x(t; \lambda_0)}{\lambda_0^* - \lambda_0} \\
= \; & (\frac{\partial}{\partial \lambda_0} x_0(\lambda_0) + r_0) + \int_0^t \Big(\Big(\frac{\partial}{\partial x} f(s, x(s; \lambda_0), \lambda_0) + r_1\Big)\frac{x(s; \lambda_0^*) - x(s; \lambda_0)}{\lambda_0^* - \lambda_0} \\
& + (\frac{\partial}{\partial \lambda_0} f(s, x(s; \lambda_0), \lambda_0) + r_2)\Big)ds,
\end{aligned} \tag{2.105}$$

where r_k, $k = 0, 1, 2$ have the property: $r_k \to 0$, if $\lambda_0^* \to \lambda_0$.

Consider the initial value problem

$$\begin{cases}
\dfrac{dy_0(t)}{dt} = \dfrac{\partial}{\partial x} f(t, x(t; \lambda_0), \lambda_0) \cdot y_0(t) + \dfrac{\partial}{\partial \lambda_0} f(t, x(t; \lambda_0), \lambda_0), \\
y_0(0) = \dfrac{\partial}{\partial \lambda_0} x_0(\lambda_0).
\end{cases} \tag{2.106}$$

Then

$$
\begin{aligned}
y_0(t) &= \frac{\partial}{\partial \lambda_0} x_0(\lambda_0) \\
&+ \int_0^t \left(\frac{\partial}{\partial x} f(s, x(s; \lambda_0), \lambda_0) y_0(s) + \frac{\partial}{\partial \lambda_0} f(s, x(s; \lambda_0), \lambda_0) \right) ds.
\end{aligned} \tag{2.107}
$$

From (2.105) and (2.107), we have

$$
\begin{aligned}
&\frac{x(t; \lambda_0^*) - x(t; \lambda_0)}{\lambda_0^* - \lambda_0} - y_0(t) \\
&= \left(\frac{\partial}{\partial \lambda_0} x_0(\lambda_0) + r_0\right) - \frac{\partial}{\partial \lambda_0} x_0(\lambda_0) + \int_0^t \left(\left(\frac{\partial}{\partial x} f(s, x(s; \lambda_0), \lambda_0) + r_1\right) \right. \\
&\quad \left. \times \frac{x(s; \lambda_0^*) - x(s; \lambda_0)}{\lambda_0^* - \lambda_0} + \left(\frac{\partial}{\partial \lambda_0} f(s, x(s; \lambda_0), \lambda_0) + r_2\right) \right) ds \\
&\quad - \int_0^t \left(\frac{\partial}{\partial x} f(s, x(s; \lambda_0), \lambda_0) y_0(s) + \frac{\partial}{\partial \lambda_0} f(s, x(s; \lambda_0), \lambda_0) \right) ds \\
&= r_0 + \int_0^t (r_2 + r_1 y_0(s)) ds \\
&\quad + \int_0^t \left(\frac{\partial}{\partial x} f(s, x(s; \lambda_0), \lambda_0) + r_1 \right) \left(\frac{x(s; \lambda_0^*) - x(s; \lambda_0)}{\lambda_0^* - \lambda_0} - y_0(s) \right) ds.
\end{aligned}
$$

Therefore,

$$
\begin{aligned}
&\left\| \frac{x(t; \lambda_0^*) - x(t; \lambda_0)}{\lambda_0^* - \lambda_0} - y_0(t) \right\| \\
&\leqslant \|r_0\| + \int_0^t \|r_2 + r_1 y_0(s)\| ds \\
&\quad + \int_0^t \left\| \frac{\partial}{\partial x} f(s, x(s; \lambda_0), \lambda_0) + r_1 \right\| \left\| \frac{x(s; \lambda_0^*) - x(s; \lambda_0)}{\lambda_0^* - \lambda_0} - y_0(s) \right\| ds \\
&\leqslant \left(\|r_0\| + \int_0^t (\|r_2\| + \|r_1\| Y_0) ds \right) e^{(M + \|r_1\|) t_1},
\end{aligned}
$$

where Y_0 is a positive constant such that $\|y_0(t)\| \leqslant Y_0$ for $t \in [0, t_1]$.

Since $\lambda_0^* \to \lambda_0$, then $r_0 \to 0$, $r_1 \to 0$, $r_2 \to 0$, we obtain

$$\lim_{\lambda_0^* \to \lambda_0} \left\| \frac{x(t; \lambda_0^*) - x(t; \lambda_0)}{\lambda_0^* - \lambda_0} - y_0(t) \right\| = 0,$$

that is

$$\lim_{\lambda_0^* \to \lambda_0} \frac{x(t; \lambda_0^*) - x(t; \lambda_0)}{\lambda_0^* - \lambda_0} = y_0(t).$$

For $t \in [0, t_1]$, there exists a continuous derivative

$$\frac{\partial}{\partial \lambda_0} x(t; \lambda_0) = y_0(t),$$

which satisfies the problem

$$\begin{cases} \dfrac{d}{dt}(\dfrac{\partial}{\partial \lambda_0} x(t; \lambda_0)) &= \left(\dfrac{\partial}{\partial x} f(t, x(t; \lambda_0), \lambda_0)\right)\left(\dfrac{\partial}{\partial \lambda_0} x(t; \lambda_0)\right) + \dfrac{\partial}{\partial \lambda_0} f(t, x(t; \lambda_0), \lambda_0), \\ \dfrac{\partial}{\partial \lambda_0} x(0; \lambda_0) &= \dfrac{\partial}{\partial \lambda_0} x_0(\lambda_0). \end{cases}$$

For $t \in (t_1, s_1]$, we get

$$\begin{aligned} & x(t; \lambda_0, \lambda_{t_1}^*) - x(t; \lambda_0, \lambda_{t_1}) \\ = \; & g_1(t, x(t_1^-; \lambda_0), \lambda_{t_1}^*) - g_1(t, x(t_1^-; \lambda_0), \lambda_{t_1}) \\ = \; & \frac{\partial}{\partial \lambda_{t_1}} g_1(t, x(t_1^-; \lambda_0), \lambda_{t_1} + \theta(\lambda_{t_1}^* - \lambda_{t_1}))(\lambda_{t_1}^* - \lambda_{t_1}), \end{aligned} \tag{2.108}$$

where $\theta \in (0, 1)$. Note

$$\frac{x(t; \lambda_0, \lambda_{t_1}^*) - x(t; \lambda_0, \lambda_{t_1})}{\lambda_{t_1}^* - \lambda_{t_1}} = \frac{\partial}{\partial \lambda_{t_1}} g_1(t, x(t_1^-; \lambda_0), \lambda_{t_1}) + r_3,$$

$r_3 \to 0$, if $\lambda_{t_1}^* \to \lambda_{t_1}$, and

$$\lim_{\lambda_{t_1}^* \to \lambda_{t_1}} \frac{x(t; \lambda_0, \lambda_{t_1}^*) - x(t; \lambda_0, \lambda_{t_1})}{\lambda_{t_1}^* - \lambda_{t_1}} = \frac{\partial}{\partial \lambda_{t_1}} g_1(t, x(t_1^-; \lambda_0), \lambda_{t_1}),$$

that is

$$\frac{\partial}{\partial \lambda_{t_1}} x(t; \lambda_0, \lambda_{t_1}) = \frac{\partial}{\partial \lambda_{t_1}} g_1(t, x(t_1^-; \lambda_0), \lambda_{t_1}).$$

For $t \in (s_1, t_2]$, note

$$
\begin{aligned}
& x(t;\lambda_0,\lambda_{t_1},\lambda_{s_1}^*) - x(t;\lambda_0,\lambda_{t_1},\lambda_{s_1}) \\
= \ & h(s_1;\lambda_{s_1}^*) - h(s_1;\lambda_{s_1}) \\
& + \int_{s_1}^{t} \Big(f(s, x(s;\lambda_0,\lambda_{t_1},\lambda_{s_1}^*),\lambda_{s_1}^*) - f(s, x(s;\lambda_0,\lambda_{t_1},\lambda_{s_1}),\lambda_{s_1})\Big)ds \\
= \ & \left(\frac{\partial}{\partial\lambda_{s_1}}h(s_1;\lambda_{s_1}) + r_4\right)(\lambda_{s_1}^* - \lambda_{s_1}) + \int_{s_1}^{t}\left(\left(\frac{\partial}{\partial x}f(s, x(s;\lambda_0,\lambda_{t_1},\lambda_{s_1}),\lambda_{s_1}) + r_5\right)\right. \\
& \times (x(s;\lambda_0,\lambda_{t_1},\lambda_{s_1}^*) - x(s;\lambda_0,\lambda_{t_1},\lambda_{s_1})) \\
& \left. + \left(\frac{\partial}{\partial\lambda_{s_1}}f(s, x(s;\lambda_0,\lambda_{t_1},\lambda_{s_1}),\lambda_{s_1}) + r_6\right)(\lambda_{s_1}^* - \lambda_{s_1})\right)ds,
\end{aligned}
\tag{2.109}
$$

where r_k, $k = 4, 5, 6$ have the property: $r_k \to 0$, if $\lambda_{s_1}^* \to \lambda_{s_1}$.

Therefore,

$$
\begin{aligned}
& \frac{x(t;\lambda_0,\lambda_{t_1},\lambda_{s_1}^*) - x(t;\lambda_0,\lambda_{t_1},\lambda_{s_1})}{\lambda_{s_1}^* - \lambda_{s_1}} \\
= \ & \left(\frac{\partial}{\partial\lambda_{s_1}}h(s_1;\lambda_{s_1}) + r_4\right) + \int_{s_1}^{t}\left(\left(\frac{\partial}{\partial x}f(s, x(s;\lambda_0,\lambda_{t_1},\lambda_{s_1}),\lambda_{s_1}) + r_5\right)\right. \\
& \times \frac{x(s;\lambda_0,\lambda_{t_1},\lambda_{s_1}^*) - x(s;\lambda_0,\lambda_{t_1},\lambda_{s_1})}{\lambda_{s_1}^* - \lambda_{s_1}} \\
& \left. + \left(\frac{\partial}{\partial\lambda_{s_1}}f(s, x(s;\lambda_0,\lambda_{t_1},\lambda_{s_1}),\lambda_{s_1}) + r_6\right)\right)ds.
\end{aligned}
\tag{2.110}
$$

Consider the initial value problem

$$
\begin{cases}
\dfrac{dy_{s_1}(t)}{dt} = \dfrac{\partial}{\partial x}f(t, x(t;\lambda_0,\lambda_{t_1},\lambda_{s_1}),\lambda_{s_1})\cdot y_{s_1}(t) + \dfrac{\partial}{\partial\lambda_{s_1}}f(t, x(t;\lambda_0,\lambda_{t_1},\lambda_{s_1}),\lambda_{s_1}), \\
y_{s_1}(s_1) = \dfrac{\partial}{\partial\lambda_{s_1}}h(s_1;\lambda_{s_1}),
\end{cases}
$$

note

$$
\begin{aligned}
y_{s_1}(t) = \ & \frac{\partial}{\partial\lambda_{s_1}}h(s_1;\lambda_{s_1}) + \int_{s_1}^{t}\left(\frac{\partial}{\partial x}f(s, x(s;\lambda_0,\lambda_{t_1},\lambda_{s_1}),\lambda_{s_1})\cdot y_{s_1}(s)\right. \\
& \left. + \frac{\partial}{\partial\lambda_{s_1}}f(s, x(s;\lambda_0,\lambda_{t_1},\lambda_{s_1}),\lambda_{s_1})\right)ds.
\end{aligned}
\tag{2.111}
$$

From (2.110) and (2.111), we have

$$
\begin{aligned}
&\left\| \frac{x(t; \lambda_0, \lambda_{t_1}, \lambda_{s_1}^*) - x(t; \lambda_0, \lambda_{t_1}, \lambda_{s_1})}{\lambda_{s_1}^* - \lambda_{s_1}} - y_{s_1}(t) \right\| \\
&\leqslant \|r_4\| + \int_{s_1}^{t} \|r_6 + r_5 y_{s_1}(s)\| ds + \int_{s_1}^{t} \left\| \frac{\partial}{\partial x} f(s, x(s; \lambda_0, \lambda_{t_1}, \lambda_{s_1}), \lambda_{s_1}) + r_5 \right\| \\
&\times \left\| \frac{x(s; \lambda_0, \lambda_{t_1}, \lambda_{s_1}^*) - x(s; \lambda_0, \lambda_{t_1}, \lambda_{s_1})}{\lambda_{s_1}^* - \lambda_{s_1}} - y_{s_1}(s) \right\| ds \\
&\leqslant \left(\|r_4\| + \int_{s_1}^{t} (\|r_6\| + \|r_5\| Y_{s_1}) ds \right) e^{(M + \|r_5\|)(t_2 - s_1)},
\end{aligned}
$$

where Y_{s_1} is a positive constant such that $\|y_{s_1}(t)\| \leqslant Y_{s_1}$ for $t \in (s_1, t_2]$.

Since $\lambda_{s_1}^* \to \lambda_{s_1}$, then $r_4 \to 0$, $r_5 \to 0$, $r_6 \to 0$, we obtain

$$
\lim_{\lambda_{s_1}^* \to \lambda_{s_1}} \left\| \frac{x(t; \lambda_0, \lambda_{t_1}, \lambda_{s_1}^*) - x(t; \lambda_0, \lambda_{t_1}, \lambda_{s_1})}{\lambda_{s_1}^* - \lambda_{s_1}} - y_{s_1}(t) \right\| = 0,
$$

that is

$$
\lim_{\lambda_{s_1}^* \to \lambda_{s_1}} \frac{x(t; \lambda_0, \lambda_{t_1}, \lambda_{s_1}^*) - x(t; \lambda_0, \lambda_{t_1}, \lambda_{s_1})}{\lambda_{s_1}^* - \lambda_{s_1}} = y_{s_1}(t).
$$

For $t \in (s_1, t_2]$, there exists a continuous derivative

$$
\frac{\partial}{\partial \lambda_{s_1}} x(t; \lambda_0, \lambda_{t_1}, \lambda_{s_1}) = y_{s_1}(t),
$$

which satisfies

$$
\begin{cases}
\dfrac{d}{dt}\left(\dfrac{\partial}{\partial \lambda_{s_1}} x(t; \lambda_0, \lambda_{t_1}, \lambda_{s_1}) \right) = \left(\dfrac{\partial}{\partial x} f(t, x(t; \lambda_0, \lambda_{t_1}, \lambda_{s_1}), \lambda_{s_1}) \right) \left(\dfrac{\partial}{\partial \lambda_{s_1}} x(t; \lambda_0, \lambda_{t_1}, \lambda_{s_1}) \right) \\
\qquad + \dfrac{\partial}{\partial \lambda_{s_1}} f(t, x(t; \lambda_0, \lambda_{t_1}, \lambda_{s_1}), \lambda_{s_1}), \\
\dfrac{\partial}{\partial \lambda_{s_1}} x(s_1; \lambda_0, \lambda_{t_1}, \lambda_{s_1}) = \dfrac{\partial}{\partial \lambda_{s_1}} h(s_1; \lambda_{s_1}).
\end{cases}
$$

Similarly (analogue of the argument in (2.108)) for $t \in (t_i, s_i]$, $i \in \mathbb{N}$, we have

$$
\begin{aligned}
& x(t; \lambda_0, \lambda_{t_1}, \ldots, \lambda_{s_{i-1}}, \lambda_{t_i}^*) - x(t; \lambda_0, \lambda_{t_1}, \ldots, \lambda_{s_{i-1}}, \lambda_{t_i}) \\
= \; & g_i(t, x(t_i^-; \lambda_0, \lambda_{t_1}, \ldots, \lambda_{s_{i-1}}), \lambda_{t_i}^*) - g_i(t, x(t_i^-; \lambda_0, \lambda_{t_1}, \ldots, \lambda_{s_{i-1}}), \lambda_{t_i}) \\
= \; & \frac{\partial}{\partial \lambda_{t_i}} g_i(t, x(t_i^-; \lambda_0, \lambda_{t_1}, \ldots, \lambda_{s_{i-1}}), \lambda_{t_1} + \theta(\lambda_{t_i}^* - \lambda_{t_i}))(\lambda_{t_i}^* - \lambda_{t_i}),
\end{aligned}
$$

where $\theta \in (0, 1)$. Note

$$\begin{aligned}
&\frac{x(t; \lambda_0, \lambda_{t_1}, \ldots, \lambda_{s_{i-1}}, \lambda_{t_i}^*) - x(t; \lambda_0, \lambda_{t_1}, \ldots, \lambda_{s_{i-1}}, \lambda_{t_i})}{\lambda_{t_i}^* - \lambda_{t_i}} \\
&= \frac{\partial}{\partial \lambda_{t_i}} g_i(t, x(t_1^-; \lambda_0, \lambda_{t_1}, \ldots, \lambda_{s_{i-1}}), \lambda_{t_i}) + r_{k-1},
\end{aligned}$$

$r_{k-1} \to 0$, if $\lambda_{t_i}^* \to \lambda_{t_i}$, and

$$\begin{aligned}
&\lim_{\lambda_{t_i}^* \to \lambda_{t_i}} \frac{x(t; \lambda_0, \lambda_{t_1}, \ldots, \lambda_{s_{i-1}}, \lambda_{t_i}^*) - x(t; \lambda_0, \lambda_{t_1}, \ldots, \lambda_{s_{i-1}}, \lambda_{t_i})}{\lambda_{t_i}^* - \lambda_{t_i}} \\
&= \frac{\partial}{\partial \lambda_{t_i}} g_i(t, x(t_1^-; \lambda_0, \lambda_{t_1}, \ldots, \lambda_{s_{i-1}}), \lambda_{t_i}),
\end{aligned}$$

that is

$$\frac{\partial}{\partial \lambda_{t_i}} x(t; \lambda_0, \lambda_{t_1}, \ldots, \lambda_{s_{i-1}}), \lambda_{t_i}) = \frac{\partial}{\partial \lambda_{t_i}} g_i(t, x(t_1^-; \lambda_0, \lambda_{t_1}, \ldots, \lambda_{s_{i-1}}), \lambda_{t_i}). \tag{2.112}$$

Now (analogue of the argument in (2.109)), for $t \in (s_i, t_{i+1}]$, we have

$$\begin{aligned}
&x(t; \lambda_0, \lambda_{t_1}, \ldots, \lambda_{t_i}, \lambda_{s_i}^*) - x(t; \lambda_0, \lambda_{t_1}, \ldots, \lambda_{t_i}, \lambda_{s_i}) \\
&= h(s_i; \lambda_{s_i}^*) - h(s_i; \lambda_{s_i}) + \int_{s_i}^{t} \Big(f(s, x(s; \lambda_0, \lambda_{t_1}, \ldots, \lambda_{t_i}, \lambda_{s_i}^*), \lambda_{s_i}^*) \\
&\quad - f(s, x(s; \lambda_0, \lambda_{t_1}, ,\ldots, \lambda_{t_i}, \lambda_{s_i}), \lambda_{s_i})\Big) ds \\
&= (\frac{\partial}{\partial \lambda_{s_i}} h(s_i; \lambda_{s_i}) + r_k)(\lambda_{s_i}^* - \lambda_{s_i}) + \int_{s_i}^{t} \left(\left(\frac{\partial}{\partial x} f(s, x(s; \lambda_0, \lambda_{t_1}, \ldots, \lambda_{t_i}, \lambda_{s_i}), \lambda_{s_i}) + r_{k+1} \right) \right. \\
&\quad \times (x(s; \lambda_0, \lambda_{t_1}, \ldots, \lambda_{t_i}, \lambda_{s_i}^*) - x(s; \lambda_0, \lambda_{t_1}, \ldots, \lambda_{t_i}, \lambda_{s_i})) \\
&\quad \left. + \left(\frac{\partial}{\partial \lambda_{s_i}} f(s, x(s; \lambda_0, \lambda_{t_1}, \ldots, \lambda_{t_i}, \lambda_{s_i}), \lambda_{s_i}) + r_{k+2} \right) (\lambda_{s_i}^* - \lambda_{s_i}) \right) ds,
\end{aligned}$$

where r_k, r_{k+1}, r_{k+2} have the property: $r_k \to 0$, $r_{k+1} \to 0$, $r_{k+2} \to 0$, if $\lambda_{s_i}^* \to \lambda_{s_i}$.

Therefore,

$$
\begin{aligned}
&\frac{x(t;\lambda_0,\lambda_{t_1},\ldots,\lambda_{t_i},\lambda^*_{s_i})-x(t;\lambda_0,\lambda_{t_1},\ldots,\lambda_{t_i},\lambda_{s_i})}{\lambda^*_{s_i}-\lambda_{s_i}}\\
&=\left(\frac{\partial}{\partial\lambda_{s_i}}h(s_i;\lambda_{s_i})+r_k\right)+\int_{s_i}^{t}\left(\left(\frac{\partial}{\partial x}f(s,x(s;\lambda_0,\lambda_{t_1},\ldots,\lambda_{t_i},\lambda_{s_i}),\lambda_{s_i})+r_{k+1}\right)\right.\\
&\quad\times\frac{x(s;\lambda_0,\lambda_{t_1},\ldots,\lambda_{t_i},\lambda^*_{s_i})-x(s;\lambda_0,\lambda_{t_1},\ldots,\lambda_{t_i},\lambda_{s_i})}{\lambda^*_{s_i}-\lambda_{s_i}}\\
&\quad\left.+\left(\frac{\partial}{\partial\lambda_{s_i}}f(s,x(s;\lambda_0,\lambda_{t_1},\ldots,\lambda_{t_i},\lambda_{s_i}),\lambda_{s_i})+r_{k+2}\right)\right)ds,
\end{aligned}
\tag{2.113}
$$

Consider the initial value problem

$$
\begin{cases}
\dfrac{dy_{s_i}(t)}{dt} = \dfrac{\partial}{\partial x}f(t,x(t;\lambda_0,\lambda_{t_1},\ldots,\lambda_{t_i},\lambda_{s_i}),\lambda_{s_i})\cdot y_{s_i}(t)\\
\qquad\qquad +\dfrac{\partial}{\partial\lambda_{s_i}}f(t,x(t;\lambda_0,\lambda_{t_1},\ldots,\lambda_{t_i},\lambda_{s_i}),\lambda_{s_i}),\\
y_{s_i}(s_i) = \dfrac{\partial}{\partial\lambda_{s_i}}h(s_i;\lambda_{s_i}),
\end{cases}
$$

and note

$$
\begin{aligned}
y_{s_i}(t) &= \frac{\partial}{\partial\lambda_{s_i}}h(s_i;\lambda_{s_i})+\int_{s_i}^{t}\left(\frac{\partial}{\partial x}f(s,x(s;\lambda_0,\lambda_{t_1},\ldots,\lambda_{t_i},\lambda_{s_i}),\lambda_{s_i})\cdot y_{s_i}(s)\right.\\
&\quad\left.+\frac{\partial}{\partial\lambda_{s_i}}f(s,x(s;\lambda_0,\lambda_{t_1},\ldots,\lambda_{t_i},\lambda_{s_i}),\lambda_{s_i})\right)ds.
\end{aligned}
\tag{2.114}
$$

From (2.113) and (2.114), we have

$$
\begin{aligned}
&\left\|\frac{x(t;\lambda_0,\lambda_{t_1},\ldots,\lambda_{t_i},\lambda^*_{s_i})-x(t;\lambda_0,\lambda_{t_1},\ldots,\lambda_{t_i},\lambda_{s_i})}{\lambda^*_{s_i}-\lambda_{s_i}}-y_{s_i}(t)\right\|\\
&\leqslant \|r_k\|+\int_{s_i}^{t}\left\|r_{k+2}+r_{k+1}y_{s_i}(s)\right\|ds\\
&\quad+\int_{s_i}^{t}\left\|\frac{\partial}{\partial x}f(s,x(s;\lambda_0,\lambda_{t_1},\ldots,\lambda_{t_i},\lambda_{s_i}),\lambda_{s_i})+r_{k+1}\right\|\\
&\quad\times\left\|\frac{x(s;\lambda_0,\lambda_{t_1},\ldots,\lambda_{t_i},\lambda^*_{s_i})-x(s;\lambda_0,\lambda_{t_1},\ldots,\lambda_{t_i},\lambda_{s_i})}{\lambda^*_{s_i}-\lambda_{s_i}}-y_{s_i}(s)\right\|ds\\
&\leqslant\left(\|r_k\|+\int_{s_i}^{t}(\|r_{k+2}\|+\|r_{k+1}\|Y_{s_i})ds\right)e^{(M+\|r_{k+1}\|)(t_{i+1}-s_i)},
\end{aligned}
$$

where Y_{s_i} is a positive constant such that $\left\|y_{s_i}(t)\right\| \leqslant Y_{s_i}$ for $t \in (s_i, t_{i+1}]$.

Since $\lambda_{s_i}^* \to \lambda_{s_i}$, then $r_k \to 0$, $r_{k+1} \to 0$, $r_{k+2} \to 0$, we obtain

$$\lim_{\lambda_{s_i}^* \to \lambda_{s_i}} \left\| \frac{x(t; \lambda_0, \lambda_{t_1}, \ldots, \lambda_{t_i}, \lambda_{s_i}^*) - x(t; \lambda_0, \lambda_{t_1}, \ldots, \lambda_{t_i}, \lambda_{s_i})}{\lambda_{s_i}^* - \lambda_{s_i}} - y_{s_i}(t) \right\| = 0,$$

that is

$$\lim_{\lambda_{s_1}^* \to \lambda_{s_1}} \frac{x(t; \lambda_0, \lambda_{t_1}, \ldots, \lambda_{t_i}, \lambda_{s_i}^*) - x(t; \lambda_0, \lambda_{t_1}, \ldots, \lambda_{t_i}, \lambda_{s_i})}{\lambda_{s_i}^* - \lambda_{s_i}} = y_{s_i}(t).$$

For $t \in (s_i, t_{i+1}]$, there exists a continuous derivative

$$\frac{\partial}{\partial \lambda_{s_i}} x(t; \lambda_0, \lambda_{t_1}, \ldots, \lambda_{t_i}, \lambda_{s_i}) = y_{s_i}(t),$$

which satisfies

$$\begin{cases} \dfrac{d}{dt}\left(\dfrac{\partial}{\partial \lambda_{s_i}} x(t; \lambda_0, \lambda_{t_1}, \ldots, \lambda_{t_i}, \lambda_{s_i})\right) \\ = \left(\dfrac{\partial}{\partial x} f(t, x(t; \lambda_0, \lambda_{t_1}, \ldots, \lambda_{t_i}, \lambda_{s_i}), \lambda_{s_i})\right)\left(\dfrac{\partial}{\partial \lambda_{s_i}} x(t; \lambda_0, \lambda_{t_1}, \ldots, \lambda_{t_i}, \lambda_{s_i})\right) \\ +\dfrac{\partial}{\partial \lambda_{s_i}} f(t, x(t; \lambda_0, \lambda_{t_1}, \ldots, \lambda_{t_i}, \lambda_{s_i}), \lambda_{s_i}), \\ \dfrac{\partial}{\partial \lambda_{s_i}} x(s_i; \lambda_0, \lambda_{t_1}, \ldots, \lambda_{t_i}, \lambda_{s_i}) = \dfrac{\partial}{\partial \lambda_{s_i}} h(s_i; \lambda_{s_i}), \end{cases}$$

The proof is completed. □

We first present continuous dependence and differentiability of solution to (2.97).

Theorem 2.3.7. *Assume* $[H_1] - [H_4]$ *hold. Then, the solution of problem (2.97) depends continuously about the order* α*, the initial parameter* λ_0*, impulsive parameters* λ_{t_i} *and junction parameters* λ_{s_i}*,* $i \in \mathbb{N}$*, respectively.*

Proof. For $t \in [0, t_1]$, $\lambda_0^* \in \mathbb{R}$ and β is the disturbance of order α, $\alpha, \beta \in (0, 1)$, $x(t; \lambda_0^*)$ and $x(t; \lambda_0)$ are the solutions of the following equations, respectively

$$\begin{cases} {}^c\mathbf{D}_{0,t}^{\beta} x(t) = f(t, x(t), \lambda_0^*), \\ x(0) = x_0(\lambda_0^*), \end{cases}$$

and

$$\begin{cases} {}^c\mathbf{D}_{0,t}^{\alpha}x(t) = f(t, x(t), \lambda_0), \\ x(0) = x_0(\lambda_0). \end{cases}$$

Then, we have

$$x(t; \lambda_0^*) = x_0(\lambda_0^*) + \frac{1}{\Gamma(\beta)}\int_0^t (t-s)^{\beta-1}f(s, x(s; \lambda_0^*), \lambda_0^*)ds,$$

and

$$x(t; \lambda_0) = x_0(\lambda_0) + \frac{1}{\Gamma(\alpha)}\int_0^t (t-s)^{\alpha-1}f(s, x(s; \lambda_0), \lambda_0)ds.$$

Therefore,

$$\begin{aligned}
&\|x(t; \lambda_0^*) - x(t; \lambda_0)\| \\
\leqslant\; & \left\|x_0(\lambda_0^*) - x_0(\lambda_0)\right\| + \int_0^t \left(\frac{1}{\Gamma(\beta)}(t-s)^{\beta-1}f(s, x(s; \lambda_0^*), \lambda_0^*)\right. \\
&\left. -\frac{1}{\Gamma(\alpha)}(t-s)^{\alpha-1}f(s, x(s; \lambda_0), \lambda_0)\right)ds \\
\leqslant\; & \left\|x_0(\lambda_0^*) - x_0(\lambda_0)\right\| + \left(\frac{1}{\Gamma(\beta)} - \frac{1}{\Gamma(\alpha)}\right)\int_0^t (t-s)^{\beta-1}\left\|f(s, x(s; \lambda_0^*), \lambda_0^*)\right\|ds \\
&+\frac{1}{\Gamma(\alpha)}\int_0^t |(t-s)^{\beta-1} - (t-s)^{\alpha-1}|\left\|f(s, x(s; \lambda_0^*), \lambda_0^*)\right\|ds \\
&+\frac{1}{\Gamma(\alpha)}\int_0^t (t-s)^{\alpha-1}\left\|f(s, x(s; \lambda_0^*), \lambda_0^*) - f(s, x(s; \lambda_0), \lambda_0)\right\|ds \\
\leqslant\; & \left\|x_0(\lambda_0^*) - x_0(\lambda_0)\right\| + \left(\frac{1}{\Gamma(\beta)} - \frac{1}{\Gamma(\alpha)}\right)M\frac{t_1^\beta}{\beta} + \frac{M}{\Gamma(\alpha)}\left|\frac{t_1^\beta}{\beta} - \frac{t_1^\alpha}{\alpha}\right| \\
&+\frac{L_f}{\Gamma(\alpha)}\int_0^t (t-s)^{\alpha-1}(\|x(s; \lambda_0^*) - x(s; \lambda_0)\| + |\lambda_0^* - \lambda_0|)ds \\
\leqslant\; & \left\|x_0(\lambda_0^*) - x_0(\lambda_0)\right\| + \left(\frac{1}{\Gamma(\beta)} - \frac{1}{\Gamma(\alpha)}\right)M\frac{t_1^\beta}{\beta} + \frac{M}{\Gamma(\alpha)}\left|\frac{t_1^\beta}{\beta} - \frac{t_1^\alpha}{\alpha}\right| \\
&+\frac{L_f}{\Gamma(\alpha+1)}|\lambda_0^* - \lambda_0|t_1^\alpha + \frac{L_f}{\Gamma(\alpha)}\int_0^t (t-s)^{\alpha-1}\left\|x(s; \lambda_0^*) - x(s; \lambda_0)\right\|ds,
\end{aligned}$$

use [18, corollary 2], we obtain

$$\begin{aligned}
\|x(t; \lambda_0^*) - x(t; \lambda_0)\| \leqslant\; & \left(\left\|x_0(\lambda_0^*) - x_0(\lambda_0)\right\| + \left(\frac{1}{\Gamma(\beta)} - \frac{1}{\Gamma(\alpha)}\right)M\frac{t_1^\beta}{\beta}\right. \\
&\left.+\frac{M}{\Gamma(\alpha)}\left|\frac{t_1^\beta}{\beta} - \frac{t_1^\alpha}{\alpha}\right| + \frac{L_f t_1^\alpha}{\Gamma(\alpha+1)}|\lambda_0^* - \lambda_0|\right)E_\alpha(L_f t_1^\alpha),
\end{aligned}$$

therefore,

$$\lim_{\substack{\beta\to\alpha \\ \lambda_0^*\to\lambda_0}} \left\| x(t; \lambda_0^*) - x(t; \lambda_0) \right\| = 0, \ \ t \in [0, t_1]. \tag{2.115}$$

For $t \in (t_1, s_1]$, we have the same form as in (2.100), we have

$$\lim_{\substack{\beta\to\alpha \\ \lambda_0^*\to\lambda_0 \\ \lambda_{t_1}^*\to\lambda_{t_1}}} \left\| x(t; \lambda_0^*, \lambda_{t_1}^*) - x(t; \lambda_0, \lambda_{t_1}) \right\| = 0, \ \ t \in (t_1, s_1].$$

For $t \in (s_1, t_2]$, we have

$$\begin{aligned} x(t; \lambda_0^*, \lambda_{t_1}^*, \lambda_{s_1}^*) &= g_1(s_1, x(t_1^-; \lambda_0^*), \lambda_{t_1}^*) + h(s_1; \lambda_{s_1}^*) \\ &\quad + \frac{1}{\Gamma(\beta)} \int_{s_1}^t (t-s)^{\beta-1} f(s, x(s; \lambda_0^*, \lambda_{t_1}^*, \lambda_{s_1}^*), \lambda_{s_1}^*) ds, \end{aligned}$$

and

$$\begin{aligned} x(t; \lambda_0, \lambda_{t_1}, \lambda_{s_1}) &= g_1(s_1, x(t_1^-; \lambda_0), \lambda_{t_1}) + h(s_1; \lambda_{s_1}) \\ &\quad + \frac{1}{\Gamma(\alpha)} \int_{s_1}^t (t-s)^{\alpha-1} f(s, x(s; \lambda_0, \lambda_{t_1}, \lambda_{s_1}), \lambda_{s_1}) ds. \end{aligned}$$

Then

$$\begin{aligned} &\left\| x(t; \lambda_0^*, \lambda_{t_1}^*, \lambda_{s_1}^*) - x(t; \lambda_0, \lambda_{t_1}, \lambda_{s_1}) \right\| \\ \leqslant\ & L_{g_1}\left(\left\| x(t_1^-; \lambda_0^*) - x(t_1^-; \lambda_0) \right\| + |\lambda_{t_1}^* - \lambda_{t_1}| \right) + L_h |\lambda_{s_1}^* - \lambda_{s_1}| \\ &+ \left(\frac{1}{\Gamma(\beta)} - \frac{1}{\Gamma(\alpha)} \right) \int_{s_1}^t (t-s)^{\beta-1} \left\| f(s, x(s; \lambda_0^*, \lambda_{t_1}^*, \lambda_{s_1}^*), \lambda_{s_1}^*) \right\| ds \\ &+ \frac{1}{\Gamma(\alpha)} \int_{s_1}^t |(t-s)^{\beta-1} - (t-s)^{\alpha-1}| \left\| f(s, x(s; \lambda_0^*, \lambda_{t_1}^*, \lambda_{s_1}^*), \lambda_{s_1}^*) \right\| ds \\ &+ \frac{1}{\Gamma(\alpha)} \int_{s_1}^t (t-s)^{\alpha-1} \left\| f(s, x(s; \lambda_0^*, \lambda_{t_1}^*, \lambda_{s_1}^*), \lambda_{s_1}^*) - f(s, x(s; \lambda_0, \lambda_{t_1}, \lambda_{s_1}), \lambda_{s_1}) \right\| ds \\ \leqslant\ & L_{g_1}\left(\left\| x(t_1^-; \lambda_0^*) - x(t_1^-; \lambda_0) \right\| + |\lambda_{t_1}^* - \lambda_{t_1}| \right) + L_h |\lambda_{s_1}^* - \lambda_{s_1}| \\ &+ \left(\frac{1}{\Gamma(\beta)} - \frac{1}{\Gamma(\alpha)} \right) M \frac{(t_2 - s_1)^\beta}{\beta} + \frac{M}{\Gamma(\alpha)} \left| \frac{(t_2 - s_1)^\beta}{\beta} - \frac{(t_2 - s_1)^\alpha}{\alpha} \right| \\ &+ \frac{L_f}{\Gamma(\alpha+1)} |\lambda_{s_1}^* - \lambda_{s_1}| (t_2 - s_1)^\alpha \\ &+ \frac{L_f}{\Gamma(\alpha)} \int_{s_1}^t (t-s)^{\alpha-1} \left\| x(s; \lambda_0^*, \lambda_{t_1}^*, \lambda_{s_1}^*) - x(s; \lambda_0, \lambda_{t_1}, \lambda_{s_1}) \right\| ds, \end{aligned}$$

apply [18, corollary 2], we obtain that

$$\left\|x(t;\lambda_0^*,\lambda_{t_1}^*,\lambda_{s_1}^*)-x(t;\lambda_0,\lambda_{t_1},\lambda_{s_1})\right\|$$

$$\leqslant \left(L_{g_1}(\left\|x(t_1^-;\lambda_0^*)-x(t_1^-;\lambda_0)\right\|+|\lambda_{t_1}^*-\lambda_{t_1}|)+\left(\frac{1}{\Gamma(\beta)}-\frac{1}{\Gamma(\alpha)}\right)M\frac{(t_2-s_1)^\beta}{\beta}\right.$$

$$\left.+\frac{M}{\Gamma(\alpha)}\left|\frac{(t_2-s_1)^\beta}{\beta}-\frac{(t_2-s_1)^\alpha}{\alpha}\right|+\left(\frac{L_f(t_2-s_1)^\alpha}{\Gamma(\alpha+1)}+L_h\right)|\lambda_{s_1}^*-\lambda_{s_1}|\right)E_\alpha(L_f(t_2-s_1)^\alpha).$$

Note (2.115), then

$$\lim_{\substack{\beta\to\alpha\\ \lambda_0^*\to\lambda_0\\ \lambda_{t_1}^*\to\lambda_{t_1}\\ \lambda_{s_1}^*\to\lambda_{s_1}}}\left\|x(t;\lambda_0^*,\lambda_{t_1}^*,\lambda_{s_1}^*)-x(t;\lambda_0,\lambda_{t_1},\lambda_{s_1})\right\|=0,\ \ t\in(s_1,t_2]. \tag{2.116}$$

For $t\in(t_2,s_2]$, we have the same form as in (2.100), we have

$$\lim_{\substack{\beta\to\alpha\\ \lambda_{s_1}^*\to\lambda_{s_1}\\ \lambda_{t_2}^*\to\lambda_{t_2}}} x(t;\lambda_0,\lambda_{t_1},\lambda_{s_1}^*,\lambda_{t_2}^*)=x(t;\lambda_0,\lambda_{t_1},\lambda_{s_1},\lambda_{t_2}),\ \ t\in(t_2,s_2].$$

For $t\in(s_2,t_3]$, we have

$$\begin{aligned}x(t;\lambda_0,\lambda_{t_1},\lambda_{s_1}^*,\lambda_{t_2}^*,\lambda_{s_2}^*) &= g_2(s_2,x(t_2^-;\lambda_0,\lambda_{t_1},\lambda_{s_1}^*),\lambda_{t_2}^*)+h(s_2;\lambda_{s_2}^*)\\ &\quad+\frac{1}{\Gamma(\beta)}\int_{s_2}^t(t-s)^{\beta-1}f(s,x(s;\lambda_0,\lambda_{t_1},\lambda_{s_1}^*,\lambda_{t_2}^*,\lambda_{s_2}^*),\lambda_{s_2}^*)ds,\end{aligned}$$

and

$$\begin{aligned}x(t;\lambda_0,\lambda_{t_1},\lambda_{s_1},\lambda_{t_2},\lambda_{s_2}) &= g_2(s_2,x(t_2^-;\lambda_0,\lambda_{t_1},\lambda_{s_1}),\lambda_{t_2})\\ &\quad+\frac{1}{\Gamma(\alpha)}\int_{s_2}^t(t-s)^{\alpha-1}f(s,x(s;\lambda_0,\lambda_{t_1},\lambda_{s_1},\lambda_{t_2},\lambda_{s_2}),\lambda_{s_2})ds.\end{aligned}$$

Therefore,

$$
\begin{aligned}
&\left\|x(t;\lambda_0,\lambda_{t_1},\lambda^*_{s_1},\lambda^*_{t_2},\lambda^*_{s_2})-x(t;\lambda_0,\lambda_{t_1},\lambda_{s_1},\lambda_{t_2},\lambda_{s_2})\right\|\\
\leqslant\ & L_{g_2}(\left\|x(t_2^-;\lambda_0,\lambda_{t_1},\lambda^*_{s_1})-x(t_2^-;\lambda_0,\lambda_{t_1},\lambda_{s_1})\right\|+|\lambda^*_{t_2}-\lambda_{t_2}|)+L_h|\lambda^*_{s_2}-\lambda_{s_2}|\\
&+\left(\frac{1}{\Gamma(\beta)}-\frac{1}{\Gamma(\alpha)}\right)\int_{s_2}^{t}(t-s)^{\beta-1}\left\|f(s,x(s;\lambda_0,\lambda_{t_1},\lambda^*_{s_1},\lambda^*_{t_2},\lambda^*_{s_2}),\lambda^*_{s_2})\right\|ds\\
&+\frac{1}{\Gamma(\alpha)}\int_{s_2}^{t}|(t-s)^{\beta-1}-(t-s)^{\alpha-1}|\left\|f(s,x(s;\lambda_0,\lambda_{t_1},\lambda^*_{s_1},\lambda^*_{t_2},\lambda^*_{s_2}),\lambda^*_{s_2})\right\|ds\\
&+\frac{1}{\Gamma(\alpha)}\int_{s_2}^{t}(t-s)^{\alpha-1}\Big\|f(s,x(s;\lambda_0,\lambda_{t_1},\lambda^*_{s_1},\lambda^*_{t_2},\lambda^*_{s_2}),\lambda^*_{s_2})\\
&-f(s,x(s;\lambda_0,\lambda_{t_1},\lambda_{s_1},\lambda_{t_2},\lambda_{s_2}),\lambda_{s_2})\Big\|ds\\
\leqslant\ & L_{g_2}(\left\|x(t_2^-;\lambda_0,\lambda_{t_1},\lambda^*_{s_1})-x(t_2^-;\lambda_0,\lambda_{t_1},\lambda_{s_1})\right\|+|\lambda^*_{t_2}-\lambda_{t_2}|)\\
&+L_h|\lambda^*_{s_2}-\lambda_{s_2}|+\left(\frac{1}{\Gamma(\beta)}-\frac{1}{\Gamma(\alpha)}\right)M\frac{(t_3-s_2)^{\beta}}{\beta}\\
&+\frac{M}{\Gamma(\alpha)}\left|\frac{(t_3-s_2)^{\beta}}{\beta}-\frac{(t_3-s_2)^{\alpha}}{\alpha}\right|+\frac{L_f}{\Gamma(\alpha+1)}|\lambda^*_{s_2}-\lambda_{s_2}|(t_3-s_2)^{\alpha}\\
&+\frac{L_f}{\Gamma(\alpha)}\int_{s_2}^{t}(t-s)^{\alpha-1}\left\|x(s;\lambda_0,\lambda_{t_1},\lambda^*_{s_1},\lambda^*_{t_2},\lambda^*_{s_2})-x(s;\lambda_0,\lambda_{t_1},\lambda_{s_1},\lambda_{t_2},\lambda_{s_2})\right\|ds\\
\leqslant\ & \Big(L_{g_2}(\left\|x(t_2^-;\lambda_0,\lambda_{t_1},\lambda^*_{s_1})-x(t_2^-;\lambda_0,\lambda_{t_1},\lambda_{s_1})\right\|+|\lambda^*_{t_2}-\lambda_{t_2}|)\\
&+\left(\frac{1}{\Gamma(\beta)}-\frac{1}{\Gamma(\alpha)}\right)M\frac{(t_3-s_2)^{\beta}}{\beta}+\frac{M}{\Gamma(\alpha)}\left|\frac{(t_3-s_2)^{\beta}}{\beta}-\frac{(t_3-s_2)^{\alpha}}{\alpha}\right|\\
&+\left(\frac{L_f(t_3-s_2)^{\alpha}}{\Gamma(\alpha+1)}+L_h\right)|\lambda^*_{s_2}-\lambda_{s_2}|\Big)E_\alpha(L_f(t_3-s_2)^{\alpha}).
\end{aligned}
$$

Use (2.115), we obtain

$$
\lim_{\substack{\beta\to\alpha\\ \lambda^*_{s_1}\to\lambda_{s_1}\\ \lambda^*_{t_2}\to\lambda_{t_2}\\ \lambda^*_{s_2}\to\lambda_{s_2}}}\left\|x(t;\lambda_0,\lambda_{t_1},\lambda^*_{s_1},\lambda^*_{t_2},\lambda^*_{s_2})-x(t;\lambda_0,\lambda_{t_1},\lambda_{s_1},\lambda_{t_2},\lambda_{s_2})\right\|=0,\ \ t\in(s_2,t_3].
$$

Repeating the above procedures, we obtain

$$
\lim_{\substack{\beta\to\alpha\\ \lambda^*_{s_{i-1}}\to\lambda_{s_{i-1}}\\ \lambda^*_{t_i}\to\lambda_{t_i}}}\left\|x(t;\lambda_0,\lambda_{t_1},\ldots,\lambda^*_{s_{i-1}},\lambda^*_{t_i})-x(t;\lambda_0,\lambda_{t_1},\ldots,\lambda_{s_{i-1}},\lambda_{t_i})\right\|=0,\ \ t\in(t_i,s_i],\ \ i\in\mathbb{N},
$$

and

$$\lim_{\substack{\beta\to\alpha \\ \lambda^*_{s_{i-1}}\to\lambda_{s_{i-1}} \\ \lambda^*_{t_i}\to\lambda_{t_i} \\ \lambda^*_{s_i}\to\lambda_{s_i}}} \left\| x(t; \lambda_0, \lambda_{t_1}, \ldots, \lambda^*_{s_{i-1}}, \lambda^*_{t_i}, \lambda^*_{s_i}) - x(t; \lambda_0, \lambda_{t_1}, \ldots, \lambda_{s_{i-1}}, \lambda_{t_i}, \lambda_{s_i}) \right\| = 0,$$

for $t \in (s_i, t_{i+1}]$, $i \in \mathbb{N}$. The proof is completed. □

Now, we consider the differentiability of solutions to (2.97).

Theorem 2.3.8. *Assume* $[H_1] - [H_8]$ *hold. Then, the solution of problem (2.97) is differentiable with respect to parameters* λ_0, λ_{t_i} *and* λ_{s_i}, $i \in \mathbb{N}$, *respectively.*

Proof. For $t \in [0, t_1]$, we have

$$x(t; \lambda_0^*) = x_0(\lambda_0^*) + \frac{1}{\Gamma(\alpha)} \int_0^t (t-s)^{\alpha-1} f(s, x(s; \lambda_0^*), \lambda_0^*) ds,$$

and

$$x(t; \lambda_0) = x_0(\lambda_0) + \frac{1}{\Gamma(\alpha)} \int_0^t (t-s)^{\alpha-1} f(s, x(s; \lambda_0), \lambda_0) ds,$$

then

$$\begin{aligned}
& x(t; \lambda_0^*) - x(t; \lambda_0) \\
= \; & x_0(\lambda_0^*) - x_0(\lambda_0) + \frac{1}{\Gamma(\alpha)} \int_0^t (t-s)^{\alpha-1} (f(s, x(s; \lambda_0^*), \lambda_0^*) - f(s, x(s; \lambda_0), \lambda_0)) ds \\
= \; & \frac{\partial}{\partial \lambda_0} x_0(\lambda_0 + \theta_0(\lambda_0^* - \lambda_0))(\lambda_0^* - \lambda_0) + \frac{1}{\Gamma(\alpha)} \int_0^t (t-s)^{\alpha-1} \\
& \left(\frac{\partial}{\partial x} f(s, x(s; \lambda_0) + \theta_1(x(s; \lambda_0^*) - x(s; \lambda_0)), \lambda_0^*)(x(s; \lambda_0^*) - x(s; \lambda_0)) \right. \\
& \left. + \frac{\partial}{\partial \lambda_0} f(s, x(s; \lambda_0), \lambda_0 + \theta_2(\lambda_0^* - \lambda_0))(\lambda_0^* - \lambda_0) \right) ds,
\end{aligned}$$

where $\theta_0, \theta_1, \theta_2 \in (0, 1)$. Note that

$$\begin{aligned}
& \frac{x(t; \lambda_0^*) - x(t; \lambda_0)}{\lambda_0^* - \lambda_0} \\
= \; & (\frac{\partial}{\partial \lambda_0} x_0(\lambda_0) + r_0) + \frac{1}{\Gamma(\alpha)} \int_0^t (t-s)^{\alpha-1} \left((\frac{\partial}{\partial x} f(s, x(s; \lambda_0), \lambda_0) + r_1) \right. \\
& \left. \times \frac{x(s; \lambda_0^*) - x(s; \lambda_0)}{\lambda_0^* - \lambda_0} + (\frac{\partial}{\partial \lambda_0} f(s, x(s; \lambda_0), \lambda_0) + r_2) \right) ds,
\end{aligned} \tag{2.117}$$

where r_k, $k = 0, 1, 2$ have the property: $r_k \to 0$, if $\lambda_0^* \to \lambda_0$.

Consider

$$\begin{cases} {}^c\mathbf{D}_{0,t}^{\alpha} y_0(t) = \frac{\partial}{\partial x} f(t, x(t; \lambda_0), \lambda_0) \cdot y_0(t) + \frac{\partial}{\partial \lambda_0} f(t, x(t; \lambda_0), \lambda_0), \\ y_0(0) = \frac{\partial}{\partial \lambda_0} x_0(\lambda_0). \end{cases} \tag{2.118}$$

Then we have the integral form of the problem (2.118),

$$\begin{aligned} y_0(t) \; = \; & \frac{\partial}{\partial \lambda_0} x_0(\lambda_0) + \frac{1}{\Gamma(\alpha)} \int_0^t (t-s)^{\alpha-1} \bigg(\frac{\partial}{\partial x} f(s, x(s; \lambda_0), \lambda_0) \cdot y_0(s) \\ & + \frac{\partial}{\partial \lambda_0} f(s, x(s; \lambda_0), \lambda_0) \bigg) ds. \end{aligned} \tag{2.119}$$

From (2.117) and (2.119), we have

$$\begin{aligned} & \frac{x(t; \lambda_0^*) - x(t; \lambda_0)}{\lambda_0^* - \lambda_0} - y_0(t) \\ = \; & r_0 + \frac{1}{\Gamma(\alpha)} \int_0^t (t-s)^{\alpha-1} \bigg((\frac{\partial}{\partial x} f(s, x(s; \lambda_0), \lambda_0) + r_1) \\ & \times \frac{x(s; \lambda_0^*) - x(s; \lambda_0)}{\lambda_0^* - \lambda_0} + r_2 - \frac{\partial}{\partial x} f(s, x(s; \lambda_0), \lambda_0) y_0(s) \bigg) ds \\ = \; & r_0 + \frac{1}{\Gamma(\alpha)} \int_0^t (t-s)^{\alpha-1} (r_2 + r_1 y_0(s)) ds \\ & + \frac{1}{\Gamma(\alpha)} \int_0^t (t-s)^{\alpha-1} \bigg(\frac{\partial}{\partial x} f(s, x(s; \lambda_0), \lambda_0) + r_1 \bigg) \bigg(\frac{x(s; \lambda_0^*) - x(s; \lambda_0)}{\lambda_0^* - \lambda_0} - y_0(s) \bigg) ds. \end{aligned}$$

Therefore,

$$\begin{aligned} & \left\| \frac{x(t; \lambda_0^*) - x(t; \lambda_0)}{\lambda_0^* - \lambda_0} - y_0(t) \right\| \\ \leqslant \; & \|r_0\| + \frac{1}{\Gamma(\alpha)} \int_0^t (t-s)^{\alpha-1} \left\| r_2 + r_1 y_0(s) \right\| ds \\ & + \frac{1}{\Gamma(\alpha)} \int_0^t (t-s)^{\alpha-1} \left\| \frac{\partial}{\partial x} f(s, x(s; \lambda_0), \lambda_0) + r_1 \right\| \left\| \frac{x(s; \lambda_0^*) - x(s; \lambda_0)}{\lambda_0^* - \lambda_0} - y_0(s) \right\| ds \\ \leqslant \; & \left(\|r_0\| + \frac{1}{\Gamma(\alpha)} \int_0^t (t-s)^{\alpha-1} (\|r_2\| + \|r_1\| Y_0) ds \right) E_\alpha((M + \|r_1\|) t_1^\alpha), \\ \leqslant \; & \left(\|r_0\| + \frac{t_1^\alpha}{\Gamma(\alpha+1)} (\|r_2\| + \|r_1\| Y_0) \right) E_\alpha((M + \|r_1\|) t_1^\alpha), \end{aligned}$$

where we use [18, corollary 2], Y_0 is a positive constant such that $\|y_0(t)\| \leqslant Y_0$, for $t \in [0, t_1]$ and E_α denotes the classical Mittag-Leffler function.

Since $\lambda_0^* \to \lambda_0$, then $r_0 \to 0$, $r_1 \to 0$, $r_2 \to 0$, we get

$$\lim_{\lambda_0^* \to \lambda_0} \left\| \frac{x(t; \lambda_0^*) - x(t; \lambda_0)}{\lambda_0^* - \lambda_0} - y_0(t) \right\| = 0,$$

that is

$$\lim_{\lambda_0^* \to \lambda_0} \frac{x(t; \lambda_0^*) - x(t; \lambda_0)}{\lambda_0^* - \lambda_0} = y_0(t).$$

For $t \in [0, t_1]$, there exists a continuous derivative

$$\frac{\partial}{\partial \lambda_0} x(t; \lambda_0) = y_0(t),$$

which satisfies the problem

$$\begin{cases} {}^c\mathbf{D}_{0,t}^{\alpha}\left(\frac{\partial}{\partial \lambda_0} x(t; \lambda_0)\right) = \left(\frac{\partial}{\partial x} f(t, x(t; \lambda_0), \lambda_0)\right)\left(\frac{\partial}{\partial \lambda_0} x(t; \lambda_0)\right) + \frac{\partial}{\partial \lambda_0} f(t, x(t; \lambda_0), \lambda_0), \\ \frac{\partial}{\partial \lambda_0} x(0; \lambda_0) = \frac{\partial}{\partial \lambda_0} x_0(\lambda_0). \end{cases}$$

For $t \in (t_1, s_1]$, we have the same conclusion as in (2.105).

For $t \in (s_1, t_2]$, note

$$\begin{aligned}
& x(t; \lambda_0, \lambda_{t_1}, \lambda_{s_1}^*) - x(t; \lambda_0, \lambda_{t_1}, \lambda_{s_1}) \\
= \ & h(s_1; \lambda_{s_1}^*) - h(s_1; \lambda_{s_1}) + \frac{1}{\Gamma(\alpha)} \int_{s_1}^{t} (t - s)^{\alpha - 1} \\
& \left(f(s, x(s; \lambda_0, \lambda_{t_1}, \lambda_{s_1}^*), \lambda_{s_1}^*) - f(s, x(s; \lambda_0, \lambda_{t_1}, \lambda_{s_1}), \lambda_{s_1})\right) ds \\
= \ & \left(\frac{\partial}{\partial \lambda_{s_1}} h(s_1; \lambda_{s_1}) + r_4\right)(\lambda_{s_1}^* - \lambda_{s_1}) \\
& + \frac{1}{\Gamma(\alpha)} \int_{s_1}^{t} (t - s)^{\alpha - 1} \left(\left(\frac{\partial}{\partial x} f(s, x(s; \lambda_0, \lambda_{t_1}, \lambda_{s_1}), \lambda_{s_1}) + r_5\right) \right. \\
& \times (x(s; \lambda_0, \lambda_{t_1}, \lambda_{s_1}^*) - x(s; \lambda_0, \lambda_{t_1}, \lambda_{s_1})) \\
& \left. + \left(\frac{\partial}{\partial \lambda_{s_1}} f(s, x(s; \lambda_0, \lambda_{t_1}, \lambda_{s_1}), \lambda_{s_1}) + r_6\right)(\lambda_{s_1}^* - \lambda_{s_1}) \right) ds,
\end{aligned} \tag{2.120}$$

where r_k, $k = 4, 5, 6$ have the property: $r_k \to 0$, if $\lambda_{s_1}^* \to \lambda_{s_1}$.

Therefore,

$$\frac{x(t;\lambda_0,\lambda_{t_1},\lambda_{s_1}^*)-x(t;\lambda_0,\lambda_{t_1},\lambda_{s_1})}{\lambda_{s_1}^*-\lambda_{s_1}}$$
$$=(\frac{\partial}{\partial\lambda_{s_1}}h(s_1;\lambda_{s_1})+r_4)+\frac{1}{\Gamma(\alpha)}\int_{s_1}^{t}(t-s)^{\alpha-1}\Big((\frac{\partial}{\partial x}f(s,x(s;\lambda_0,\lambda_{t_1},\lambda_{s_1}),\lambda_{s_1})+r_5) \tag{2.121}$$
$$\times\frac{x(s;\lambda_0,\lambda_{t_1},\lambda_{s_1}^*)-x(s;\lambda_0,\lambda_{t_1},\lambda_{s_1})}{\lambda_{s_1}^*-\lambda_{s_1}}+(\frac{\partial}{\partial\lambda_{s_1}}f(s,x(s;\lambda_0,\lambda_{t_1},\lambda_{s_1}),\lambda_{s_1})+r_6)\Big)ds.$$

Consider the initial value problem

$$\begin{cases}{}^{c}\mathbf{D}_{s_1,t}^{\alpha}y_{s_1}(t)=\frac{\partial}{\partial x}f(t,x(t;\lambda_0,\lambda_{t_1},\lambda_{s_1}),\lambda_{s_1})\cdot y_{s_1}(t)+\frac{\partial}{\partial\lambda_{s_1}}f(t,x(t;\lambda_0,\lambda_{t_1},\lambda_{s_1}),\lambda_{s_1}),\\ y_{s_1}(s_1)=\frac{\partial}{\partial\lambda_{s_1}}h(s_1;\lambda_{s_1}),\end{cases}$$

and note

$$y_{s_1}(t)=\frac{\partial}{\partial\lambda_{s_1}}h(s_1;\lambda_{s_1})+\frac{1}{\Gamma(\alpha)}\int_{s_1}^{t}(t-s)^{\alpha-1}\Big(\frac{\partial}{\partial x}f(s,x(s;\lambda_0,\lambda_{t_1},\lambda_{s_1}),\lambda_{s_1})\cdot y_{s_1}(s)$$
$$+\frac{\partial}{\partial\lambda_{s_1}}f(s,x(s;\lambda_0,\lambda_{t_1},\lambda_{s_1}),\lambda_{s_1})\Big)ds. \tag{2.122}$$

From (2.121) and (2.122) and apply [18, corollary 2], we have

$$\left\|\frac{x(t;\lambda_0,\lambda_{t_1},\lambda_{s_1}^*)-x(t;\lambda_0,\lambda_{t_1},\lambda_{s_1})}{\lambda_{s_1}^*-\lambda_{s_1}}-y_{s_1}(t)\right\|$$
$$\leqslant\left\|r_4\right\|+\frac{1}{\Gamma(\alpha)}\int_{s_1}^{t}(t-s)^{\alpha-1}\left\|r_6+r_5y_{s_1}(s)\right\|ds$$
$$+\frac{1}{\Gamma(\alpha)}\int_{s_1}^{t}(t-s)^{\alpha-1}\left\|\frac{\partial}{\partial x}f(s,x(s;\lambda_0,\lambda_{t_1},\lambda_{s_1}),\lambda_{s_1})+r_5\right\|$$
$$\times\left\|\frac{x(s;\lambda_0,\lambda_{t_1},\lambda_{s_1}^*)-x(s;\lambda_0,\lambda_{t_1},\lambda_{s_1})}{\lambda_{s_1}^*-\lambda_{s_1}}-y_{s_1}(s)\right\|ds$$
$$\leqslant\left(\left\|r_4\right\|+\frac{1}{\Gamma(\alpha)}\int_{s_1}^{t}(t-s)^{\alpha-1}(\left\|r_6\right\|+\left\|r_5\right\|Y_{s_1})ds\right)E_\alpha((M+\|r_5\|)(t_2-s_1)^\alpha)$$
$$\leqslant\left(\left\|r_4\right\|+\frac{(t_2-s_1)^\alpha}{\Gamma(\alpha+1)}(\left\|r_6\right\|+\left\|r_5\right\|Y_{s_1})\right)E_\alpha((M+\|r_5\|)(t_2-s_1)^\alpha),$$

where Y_{s_1} is a positive constant such that $\left\|y_{s_1}(t)\right\|\leqslant Y_{s_1}$, for $t\in(s_1,t_2]$.

Since $\lambda^*_{s_1} \to \lambda_{s_1}$, then $r_4 \to 0$, $r_5 \to 0$, $r_6 \to 0$, we obtain

$$\lim_{\lambda^*_{s_1} \to \lambda_{s_1}} \frac{x(t; \lambda_0, \lambda_{t_1}, \lambda^*_{s_1}) - x(t; \lambda_0, \lambda_{t_1}, \lambda_{s_1})}{\lambda^*_{s_1} - \lambda_{s_1}} = y_{s_1}(t).$$

For $t \in (s_1, t_2]$, there exists a continuous derivative

$$\frac{\partial}{\partial \lambda_{s_1}} x(t; \lambda_0, \lambda_{t_1}, \lambda_{s_1}) = y_{s_1}(t),$$

which satisfies

$$\begin{cases} {}^c\mathbf{D}^\alpha_{s_1,t}\left(\frac{\partial}{\partial \lambda_{s_1}} x(t; \lambda_0, \lambda_{t_1}, \lambda_{s_1})\right) &= \left(\frac{\partial}{\partial x} f(t, x(t; \lambda_0, \lambda_{t_1}, \lambda_{s_1}), \lambda_{s_1})\right)\left(\frac{\partial}{\partial \lambda_{s_1}} x(t; \lambda_0, \lambda_{t_1}, \lambda_{s_1})\right) \\ & \quad + \frac{\partial}{\partial \lambda_{s_1}} f(t, x(t; \lambda_0, \lambda_{t_1}, \lambda_{s_1}), \lambda_{s_1}), \\ \frac{\partial}{\partial \lambda_{s_1}} x(s_1; \lambda_0, \lambda_{t_1}, \lambda_{s_1}) &= \frac{\partial}{\partial \lambda_{s_1}} h(s_1; \lambda_{s_1}). \end{cases}$$

For $t \in (t_i, s_i]$, $i \in \mathbb{N}$, we have the same result as in (2.112).

Repeating the procedure as in (2.120), for $t \in (s_i, t_{i+1}]$, $i \in \mathbb{N}$, we have

$$\begin{aligned} & \frac{x(t; \lambda_0, \lambda_{t_1}, \ldots, \lambda_{t_i}, \lambda^*_{s_i}) - x(t; \lambda_0, \lambda_{t_1}, \ldots, \lambda_{t_i}, \lambda_{s_i})}{\lambda^*_{s_i} - \lambda_{s_i}} \\ = & \left(\frac{\partial}{\partial \lambda_{s_i}} h(s_i; \lambda_{s_i}) + r_k\right) \\ & + \frac{1}{\Gamma(\alpha)} \int_{s_i}^t (t-s)^{\alpha-1} \Bigg(\left(\frac{\partial}{\partial x} f(s, x(s; \lambda_0, \lambda_{t_1}, \ldots, \lambda_{t_i}, \lambda_{s_i}), \lambda_{s_i}) + r_{k+1}\right) \\ & \times \frac{x(s; \lambda_0, \lambda_{t_1}, \ldots, \lambda_{t_i}, \lambda^*_{s_i}) - x(s; \lambda_0, \lambda_{t_1}, \ldots, \lambda_{t_i}, \lambda_{s_i})}{\lambda^*_{s_i} - \lambda_{s_i}} \\ & + \left(\frac{\partial}{\partial \lambda_{s_i}} f(s, x(s; \lambda_0, \lambda_{t_1}, \ldots, \lambda_{t_i}, \lambda_{s_i}), \lambda_{s_i}) + r_{k+2}\right) \Bigg) ds, \end{aligned} \tag{2.123}$$

where r_k, r_{k+1}, r_{k+2} have the property: $r_k \to 0$, $r_{k+1} \to 0$, $r_{k+2} \to 0$, if $\lambda^*_{s_i} \to \lambda_{s_i}$.

Consider

$$\begin{cases} {}^c\mathbf{D}^\alpha_{s_i,t} y_{s_i}(t) &= \frac{\partial}{\partial x} f(t, x(t; \lambda_0, \lambda_{t_1}, \ldots, \lambda_{t_i}, \lambda_{s_i}), \lambda_{s_i}) \cdot y_{s_i}(t) \\ & \quad + \frac{\partial}{\partial \lambda_{s_i}} f(t, x(t; \lambda_0, \lambda_{t_1}, \ldots, \lambda_{t_i}, \lambda_{s_i}), \lambda_{s_i}), \\ y_{s_i}(s_i) &= \frac{\partial}{\partial \lambda_{s_i}} h(s_i; \lambda_{s_i}), \end{cases}$$

and note

$$
\begin{aligned}
y_{s_i}(t) &= \frac{\partial}{\partial \lambda_{s_i}} h(s_i; \lambda_{s_i}) \\
&+ \frac{1}{\Gamma(\alpha)} \int_{s_i}^{t} (t-s)^{\alpha-1} \Big(\frac{\partial}{\partial x} f(s, x(s; \lambda_0, \lambda_{t_1}, \ldots, \lambda_{t_i}, \lambda_{s_i}), \lambda_{s_i}) \cdot y_{s_i}(s) \\
&+ \frac{\partial}{\partial \lambda_{s_i}} f(s, x(s; \lambda_0, \lambda_{t_1}, \ldots, \lambda_{t_i}, \lambda_{s_i}), \lambda_{s_i}) \Big) ds.
\end{aligned} \tag{2.124}
$$

From (2.123) and (2.124) and use [18, corollary 2], we have

$$
\begin{aligned}
&\left\| \frac{x(t; \lambda_0, \lambda_{t_1}, \ldots, \lambda_{t_i}, \lambda_{s_i}^*) - x(t; \lambda_0, \lambda_{t_1}, \ldots, \lambda_{t_i}, \lambda_{s_i})}{\lambda_{s_i}^* - \lambda_{s_i}} - y_{s_i}(t) \right\| \\
\leqslant\; & \left\| r_k \right\| + \frac{1}{\Gamma(\alpha)} \int_{s_i}^{t} (t-s)^{\alpha-1} \left\| r_{k+2} + r_{k+1} y_{s_i}(s) \right\| ds \\
&+ \frac{1}{\Gamma(\alpha)} \int_{s_i}^{t} (t-s)^{\alpha-1} \left\| \frac{\partial}{\partial x} f(s, x(s; \lambda_0, \lambda_{t_1}, \ldots, \lambda_{t_i}, \lambda_{s_i}), \lambda_{s_i}) + r_{k+1} \right\| \\
&\times \left\| \frac{x(s; \lambda_0, \lambda_{t_1}, \ldots, \lambda_{t_i}, \lambda_{s_i}^*) - x(s; \lambda_0, \lambda_{t_1}, \ldots, \lambda_{t_i}, \lambda_{s_i})}{\lambda_{s_i}^* - \lambda_{s_i}} - y_{s_i}(s) \right\| ds \\
\leqslant\; & \left(\left\| r_k \right\| + \frac{1}{\Gamma(\alpha)} \int_{s_i}^{t} (t-s)^{\alpha-1} \left(\left\| r_{k+2} \right\| + \left\| r_{k+1} \right\| Y_{s_i} \right) ds \right) E_\alpha \\
&\left((M + \|r_{k+1}\|)(t_{i+1} - s_i)^\alpha \right) \\
\leqslant\; & \left(\left\| r_k \right\| + \frac{(t_{i+1} - s_i)^\alpha}{\Gamma(\alpha+1)} \left(\left\| r_{k+2} \right\| + \left\| r_{k+1} \right\| Y_{s_i} \right) \right) E_\alpha \left((M + \|r_{k+1}\|)(t_{i+1} - s_i)^\alpha \right),
\end{aligned}
$$

where Y_{s_i} is a positive constant such that $\left\| y_{s_i}(t) \right\| \leqslant Y_{s_i}$, for $t \in (s_i, t_{i+1}]$.

Since $\lambda_{s_i}^* \to \lambda_{s_i}$, then $r_k \to 0$, $r_{k+1} \to 0$, $r_{k+2} \to 0$, we obtain

$$
\lim_{\lambda_{s_i}^* \to \lambda_{s_i}} \frac{x(t; \lambda_0, \lambda_{t_1}, \ldots, \lambda_{t_i}, \lambda_{s_i}^*) - x(t; \lambda_0, \lambda_{t_1}, \ldots, \lambda_{t_i}, \lambda_{s_i})}{\lambda_{s_i}^* - \lambda_{s_i}} = y_{s_i}(t).
$$

For $t \in (s_i, t_{i+1}]$, there exists a continuous derivative

$$
\frac{\partial}{\partial \lambda_{s_i}} x(t; \lambda_0, \lambda_{t_1}, \ldots, \lambda_{t_i}, \lambda_{s_i}) = y_{s_i}(t),
$$

which satisfies

$$\begin{cases} {}^c\mathbf{D}^{\alpha}_{s_i,t}\left(\frac{\partial}{\partial\lambda_{s_i}}x(t;\lambda_0,\lambda_{t_1},\ldots,\lambda_{t_i},\lambda_{s_i})\right) \\ =\left(\frac{\partial}{\partial x}f(t,x(t;\lambda_0,\lambda_{t_1},\ldots,\lambda_{t_i},\lambda_{s_i}),\lambda_{s_i})\right)\left(\frac{\partial}{\partial\lambda_{s_i}}x(t;\lambda_0,\lambda_{t_1},\ldots,\lambda_{t_i},\lambda_{s_i})\right) \\ +\frac{\partial}{\partial\lambda_{s_i}}f(t,x(t;\lambda_0,\lambda_{t_1},\ldots,\lambda_{t_i},\lambda_{s_i}),\lambda_{s_i}), \\ \frac{\partial}{\partial\lambda_{s_i}}x(s_i;\lambda_0,\lambda_{t_1},\ldots,\lambda_{t_i},\lambda_{s_i}) = \frac{\partial}{\partial\lambda_{s_i}}h(s_i;\lambda_{s_i}). \end{cases}$$

The proof is completed. □

2.3.4 Examples

In this section, we extend the example in [24, section 4] to two classes of non-instantaneous impulsive logistic equations.

Example 2.3.9. *Consider the following integer order non-instantaneous impulsive logistic equations*

$$\begin{cases} x'(t) = \frac{r}{\kappa}((x+\lambda)(\kappa-(x+\lambda))), \ t\in(s_i,t_{i+1}], \ i\in\{0\}\bigcup\mathbb{N}, \\ x(t_i^+) = x(t_i^-) - p_i(x(t_i^-)+\lambda), \ i\in\mathbb{N}, \\ x(t) = x(t_i^-) - p_i(x(t_i^-)+\lambda) - |\sin(t-\lambda)|, \ t\in(t_i,s_i], \ i\in\mathbb{N}, \\ x(0) = \lambda_0, \end{cases} \tag{2.125}$$

where $x(t)$ denotes the quantity of living organisms at time $t \geqslant 0$, $\kappa > 0$ denotes the capacity of environment, $r > 0$ denotes the reproductive potential of population. Moreover, $\lambda_{s_i} = \lambda = \lambda_{t_i} = 2i - 1$, $i\in\mathbb{R}^+$ and $0<\lambda<\kappa$; $p_i\in\mathbb{R}$ with $0<p_i<1$, $i\in\mathbb{N}$. Set $t_i = 2i-1$ and $s_i = 2i$, $i\in\mathbb{N}$, and note, $t_i < s_i$ holds.

We make some additional notes on (2.125):

The first equation describes the varying principle of living organisms at the time set $\bigcup_{i\in\{0\}\cup\mathbb{N}}(s_i,t_{i+1}]$;

The second equation describe that people take some strategy to interpose the quantity of living organisms at some fixed time points t_i^-, $i\in\mathbb{N}$;

The third equation describe that people take the strategy in the second equation and act on the first equation on a time interval $t\in(t_i,s_i]$, $i\in\mathbb{N}$.

Set $f(\cdot,x,\lambda_{s_i}) = \frac{r}{\kappa}((x+\lambda)(\kappa-(x+\lambda)))$, $g_i(\cdot,x(t_i^-),\lambda_{t_i}) := x(t_i^-) - p_i(x(t_i^-)+\lambda_{t_i})$, $h(\cdot,\lambda_{s_i}) = -|\sin(t-\lambda)|$ with $h(t_i,\lambda_{s_i}) = 0$ and $\lambda_0\in\mathbb{D} := (0,\kappa-\lambda)$.

The solution of $x' = \frac{r}{\kappa}((x+\lambda)(\kappa-(x+\lambda)))$ *with* $x(0) = \lambda_0 \in \mathbb{D}$ *satisfies the inequality:* $0 < x(\cdot) + \lambda < \kappa \Longleftrightarrow x(\cdot) \in \mathbb{D}$. *Thus,* $[H_5]$ *holds. Moreover,* $0 < x(t_i^+) = x(t_i^-) - p_i(x(t_i^-) + \lambda) < \kappa - \lambda$, $i \in \mathbb{N}$ *and* $x(t) = x(t_i^-) - p_i(x(t_i^-) + \lambda) - |\sin(t-\lambda)| < \kappa - \lambda$, $t \in (t_i, s_i]$, $i \in \mathbb{N}$. *Next,* $\left\|\frac{\partial f}{\partial x}\right\| = \frac{r}{\kappa}|\kappa - 2(x+\lambda)| \leqslant r := M$ *since* $x \in \mathbb{D}$. *Thus,* $[H_2]$ *holds. The remaining conditions* $[H_1]$, $[H_3]$, $[H_4]$, $[H_6]$, $[H_7]$, $[H_8]$ *hold directly.*

Example 2.3.10. *Consider the following fractional order non-instantaneous impulsive logistic equations*

$$\begin{cases} {}^{c}\mathbf{D}_{s_i,t}^{1/3}x(t) = -ax(t), & t \in (s_i, t_{i+1}],\ i \in \{0\} \bigcup \mathbb{N}, \\ x(t_i^+) = x(t_i^-) - c_i x(t_i^-) - d_i|\sin\lambda|, & i \in \mathbb{N}, \\ x(t) = x(t_i^-) - c_i x(t_i^-) - d_i|\sin\lambda|, & t \in (t_i, s_i],\ i \in \mathbb{N}, \\ x(0) = \lambda_0 > 0, \end{cases} \tag{2.126}$$

where $a, c_i, d_i > 0$ *and* $\lambda_{s_i} = 0$, $\lambda_{t_i} \in \mathbb{R}$ *and* $t_i = 2i - 1$ *and* $s_i = 2i$, $i \in \mathbb{N}$, *and note,* $t_i < s_i$ *holds.*

For any $t \in (s_i, t_{i+1}]$, $i \in \{0\} \cup \mathbb{N}$,

$$0 < x(t) = E_{1/3}(-at^{1/3})\lambda_0 \leqslant \lambda_0, \quad (E_{1/3}(z) \leqslant 1,\ z \leqslant 0),$$

and $x(t_i^+) = x(t_i^-) - c_i x(t_i^-) - d_i|\sin\lambda| < \lambda_0$, $i \in \mathbb{N}$, $x(t) = x(t_i^-) - c_i x(t_i^-) - d_i|\sin\lambda| < \lambda_0$, $t \in (t_i, s_i]$. *Thus, we choose* $\mathbb{D} = (0, \lambda_0 + 1)$.

Set $f(\cdot, x, \lambda_{s_i}) = -ax$, $g_i(\cdot, x(t_i^-), \lambda_{t_i}) := (1 - c_i)x(t_i^-) - d_i|\sin\lambda|$ *and* $h(\cdot, \lambda_{s_i}) = 0$. *Note,* $[H_1]$*–*$[H_8]$ *hold directly.*

From above, one can see the results in theorems 2.3.7 and 2.3.8 can be applied to examples 2.3.9 and 2.3.10, respectively.

We obtain the following facts:

- The solution of problem (2.125) and (2.126) depend continuously on the parameters λ_0 and λ, respectively.
- The solution of problem (2.125) and (2.126) is differentiable with respect to the parameters λ_0 and λ, respectively.
- If the parameters λ_0 and λ are chosen in the certain range, then the quantity of living organisms will not be changed largely by using a instantaneous impulsive test technique and a non-instantaneous impulsive control strategy. Moreover, the varying of the quantity of living organisms depend smoothly on the initial amount of the living organisms, instantaneous impulsive size and non-instantaneous impulsive process via impulsive parameters and junction parameters, which are associated with impulsive points and junction points.

2.3.5 Notes and remarks

This section transfer ideas in [25] to study asymptotic and smooth properties of solutions to nonlinear non-instantaneous impulsive differential equations involving parameters of integer order and fractional order. We introduce the concept of continuous dependence and differentiability of solutions and establish sufficient conditions to guarantee the solution depends continuously and is differentiable on the initial condition, impulsive parameters and junction parameters.

The results in section 2.3 are motivated by [27].

References

[1] Kilbas A A, Srivastava H M and Trujillo J J 2006 *Theory and Applications of Fractional Differential Equations* (Amsterdam: Elsevier Science)

[2] Hilfer R 1999 *Applications of Fractional Calculus in Physics* (Singapore: World Scientific)

[3] Machado J T, Kiryakova V and Mainardi F 2011 Recent history of fractional calculus *Commun. Nonlinear Sci. Numer. Simula.* **16** 1140–53

[4] Baleanu D, Machado J A T and Luo A C J 2012 *Fractional Dynamics and Control* (New York: Springer)

[5] Diethelm K 2010 The analysis of fractional differential equations *Lecture Notes in Mathematics* (New York: Springer)

[6] Miller K S and Ross B 1993 *An Introduction to Fractional Calculus and Differential Equations* (New York: John Wiley)

[7] Podlubny I 1999 *Fractional Differential Equations* (New York: Academic)

[8] Tarasov V E 2011 *Fractional Dynamics: Application of Fractional Calculus to Dynamics of Particles, Fields and Media* (New York: Springer)

[9] Zhou Y 2014 *Basic Theory of Fractional Differential Equations* (Singapore: World Scientific)

[10] Zhou Y 2016 *Fractional Evolution Equations and Inclusions: Analysis and Control* (New York: Academic)

[11] Metzler R and Klafter J 2004 The restaurant at the end of the random walk: recent developments in the description of anomalous transport by fractional dynamics *J. Phys. A: Math. Gen.* **37** R161–208

[12] Pagnini G 2014 Short note on the emergence of fractional kinetics *Physica* A **409** 29–34

[13] West B J 2014 Fractional calculus view of complexity: a tutorial *Rev. Mod. Phys.* **86** 1169–84

[14] Dishliev A, Dishlieva K and Nenov S 2012 *Specific Asymptotic Properties of the Solutions of Impulsive Differential Equations: Methods and Applications* (New York: Academic)

[15] Samoilenko A M, Perestyuk N A and Chapovsky Y 1995 *Impulsive Differential Equations* (New York: World Scientific)

[16] Wang J, Zhou Y and Fečkan M 2012 Nonlinear impulsive problems for fractional differential equations and Ulam stability *Comp. Math. Appl.* **64** 3389–405

[17] Bainov D D and Simeonov P S 1992 *Integral Inequalities and Applications* (Dordrecht: Kluwer)

[18] Ye H, Gao J and Ding Y 2007 A generalized Grönwall inequality and its application to a fractional differential equation *J. Math. Anal. Appl.* **328** 1075–81

[19] Wang J, Fečkan M and Zhou Y 2016 A survey on impulsive fractional differential equations *Fract. Calc. Appl. Anal.* **19** 806–31

[20] Wang J and Fečkan M 2015 A general class of impulsive evolution equations *Topol. Meth. Nonlinear Anal.* **46** 915–34

[21] Dishlieva K and Antonov A 2015 *Hausdorff Metric and Differential Equations with Variable Structure and Impulses* (Bulgaria: Technical University of Sofia)

[22] Yang D, Wang J and O'Regan D 2017 Asymptotic properties of the solutions of nonlinear non-instantaneous impulsive differential equations *J. Franklin Inst.* **354** 6978–7011

[23] Sendov B 1990 *Hausdorff Approximations* (Berlin: Springer)

[24] Agarwal R, O'Regan D and Hristova S 2016 A survey of Lyapunov functions, stability and impulsive Caputo fractional differential equations *Fract. Calc. Appl. Anal.* **19** 290–318

[25] Dishlieva K G 2011 Differentiability of solutions of impulsive differential equations with respect to the impulsive perturbations *Nonlinear Anal.: RWA* **12** 3541–51

[26] Yang D, Wang J and O'Regan D 2018 On the orbital Hausdorff dependence of differential equations with non-instantaneous impulses *C. R. Acad. Sci. Paris Ser. I* **356** 150–171

[27] Yang D, Wang J and O'Regan D 2018 Differentiability of solutions to nonlinear non-instantaneous impulsive differential equations involving parameters *Appl. Math. Comput.* **321** 654–71

Chapter 3

Semilinear evolution equations

In this chapter, we introduce two new classes of semilinear evolution equations with impulses and derive the existence and Ulam's stability results. In section 3.1 we discuss the existence and stability for first-order evolution equations by using the theory of strongly continuous semigroups. Section 3.2 is devoted to studying the existence and stability for second-order evolution equations by using the theory of a strongly continuous cosine family.

3.1 First-order evolution equations

3.1.1 Introduction

Set $J := [0, T]$. We need the Banach space $PC(J, X) := \{x: J \to X: x \in C((t_k, t_{k+1}], X),\ k = 0, 1, \ldots, m$ and there exist $x(t_k^-)$ and $x(t_k^+)$, $k = 1, \ldots, m$ with $x(t_k^-) = x(t_k)\}$ endowed either with the Chebyshev PC-norm $\|x\|_{PC} := \sup\{\|x(t)\|: t \in J\}$ or with the Bielecki PCB-norm $\|x\|_{PCB} := \sup\{\|x(t)\|e^{-\Omega t}: t \in J\}$ for some $\Omega \in \mathbb{R}$.

Many researchers have devoted their study to mild solutions for impulsive evolution equations with instantaneous impulses of the form

$$\begin{cases} x'(t) = Ax(t) + f(t, x(t)),\ t \in J' := J/\{t_1, \ldots, t_m\},\ J, \\ x(t_k^+) = x(t_k^-) + I_k(x(t_k^-)),\ k = 1, 2, \ldots, m, \end{cases} \tag{3.1}$$

where the linear unbounded operator $A: D(A) \subseteq X \to X$ is the generator of a C_0-semigroup (analytic or compact) $\{T(t),\ t \geqslant 0\}$ on a Banach space X with a norm $\|\cdot\|$, $f : J \times X \to X$ and $I_k : X \to X$ and fixed impulsive time t_k satisfy $0 = t_0 < t_1 < \cdots < t_m < t_{m+1} = T$, and the symbols $x(t_k^+) := \lim_{\epsilon\to 0^+} x(t_k + \epsilon)$ and $x(t_k^-) := \lim_{\epsilon\to 0^-} x(t_k + \epsilon)$ represent the right and left limits of $x(t)$ at $t = t_k$, respectively. Note that I_k in (3.1) is a sequence of instantaneous impulse operators and has been developed in physics, population dynamics, biotechnology and so forth.

doi:10.1088/2053-2563/aada21ch3

It is notable that Hernández and O'Regan [1] introduced a new class of impulsive evolution equations (with non-instantaneous impulses) of the form

$$\begin{cases} x'(t) = Ax(t) + f(t, x(t)), \ t \in (s_i, t_{i+1}], \ i = 0, 1, 2, \ldots, m, \\ x(t) = g_i(t, x(t)), \ t \in (t_i, s_i], \ i = 1, 2, \ldots, m, \\ x(0) = x_0 \in X, \end{cases} \tag{3.2}$$

where A and f are the same as in (3.1) and the fixed points s_i and t_i satisfy $0 = t_0 = s_0 < t_1 < s_1 < t_2 < \cdots < s_{m-1} < t_m < s_m < t_{m+1} = T$, and $g_i : [t_i, s_i] \times X \to X$ is continuous for all $i = 1, 2, \ldots, m$. Here g_i is regarded as a continuous action process.

The concepts of mild solutions and classical solutions were introduced by Hernández and O'Regan [1, definitions 2.1 and 2.2]. The existence and uniqueness results of (3.2) are presented by using the theory of strongly continuous semigroups and compact semigroups via fixed point theorems (see [1, theorems 2.1 and 2.2] and [2, theorems 2.1 and 2.2]). Next, Pierri *et al* [2] continued the work and development in [1] in which they study the existence and uniqueness of mild solutions to semilinear impulsive differential equations with non-instantaneous impulses in the fractional power space using the theory of the analytic semigroup.

Remark 3.1.1. *We revisit the condition in (3.2):*

$$x(t) = g_i(t, x(t)), \ t \in (t_i, s_i], \ i = 1, 2, \ldots, m, \tag{3.3}$$

where $g_i \in C([t_i, s_i] \times X, X)$ *and there are positive constants* L_{g_i}, $i = 1, 2, \ldots, m$ *such that*

$$\|g_i(t, u_1) - g_i(t, u_2)\| \leqslant L_{g_i}\|u_1 - u_2\| \text{ for each } t \in [t_i, s_i] \text{ and all } u_1, u_2 \in X.$$

It follows from [1, 2, theorems 2.1 and 2.2], that a necessary condition $\max\{L_{g_i} : i = 1, \ldots, m\} < 1$ *is considered. Then the Banach fixed point theorem gives a unique* $z_i \in C([t_i, s_i], X)$ *so that* $z = g_i(t, z)$ *if and only if* $z = z_i(t)$*. So (3.3) is equivalent to*

$$x(t) = z_i(t), \ t \in (t_i, s_i], \ i = 1, 2, \ldots, m, \tag{3.4}$$

which does not depend on the state x. Thus it is necessary to modify (3.3) and we recommend you consider the conditions

$$x(t) = g_i(t, x(t_i^-)), \ t \in (t_i, s_i], \ i = 1, 2, \ldots, m. \tag{3.5}$$

Compared to (3.3), the new conditions (3.5) are a better and more reasonable generalization of sudden impulses to non-instantaneous ones.

Based on remark 3.1.1, we are aiming to study a new class of impulsive evolution equations of the form

$$\begin{cases} x'(t) = Ax(t) + f(t, x(t)), & t \in (s_i, t_{i+1}), \ i = 0, 1, 2, \dots, m, \\ x(t_i^+) = g_i(t_i, x(t_i^-)), & i = 1, 2, \dots, m, \\ x(t) = g_i(t, x(t_i^-)), & t \in (t_i, s_i], \ i = 1, 2, \dots, m. \end{cases} \tag{3.6}$$

Note that we consider in (3.6) that $x \in C((t_k, t_{k+1}], X)$, $k = 0, 1, \dots, m$, and there exist $x(t_k^-)$, $x(t_k^+)$, $k = 1, \dots, m$ with $x(t_k^-) = x(t_k)$. This model is more suitable to show the dynamics of evolution processes in pharmacotherapy: the first equation denotes the health status of a patient; the second equation denotes the doctor taking some practical actions to test medicine for the patient; the third equation denotes that the tested medicine is valid for this patient and then begins to deal with the effects on the patient for some time.

The rest of this section is organized as follows. In section 3.1.2 we give an existence and uniqueness result of (3.6) with $x(0) = x_0 \in X$. In section 3.1.3 we introduce four new types of Ulam's type stability for differential equations with non-instantaneous impulses (see definitions 3.1.4–3.1.7). The Ulam problem has attracted many researchers, we refer the reader to the monographs of Cădariu [3], Hyers [4, 5], Jung [6], Rassias [7] and other mathematicians. We mainly present the generalized Ulam–Hyers–Rassias stability results for equation (3.6) on a compact interval. In section 3.1.4 we extend our study to

$$\begin{cases} x'(t) = Ax(t) + f(t, x(t)), & t \in (s_i, t_{i+1}), \ i \in \mathbb{M}, \\ x(t_i^+) = g_i(t_i, x(t_i^-)), & i \in \mathbb{M}, \\ x(t) = g_i(t, x(t_i^-)), & t \in (t_i, s_i], \ i \in \mathbb{M}, \end{cases} \tag{3.7}$$

where $t \in \mathbb{R}^+ := [0, \infty)$, the fixed points s_i and t_i satisfy $t_i < s_i < t_{i+1}$, and either $\mathbb{M} = \{1, \dots, m\}$ or $\mathbb{M} = \mathbb{N}$, with $\lim_{i\to\infty} t_i = \infty$. We also set $t_{m+1} = \infty$ for $\mathbb{M} = \{1, \dots, m\}$. Some extensions of Ulam–Hyers–Rassias stability for the case with infinite impulses are given. The existence and uniqueness result is also presented for this case.

3.1.2 An existence and uniqueness result

We recall the following concepts of mild solutions of semilinear evolution equations with non-instantaneous impulses.

Definition 3.1.2. *(See [1, definition 2.1].) A function $x \in PC(J, X)$ is called a mild solution of the problem*

$$\begin{cases} x'(t) = Ax(t) + f(t, x(t)), & t \in (s_i, t_{i+1}), \ i = 0, 1, 2, \dots, m, \\ x(t_i^+) = g_i(t_i, x(t_i^-)), & i = 1, 2, \dots, m, \\ x(t) = g_i(t, x(t_i^-)), & t \in (t_i, s_i], \ i = 1, 2, \dots, m, \\ x(0) = x_0 \in X, \end{cases} \tag{3.8}$$

if $x(0) = x_0$ *and*

$$\begin{aligned}
x(t) &= g_i(t, x(t_i^-)), \quad t \in (t_i, s_i], \quad i = 1, 2, \dots, m; \\
x(t_i^+) &= g_i(t_i, x(t_i^-)), \quad i = 1, 2, \dots, m; \\
x(t) &= T(t)x_0 + \int_0^t T(t-s)f(s, x(s))ds, \quad t \in [0, t_1]; \\
x(t) &= T(t-s_i)g_i(s_i, x(t_i^-)) + \int_{s_i}^t T(t-s)f(s, x(s))ds, \\
&\quad t \in [s_i, t_{i+1}], \quad i = 1, 2, \dots, m.
\end{aligned}$$

Note that we consider $x \in C((t_i, t_{i+1}], X)$, $i = 0, 1, 2, \dots, m$.

Considering the existence and uniqueness of solutions to problem (3.8), Hernández and O'Regan [1] initially obtain a interesting result under strong conditions via the PC-norm. Here, we give another result under weak conditions via the PCB-norm.

We introduce the following conditions.

$[H_0]$ A: $D(A) \subseteq X \to X$ is the generator of a C_0-semigroup $\{T(t), t \geqslant 0\}$ on a Banach space X with a norm $\|\cdot\|$. Then $\|T(t)\| \leqslant Me^{\omega t}$ for some $M \geqslant 1$ and $\omega \in \mathbb{R}$ [8].

$[H_1]$ $f \in C(J \times X, X)$.

$[H_2]$ There exists a positive constant L_f such that

$$\|f(t, u_1) - f(t, u_2)\| \leqslant L_f\|u_1 - u_2\| \text{ for each } t \in J \text{ and all } u_1, u_2 \in X.$$

$[H_3]$ $g_i \in C([t_i, s_i] \times X, X)$ and there are positive constants L_{g_i}, $i = 1, 2, \dots, m$ such that

$$\|g_i(t, u_1) - g_i(t, u_2)\| \leqslant L_{g_i}\|u_1 - u_2\| \text{ for each } t \in [t_i, s_i] \text{ and all } u_1, u_2 \in X.$$

Theorem 3.1.3. *Assume that $[H_0]$–$[H_3]$ are satisfied. Then equation (3.8) has a unique mild solution $x \in PC(J, X)$.*

Proof. Consider a mapping F: $PC(J, X) \to PC(J, X)$ defined by

$$\begin{aligned}
(Fx)(0) &= x_0; \\
(Fx)(t) &= g_i\Big(t, T(t_i - s_{i-1})g_{i-1}(s_{i-1}, x(t_{i-1}^-)) \\
&\quad + \int_{s_{i-1}}^{t_i} T(t_i - s)f(s, x(s))ds\Big), \quad t \in (t_i, s_i], \quad i = 1, 2, \dots, m; \\
(Fx)(t) &= T(t)x_0 + \int_0^t T(t-s)f(s, x(s))ds, \quad t \in [0, t_1]; \\
(Fx)(t) &= T(t-s_i)g_i(s_i, x(t_i^-)) + \int_{s_i}^t T(t-s)f(s, x(s))ds, \\
&\quad t \in (s_i, t_{i+1}], \quad i = 1, 2, \dots, m,
\end{aligned}$$

where we set $g_0(t, x) := x_0$ and so $L_{g_0} = 0$. Obviously, F is well defined due to $[H_1]$.

Supposing $\Omega > \omega$, for any $x, y \in PC(J, X)$ and $t \in (s_i, t_{i+1}]$, $i = 1, 2, \dots, m$, we have

$$\begin{aligned}
&\|(Fx)(t) - (Fy)(t)\| \\
&\leqslant Me^{\omega(t-s_i)}L_{g_i}\|x(t_i^-) - y(t_i^-)\| + ML_f\int_{s_i}^{t} e^{\omega(t-s)}\|x(s) - y(s)\|ds \\
&\leqslant Me^{\omega(t-s_i)+\Omega t_i}L_{g_i}\|x - y\|_{PCB} + ML_f\int_{s_i}^{t} e^{\omega(t-s)+\Omega s}ds\|x - y\|_{PCB} \\
&\leqslant M\left(e^{\omega(t-s_i)+\Omega t_i}L_{g_i} + \frac{e^{\Omega t}L_f}{\Omega - \omega}\right)\|x - y\|_{PCB},
\end{aligned}$$

which implies that

$$\begin{aligned}
e^{-\Omega t}\|(Fx)(t) - (Fy)(t)\| &\leqslant M\left(e^{\omega(t-s_i)+\Omega(t_i-t)}L_{g_i} + \frac{L_f}{\Omega - \omega}\right)\|x - y\|_{PCB} \\
&\leqslant M\left(e^{\Omega(t_i-s_i)}L_{g_i} + \frac{L_f}{\Omega - \omega}\right)\|x - y\|_{PCB}, \\
&\quad t \in (s_i, t_{i+1}].
\end{aligned}$$

Proceeding as above for $t \in [0, t_1]$, we obtain that

$$\begin{aligned}
\|(Fx)(t) - (Fy)(t)\| &\leqslant ML_f\int_0^t e^{\omega(t-s)}\|x(s) - y(s)\|ds \\
&\leqslant ML_f\int_0^t e^{\omega(t-s)+\Omega s}ds\|x - y\|_{PCB} \\
&\leqslant \frac{ML_f e^{\Omega t}}{\Omega - \omega}\|x - y\|_{PCB},
\end{aligned}$$

which implies that

$$e^{-\Omega t}\|(Fx)(t) - (Fy)(t)\| \leqslant \frac{ML_f}{\Omega - \omega}\|x - y\|_{PCB}, \quad t \in [0, t_1].$$

Using the above estimates, similarly for $t \in (t_i, s_i]$, $i = 1, 2, \dots, m$, we derive

$$\begin{aligned}
\|(Fx)(t) - (Fy)(t)\| &\leqslant ML_{g_i}\Big(e^{\omega(t_i-s_{i-1})}L_{g_{i-1}}\|x(t_{i-1}^-) - y(t_{i-1}^-)\| \\
&\quad + L_f\int_{s_{i-1}}^{t_i} e^{\omega(t_i-s)}\|x(s) - y(s)\|ds\Big) \\
&\leqslant ML_{g_i}\left(e^{\omega(t_i-s_{i-1})+\Omega t_{i-1}}L_{g_{i-1}} + \frac{e^{\Omega t_i}L_f}{\Omega - \omega}\right)\|x - y\|_{PCB},
\end{aligned}$$

which implies that

$$\begin{aligned}
&e^{-\Omega t}\|(Fx)(t) - (Fy)(t)\| \\
&\leqslant ML_{g_i}\left(e^{\omega(t_i - s_{i-1}) + \Omega(t_{i-1} - t_i)}L_{g_{i-1}} + \frac{L_f}{\Omega - \omega}\right)\|x - y\|_{PCB} \\
&\leqslant ML_{g_i}\left(e^{\Omega(t_{i-1} - s_{i-1})}L_{g_{i-1}} + \frac{L_f}{\Omega - \omega}\right)\|x - y\|_{PCB}, \\
&t \in (t_i, s_i],\ i = 1, 2, \ldots, m.
\end{aligned}$$

Summarizing the above estimates, we have that

$$\|Fx - Fy\|_{PCB} \leqslant L_F\|x - y\|_{PCB},$$

for

$$L_F := M \times \max_{1\leqslant i\leqslant m}\left\{\frac{L_f}{\Omega - \omega},\ L_{g_i}\left(e^{\Omega(t_{i-1} - s_{i-1})}L_{g_{i-1}} + \frac{L_f}{\Omega - \omega}\right),\ e^{\Omega(t_i - s_i)}L_{g_i} + \frac{L_f}{\Omega - \omega}\right\}.$$

Obviously, one can choose a sufficiently large $\Omega > \omega$ such that $L_F < 1$, and so F is a contraction mapping. Then, one can derive the results directly. The proof is complete. □

3.1.3 Concepts and results of Ulam's type stability on a compact interval

In this section, we introduce Ulam's type stability concepts for equation (3.6).

Let $\epsilon > 0$, $\psi \geqslant 0$ and $\varphi \in PC(J, \mathbb{R}^+)$ be nondecreasing.

We consider the following inequalities

$$\begin{cases}
\|y'(t) - Ay(t) - f(t, y(t))\| \leqslant \epsilon,\ t \in (s_i, t_{i+1}),\ i = 0, 1, 2, \ldots, m, \\
\|y(t_i^+) - g_i(t_i, y(t_i^-))\| \leqslant \epsilon,\ i = 1, 2, \ldots, m, \\
\|y(t) - g_i(t, y(t_i^-))\| \leqslant \epsilon,\ t \in (t_i, s_i],\ i = 1, 2, \ldots, m,
\end{cases} \tag{3.9}$$

and

$$\begin{cases}
\|y'(t) - Ay(t) - f(t, y(t))\| \leqslant \varphi(t),\ t \in (s_i, t_{i+1}), \\
\qquad i = 0, 1, 2, \ldots, m, \\
\|y(t_i^+) - g_i(t_i, y(t_i^-))\| \leqslant \psi,\ i = 1, 2, \ldots, m, \\
\|y(t) - g_i(t, y(t_i^-))\| \leqslant \psi,\ t \in (t_i, s_i],\ i = 1, 2, \ldots, m,
\end{cases} \tag{3.10}$$

and

$$\begin{cases}
\|y'(t) - Ay(t) - f(t, y(t))\| \leqslant \epsilon\varphi(t),\ t \in (s_i, t_{i+1}), \\
\qquad i = 0, 1, 2, \ldots, m, \\
\|y(t_i^+) - g_i(t_i, y(t_i^-))\| \leqslant \epsilon\psi,\ i = 1, 2, \ldots, m, \\
\|y(t) - g_i(t, y(t_i^-))\| \leqslant \epsilon\psi,\ t \in (t_i, s_i],\ i = 1, 2, \ldots, m.
\end{cases} \tag{3.11}$$

Now we set the vector space

$$Z := PC(J, X)\bigcap_{i=0}^{m} C^1((s_i, t_{i+1}), X)\bigcap_{i=0}^{m} C((s_i, t_{i+1}), D(A)).$$

The following concepts are inspired by Wang *et al* [9].

Definition 3.1.4. *Equation (3.6) is Ulam–Hyers stable if there exists a real number* $c_{f, m, g} > 0$ *such that for each* $\epsilon > 0$ *and for each solution* $y \in Z$ *of inequality (3.9), there exists a mild solution* $x \in PC(J, X)$ *of equation (3.6) with*

$$\|y(t) - x(t)\| \leqslant c_{f, m, g}\epsilon, \quad t \in J. \tag{3.12}$$

Definition 3.1.5. *Equation (3.6) is generalized Ulam–Hyers stable if there exists* $\theta_{f, m, g} \in C(\mathbb{R}^+, \mathbb{R}^+)$, $\theta_{f, m, g}(0) = 0$ *such that for each solution* $y \in Z$ *of inequality (3.9), there exists a mild solution* $x \in PC(J, X)$ *of equation (3.6) with*

$$\|y(t) - x(t)\| \leqslant \theta_{f, m, g}(\epsilon), \quad t \in J. \tag{3.13}$$

Definition 3.1.6. *Equation (3.6) is Ulam–Hyers–Rassias stable with respect to* (φ, ψ) *if there exists* $c_{f, m, g, \varphi} > 0$ *such that for each* $\epsilon > 0$ *and for each solution* $y \in Z$ *of inequality (3.11), there exists a mild solution* $x \in PC(J, X)$ *of equation (3.6) with*

$$\|y(t) - x(t)\| \leqslant c_{f, m, g, \varphi}\epsilon(\varphi(t) + \psi), \quad t \in J. \tag{3.14}$$

Definition 3.1.7. *Equation (3.6) is generalized Ulam–Hyers–Rassias stable with respect to* (φ, ψ) *if there exists* $c_{f, m, g, \varphi} > 0$ *such that for each solution* $y \in Z$ *of inequality (3.10) there exists a mild solution* $x \in PC(J, X)$ *of equation (3.6) with*

$$\|y(t) - x(t)\| \leqslant c_{f, m, g, \varphi}(\varphi(t) + \psi), \quad t \in J. \tag{3.15}$$

Remark 3.1.8. *It is clear that: (i) definition 3.1.4* $\Longrightarrow$ *definition 3.1.5; (ii) definition 3.1.6* $\Longrightarrow$ *definition 3.1.7; (iii) definition 3.1.6 for* $\varphi(t) = \psi = 1 \Longrightarrow$ *definition 3.1.4.*

Remark 3.1.9. *A function* $y \in Z$ *is a solution of inequality (3.11) if and only if there is* $G \in \bigcap_{i=0}^{m} C^1((s_i, t_{i+1}), X)\bigcap_{i=0}^{m} C((s_i, t_{i+1}), D(A))$ *and* $g \in \bigcap_{i=1}^{m} C([t_i, s_i], X)$ *(which depends on y) such that*

(i) $\|G(t)\| \leqslant \epsilon\varphi(t)$, $t \in \bigcup_{i=0}^{m}(s_i, t_{i+1})$ *and* $\|g(t)\| \leqslant \epsilon\psi$, $t \in \bigcup_{i=0}^{m}[t_i, s_i]$;

(ii) $y'(t) = Ay(t) + f(t, y(t)) + G(t)$, $t \in (s_i, t_{i+1})$, $i = 0, 1, 2, \dots, m$;

(iii) $y(t) = g_i(t, y(t_i^-)) + g(t),\ t \in (t_i, s_i],\ i = 1, 2, \dots, m;$
(iv) $y(t_i^+) = g_i(t_i, y(t_i^-)) + g(t_i),\ i = 1, 2, \dots, m.$

One can have similar remarks for inequalities (3.9) and (3.10).

Remark 3.1.10. *If $y \in Z$ is a solution of inequality (3.11) then y is a solution of the following integral inequality*

$$\begin{cases} \|y(t) - g_i(t, y(t_i^-))\| \leqslant \epsilon\psi,\ t \in (t_i, s_i],\ i = 1, 2, \dots, m, \\ \left\| y(t) - T(t)y(0) - \int_0^t T(t-s)(s, y(s))ds \right\| \leqslant \epsilon M \int_0^t e^{\omega(t-s)}\varphi(s)ds, \\ \qquad t \in [0, t_1], \\ \|y(t_i^+) - g_i(t_i, y(t_i^-))\| \leqslant \epsilon\psi,\ i = 1, 2, \dots, m, \\ \left\| y(t) - T(t-s_i)g_i(s_i, y(t_i^-)) - \int_{s_i}^t T(t-s)f(s, y(s))ds \right\| \\ \leqslant \epsilon M e^{\omega(t-s_i)}\psi + \epsilon M \int_{s_i}^t e^{\omega(t-s)}\varphi(s)ds,\ t \in [s_i, t_{i+1}],\ i = 1, 2, \dots, m. \end{cases} \tag{3.16}$$

In fact, from remark 3.1.9 we obtain

$$\begin{cases} y'(t) = Ay(t) + f(t, y(t)) + G(t),\ t \in (s_i, t_{i+1}),\ i = 1, 2, \dots, m, \\ y(t_i^+) = g_i(t_i, y(t_i^-)) + g(t_i),\ i = 1, 2, \dots, m, \\ y(t) = g_i(t, y(t_i^-)) + g(t),\ t \in (t_i, s_i],\ i = 1, 2, \dots, m. \end{cases} \tag{3.17}$$

Clearly [8, p 105], the solution $y \in Z$ of equation (3.17) is given by

$$y(t) = g_i(t, y(t_i^-)) + g(t),\ t \in (t_i, s_i],\ i = 1, 2, \dots, m;$$

$$y(t) = T(t)y(0) + \int_0^t T(t-s)(f(s, y(s)) + G(s))ds,\ t \in [0, t_1];$$

$$y(t) = T(t-s_i)(g_i(s_i, y(t_i^-)) + g(t_i)) + \int_{s_i}^t T(t-s)(f(s, y(s)) + G(s))ds,$$
$$t \in [s_i, t_{i+1}],\ i = 1, 2, \dots, m.$$

For $t \in [s_i, t_{i+1}]$, $i = 0, 1, 2, \dots, m$, we obtain

$$\begin{aligned} &\left\| y(t) - T(t-s_i)g_i(s_i, y(t_i^-)) - \int_{s_i}^t T(t-s)f(s, y(s))ds \right\| \\ &\leqslant M e^{\omega(t-s_i)}\|g(t_i)\| + \int_{s_i}^t e^{\omega(t-s)}\|G(s)\|ds \\ &\leqslant \epsilon M e^{\omega(t-s_i)}\psi + \epsilon M \int_{s_i}^t e^{\omega(t-s)}\varphi(s)ds. \end{aligned}$$

Proceeding as above, we derive that

$$\|y(t) - g_i(t, y(t_i^-))\| \leqslant \|g(t)\| \leqslant \epsilon\psi, \ t \in (t_j, s_j], \ j = 1, 2, \ldots, m;$$

$$\begin{aligned} &\left\| y(t) - T(t)y(0) - \int_0^t T(t-s)f(s, y(s))ds \right\| \\ &\leqslant M \int_0^t e^{\omega(t-s)}\|G(s)\|ds \\ &\leqslant \epsilon M \int_0^t e^{\omega(t-s)}\varphi(s)ds, \ t \in [0, t_1]. \end{aligned}$$

Similarly, one can give similar remarks for the solutions of inequalities (3.10) and (3.9).

To discuss stability, we need the following additional assumption:

$[H_4]$ Let $\varphi \in C(J, \mathbb{R}^+)$ be a nondecreasing function. There exists $c_\varphi > 0$ such that

$$\int_0^t \varphi(s)ds \leqslant c_\varphi\varphi(t), \ \text{for each } t \in J.$$

We need an impulsive Grönwall inequality which was given by Bainov and Simeonov (see [10, theorem 16.4]).

Lemma 3.1.11. *Let* $\mathbb{M}_0 := \mathbb{M} \cup \{0\}$ *and the following inequality holds*

$$u(t) \leqslant a(t) + \int_0^t b(s)u(s)ds + \sum_{0<t_k<t} \beta_k u(t_k^-), \ t \geqslant 0, \tag{3.18}$$

where $u, a, b \in PC(\mathbb{R}^+, \mathbb{R}^+)$, *a is nondecreasing and* $b(t) > 0$, $\beta_k > 0$, $k \in \mathbb{M}$.

Then, for $t \in \mathbb{R}^+$, *the following inequality is valid:*

$$u(t) \leqslant a(t)(1+\beta)^k \exp\left(\int_0^t b(s)ds\right), \ t \in (t_k, t_{k+1}], \ k \in \mathbb{M}_0, \tag{3.19}$$

where $\beta = \sup_{k\in\mathbb{M}}\{\beta_k\}$.

Now we are ready to state our main results in this section.

Theorem 3.1.12. *Assume that* $[H_1]$*–*$[H_4]$ *are satisfied. Then equation (3.6) is Ulam–Hyers–Rassias stable with respect to* (φ, ψ).

Proof. Let $y \in PC(J, D(A)) \cap C^1((s_i, t_{i+1}], X)$ be a solution of inequality (3.11). Denote by x the unique mild solution of the impulsive Cauchy problem

$$\begin{cases} x'(t) = Ax(t) + f(t, x(t)), \ t \in (s_i, t_{i+1}), \ i = 0, 1, 2, \dots, m, \\ x(t_i^+) = g_i(t_i, x(t_i^-)), \ i = 1, 2, \dots, m, \\ x(t) = g_i(t, x(t_i^-)), \ t \in (t_i, s_i], \ i = 1, 2, \dots, m, \\ x(0) = y(0). \end{cases} \tag{3.20}$$

Then we obtain

$$x(t) = \begin{cases} g_i(t, x(t_j^-)), \ t \in (t_j, s_j], \ j = 1, 2, \dots, m, \\ T(t)y(0) + \int_0^t T(t-s)f(s, x(s))ds, \ t \in [0, t_1], \\ T(t - s_i)g_i(s_i, x(t_i^-)) + \int_{s_i}^t T(t-s)f(s, x(s))ds, \\ \ t \in (s_i, t_{i+1}], \ i = 1, 2, \dots, m. \end{cases}$$

Keeping in mind (3.16), for each $t \in (s_i, t_{i+1}]$, $i = 1, 2, \dots, m$, we have

$$\begin{aligned} &\left\| y(t) - T(t - s_i)g_i(s_i, y(t_i^-)) - \int_{s_i}^t T(t-s)f(s, y(s))ds \right\| \\ &\leqslant \epsilon M e^{|\omega|T}\left(\psi + \int_0^t \varphi(s)ds\right) \\ &\leqslant \epsilon M e^{|\omega|T}\left(\psi + c_\varphi \varphi(t)\right), \end{aligned}$$

and for $t \in (t_j, s_j]$, $j = 1, 2, \dots, m$, we have

$$\|y(t) - g_i(t, y(t_i))\| \leqslant \epsilon\psi,$$

and for $t \in [0, t_1]$, we have

$$\left\| y(t) - T(t)y(0) - \int_0^t T(t-s)f(s, y(s))ds \right\| \leqslant \epsilon M e^{|\omega|T} c_\varphi \varphi(t).$$

Hence, for each $t \in (s_i, t_{i+1}]$, $i = 1, 2, \dots, m$, we obtain

$$\begin{aligned} \|y(t) - x(t)\| \leqslant & \left\| y(t) - T(t - s_i)g_i(s_i, y(t_i^-)) - \int_{s_i}^t T(t-s)f(s, y(s))ds \right\| \\ & + M e^{|\omega|T} \|g_i(s_i, y(t_i^-)) - g_i(s_i, x(t_i^-))\| \\ & + M e^{|\omega|T} \int_{s_i}^t \|f(s, y(s)) - f(s, x(s))\| ds \\ \leqslant & \ M e^{|\omega|T}\Bigg(\epsilon(1 + c_\varphi)[\psi + \varphi(t)] + L_{g_i}\|y(t_i^-) - x(t_i^-)\| \\ & + L_f \int_{s_i}^t \| y(s) - x(s)\| \, ds \Bigg) \\ \leqslant & \ M e^{|\omega|T}\Bigg(\epsilon(1 + c_\varphi)[\psi + \varphi(t)] + L_f \int_0^t \| y(s) - x(s)\| \, ds \\ & + \sum_{j=1}^{i} L_{g_j}\|y(t_i^-) - x(t_i^-)\| \Bigg). \end{aligned}$$

Further, for $t \in (t_j, s_j]$, $j = 1, 2, \ldots, m$, we have

$$\begin{aligned}
\|y(t) - x(t)\| &\leqslant \|y(t) - g_i(t, y(t_i^-))\| + \|g_i(t, y(t_i^-)) - g_i(t, x(t_i^-))\| \\
&\leqslant \epsilon\psi + \sum_{j=1}^{i} L_{g_j}\|y(t_j^-) - x(t_j^-)\| \\
&\leqslant Me^{|\omega|T}\left(\epsilon(1 + c_\varphi)[\psi + \varphi(t)] + L_f\int_0^t \| y(s) - x(s)\| \, ds \right. \\
&\quad \left. + \sum_{j=1}^{i} L_{g_j}\|y(t_i^-) - x(t_i^-)\| \right).
\end{aligned}$$

Next, for $t \in [0, t_1]$, we have

$$\begin{aligned}
\|y(t) - x(t)\| &\leqslant Me^{|\omega|T}\left(\epsilon c_\varphi \varphi(t) + L_f\int_0^t \| y(s) - x(s)\| \, ds\right) \\
&\leqslant Me^{|\omega|T}\left(\epsilon(1 + c_\varphi)[\psi + \varphi(t)] + L_f\int_0^t \| y(s) - x(s)\| \, ds\right).
\end{aligned}$$

Consequently, for $t \in (t_i, t_{i+1}]$, using lemma 3.1.11, we derive that

$$\begin{aligned}
\|y(t) - x(t)\| &\leqslant Me^{|\omega|T}(1 + c_\varphi)(1 + Me^{|\omega|T}L_g)^i e^{Me^{|\omega|T}L_f t}\epsilon(\psi + \varphi(t)) \\
&\leqslant Me^{|\omega|T}(1 + c_\varphi)(1 + Me^{|\omega|T}L_g)^m e^{Me^{|\omega|T}L_f T}\epsilon(\psi + \varphi(t)) \\
&:= c_{f,\, m,\, g,\, \varphi}\epsilon(\psi + \varphi(t)),
\end{aligned}$$

for $L_g := \sup_{i \in \mathbb{M}} L_{g_i}$, which implies that equation (3.6) is Ulam–Hyers–Rassias stable with respect to (φ, ψ). The proof is complete. □

By simply repeating the procedure in theorem 3.1.12, we establish without proof the following results.

Remark 3.1.13. *Under the assumptions of theorem 3.1.12, we consider equation (3.6) and inequality (3.10). One can repeat the same process to verify that equation (3.6) is generalized Ulam–Hyers–Rassias stable with respect to* (φ, ψ).

Remark 3.1.14. *Under the assumptions of theorem 3.1.12, we consider equation (3.6) and inequality (3.9). One can repeat the same process to verify that equation (3.6) is Ulam–Hyers stable.*

To end this section, we give an example to illustrate our abstract results above.

Example 3.1.15. *We consider one-dimensional diffusion processes with non-instantaneous changes of states. This example can explain either evolution of the temperature of a rod or the chemical concentration of a substance.*

Consider the following impulsive partial differential equation

$$\begin{cases} \dfrac{\partial}{\partial t}x(t, y) = \dfrac{\partial^2}{\partial y^2}x(t, y), \; y \in (0, 1), \; t \in [0, 1) \bigcup (2, 3], \\ \dfrac{\partial}{\partial y}x(t, 0) = \dfrac{\partial}{\partial y}x(t, 1) = 0, \; t \in [0, 1) \bigcup (2, 3], \\ x(t, y) = \lambda x(1^-, y), \; \lambda \in (-1, 1), \; t \in (1, 2], \; y \in (0, 1). \end{cases} \tag{3.21}$$

Hence $J = [0, 3]$, $0 = t_0 = s_0$, $t_1 = 1$, $s_1 = 2$ *and* $T = 3$. *Let* $X = L^2(0, 1)$. *Define* $Ax = \frac{\partial^2}{\partial y^2}x$ *for* $x \in D(A)$ *with* $D(A) = \{x \in X: \frac{\partial x}{\partial y}, \frac{\partial^2 x}{\partial y^2} \in X, \; x(0) = x(1) = 0\}$. *Then,* A *is the infinitesimal generator of a* C_0*-semigroup* $\{T(t), t \geqslant 0\}$ *in* X. *Moreover,* $\|T(t)\| \leqslant 1$ *for all* $t \geqslant 0$. *So* $M = 1$ *and* $\omega = 0$.

Denote $x(t)(y) = x(t, y)$, $f(t, x)(y) = 0$ *and* $g_1(t, x)(y) = \lambda x(y)$, *then problem (3.21) can be abstracted into*

$$\begin{cases} x'(t) = Ax(t), \; t \in (0, 1) \bigcup (2, 3), \\ x(t) = \lambda x(1^-), \; t \in (1, 2]. \end{cases} \tag{3.22}$$

Clearly, $[H_0]$–$[H_3]$ *are satisfied. Then equation (3.22) with a given initial value has a unique solution* $x \in PC(J, X)$. *Moreover, we put* $\varphi(t) = e^t$ *and* $\psi = 1$. *Then* $[H_4]$ *holds if* $c_\varphi = 1$. *Thus, from theorem 3.1.12, equation (3.21) is Ulam–Hyers–Rassias stable with respect to* $(e^t, 1)$ *on* $[0, 3]$ *and* $c_{f, \psi, g, \varphi} = 2(1 + |\lambda|)$.

3.1.4 Extension stability results on the unbounded interval

In this section we consider the case $J = \mathbb{R}^+$. Then one can replace $i = 1, \ldots, m$ by $i \in \mathbb{N}$ and $i = 0, 1, \ldots, m$ by $i \in \{0\} \bigcup \mathbb{N}$ in inequalities (3.9)–(3.11) and (3.12)–(3.15), respectively. Then, one can rewrite four parallel stability definitions like definitions 3.1.4–3.1.7 when we take

$$Z := PC(J, X) \bigcap_{i \in \mathbb{M}_0} C^1((s_i, t_{i+1}), X) \bigcap_{i \in \mathbb{M}_0} C((s_i, t_{i+1}), D(A)).$$

We state the following assumptions:

$[H_0']$ C_0-semigroup $\{T(t), t \geqslant 0\}$ is exponentially stable, that is, $\omega < 0$ in $[H_0]$.

$[H_1']$ $f \in C([0, \infty) \times X, X)$.

$[H_2']$ There exists positive constant $L_f \in C(\mathbb{R}^+, (0, \infty))$ such that

$$\|f(t, u_1) - f(t, u_2)\| \leqslant L_f\|u_1 - u_2\| \text{ for each } t \in [0, \infty) \text{ and all } u_1, u_2 \in X.$$

$[H_3']$ $g_i \in C([t_i, s_i] \times X, X)$ and there are positive constants L_{g_i}, $i \in \mathbb{N}$ such that

$$\|g_i(t, u_1) - g_i(t, u_2)\| \leqslant L_{g_i}\|u_1 - u_2\| \text{ for each } t \in [t_i, s_i] \text{ and all } u_1, u_2 \in X.$$

$[H_4']$ There exists $c_\varphi \geqslant 1$ such that

$$\int_{s_i}^{t} e^{\omega(s_i-s)}\varphi(s)ds \leqslant c_\varphi\varphi(t),$$

for any $t \in [s_i, t_{i+1}]$ and $i \in \mathbb{M}_0$.

First we have the following extension of theorem 3.1.3.

Theorem 3.1.16. *Assume that $[H_0]$, $[H'_1]$–$[H'_3]$ are satisfied, then equation (3.7) has a unique mild solution $x \in PC([0, \infty), X)$ with $x(0) = x_0 \in X$.*

Proof. Take $T \in \mathbb{R}_+' := \mathbb{R}^+/\mathbb{M}$. Then from theorem 3.1.3, equation (3.7) has a unique mild solution $x \in PC([0, T]), X)$ with $x(0) = x_0 \in X$. Taking $T \to \infty$, the proof is complete. □

Now we move on to the stability problem.

Theorem 3.1.17. *Assume that $[H'_1]$–$[H'_4]$ are satisfied. Then equation (3.7) is Ulam–Hyers–Rassias stable with respect to (φ, ψ) provided that*

$$\begin{aligned} L_g &:= \sup_{i\in\mathbb{M}} L_{g_i} < \infty; \\ \alpha &:= \inf_{i\in\mathbb{M}}(t_{i+1} - s_i) > 0; \\ &\omega + ML_f < 0; \\ \beta &:= ML_g e^{(\omega+ML_f)\alpha} < 1. \end{aligned}$$

Proof. Let $y \in Z$ be a solution of inequality (3.11). Denote by x the unique mild solution of

$$\begin{cases} x'(t) = Ax(t) + f(t, x(t)), \ t \in (s_i, t_{i+1}), \ i \in \mathbb{M}_0, \\ x(t_i^+) = g_i(t_i, x(t_i^-)), \ i \in \mathbb{M}, \\ x(t) = g_i(t, x(t_i^-)), \ t \in (t_i, s_i], \ i \in \mathbb{M}, \\ x(0) = y(0). \end{cases} \tag{3.23}$$

Then we obtain

$$x(t) = \begin{cases} g_i(t, x(t_i^-)), \ t \in (t_i, s_i], \ i \in \mathbb{M}, \\ T(t)x_0 + \int_0^t T(t-s)f(s, x(s))ds, \ t \in [0, t_1], \\ T(t-s_i)g_i(s_i, x(t_i^-)) + \int_{s_i}^t T(t-s)f(s, x(s))ds, \ t \in (s_i, t_{i+1}], \ i \in \mathbb{M}. \end{cases}$$

Here we note that we mean $[a, \infty] = [a, \infty)$ to avoid confusion. From (3.16), for each $t \in (s_i, t_{i+1}]$, $i \in \mathbb{M}$, we have

$$\left\| y(t) - T(t-s_i)g_i(s_i, y(t_i^-)) - \int_{s_i}^{t} T(t-s)f(s, y(s))ds \right\|$$
$$\leqslant \epsilon M e^{\omega(t-s_i)}\psi + \epsilon M \int_{s_i}^{t} e^{\omega(t-s)}\varphi(s)ds$$
$$\leqslant \epsilon M e^{\omega(t-s_i)} c_\varphi(\psi + \varphi(t)),$$

and for $t \in (t_i, s_i]$, $i \in \mathbb{M}$, we have

$$\|y(t) - g_i(t, y(t_i^-))\| \leqslant \epsilon\psi,$$

and for $t \in [0, t_1]$, we have

$$\left\| y(t) - T(t)y(0) - \int_0^t T(t-s)f(s, y(s))ds \right\| \leqslant \epsilon M \int_0^t e^{\omega(t-s)}\varphi(s)ds$$
$$\leqslant \epsilon M c_\varphi e^{\omega t}\varphi(t).$$

Hence, for each $t \in (s_i, t_{i+1}]$, $i \in \mathbb{M}$, we obtain

$$\|y(t) - x(t)\| \leqslant \left\| y(t) - T(t-s_i)g_i(s_i, y(t_i^-)) - \int_{s_i}^{t} T(t-s)f(s, y(s))ds \right\|$$
$$+ M e^{\omega(t-s_i)}\|g_i(s_i, y(t_i^-)) - g_i(s_i, x(t_i^-))\|$$
$$+ M \int_{s_i}^{t} e^{\omega(t-s)}\|f(s, y(s)) - f(s, x(s))\|ds$$
$$\leqslant \epsilon M c_\varphi e^{\omega(t-s_i)}(\psi + \varphi(t)) + M e^{\omega(t-s_i)} L_g\|y(t_i^-) - x(t_i^-)\|$$
$$+ M L_f \int_{s_i}^{t} e^{\omega(t-s)} \| y(s) - x(s)\| \, ds,$$

which implies for $\bar{y}(t) := e^{-\omega t}y(t)$ and $\bar{x}(t) := e^{-\omega t}x(t)$ that

$$\|\bar{y}(t) - \bar{x}(t)\| \leqslant \epsilon M c_\varphi e^{-\omega s_i}(\psi + \varphi(t)) + M e^{\omega(t_i - s_i)} L_g\|\bar{y}(t_i^-) - \bar{x}(t_i^-)\|$$
$$+ M L_f \int_{s_i}^{t} \| \bar{y}(s) - \bar{x}(s)\| \, ds,$$

for $t \in [s_i, t_{i+1}]$, $i \in \mathbb{M}$. From the Grönwall inequality (note $\varphi(t)$ is nondecreasing), we derive

$$\|\bar{y}(t) - \bar{x}(t)\|$$
$$\leqslant \left(\epsilon M c_\varphi e^{-\omega s_i}(\psi + \varphi(t)) + M e^{\omega(t_i - s_i)} L_g\|\bar{y}(t_i^-) - \bar{x}(t_i^-)\|\right) e^{ML_f(t-s_i)},$$

which gives back

$$\|y(t) - x(t)\| \leqslant M e^{(\omega + ML_f)(t-s_i)}\left(\epsilon c_\varphi(\psi + \varphi(t)) + L_g\|y(t_i^-) - x(t_i^-)\|\right), \qquad (3.24)$$

for $t \in [s_i, t_{i+1}]$, $i \in \mathbb{M}$, and in particular

$$\begin{aligned} &\|y(t_{i+1}^-) - x(t_{i+1}^-)\| \\ &\leqslant Me^{(\omega+ML_f)(t_{i+1}-s_i)}\left(\epsilon c_\varphi(\psi + \varphi(t_{i+1}^-)) + L_g\|y(t_i^-) - x(t_i^-)\|\right) \\ &\leqslant Mc_\varphi\epsilon(\psi + \varphi(t_{i+1}^-)) + \beta\ \|y(t_i^-) - x(t_i^-)\|, \end{aligned} \tag{3.25}$$

for $i \in \mathbb{M}$ when $t_{i+1} < \infty$.

Further, for $t \in (t_i, s_i]$, $i \in \mathbb{M}$, we have

$$\begin{aligned} \|y(t) - x(t)\| &\leqslant \|y(t) - g_i(t, y(t_i^-))\| + \|g_i(t, y(t_i^-)) - g_i(t, x(t_i^-))\| \\ &\leqslant \epsilon\psi + L_g\|y(t_i^-) - x(t_i^-)\|. \end{aligned} \tag{3.26}$$

Moreover, for $t \in [0, t_1]$, we have

$$\begin{aligned} \|y(t) - x(t)\| &\leqslant \epsilon M \int_0^t e^{\omega(t-s)}\varphi(s)ds + M\int_0^t e^{\omega(t-s)}L_f\ \|\ y(s) - x(s)\|\ ds \\ &\leqslant \epsilon Mc_\varphi e^{\omega t}\varphi(t) + ML_f\int_0^t e^{\omega(t-s)}\ \|\ y(s) - x(s)\|\ ds, \end{aligned}$$

which yields as above that

$$\|y(t) - x(t)\| \leqslant \epsilon Mc_\varphi e^{(\omega+ML_f)t}\varphi(t) \leqslant \epsilon Mc_\varphi\varphi(t), \tag{3.27}$$

for $t \in [0, t_1]$, and in particular

$$\|y(t_1^-) - x(t_1^-)\| \leqslant \epsilon Mc_\varphi\varphi(t_1^-). \tag{3.28}$$

Solving (3.25) and using (3.28) we derive

$$\begin{aligned} \|y(t_i^-) - x(t_i^-)\| &\leqslant M\epsilon c_\varphi\sum_{j=2}^{i}(\psi + \varphi(t_j^-))\beta^{i-j} + \beta^{i-1}\|y(t_1^-) - x(t_1^-)\| \\ &\leqslant \frac{M\epsilon c_\varphi(\psi + \varphi(t_i^-))}{1-\beta} + \epsilon Mc_\varphi\varphi(t_i^-), \end{aligned} \tag{3.29}$$

since $\varphi(\cdot)$ is nondecreasing.

Now let $t \geqslant 0$. Then either $t \in [0, t_1]$ and (3.27) gives

$$\|y(t) - x(t)\| \leqslant \epsilon Mc_\varphi\varphi(t), \tag{3.30}$$

or $t \in (t_i, t_{i+1}]$ for some $i \in \mathbb{M}_0$. Then either $t \in (t_i, s_i]$ and then (3.26) and (3.29) give

$$\begin{aligned} \|y(t) - x(t)\| &\leqslant \epsilon\psi + L_g\left(\frac{M\epsilon c_\varphi(\psi + \varphi(t_i^-))}{1-\beta} + \epsilon Mc_\varphi\varphi(t_i^-)\right) \\ &\leqslant \epsilon\left(\left(1 + \frac{ML_gc_\varphi}{1-\beta}\right)\psi + \left(\frac{ML_gc_\varphi}{1-\beta} + Mc_\varphi\right)\varphi(t)\right), \end{aligned} \tag{3.31}$$

or $t \in (s_i, t_{i+1}]$ and then (3.24) and (3.29) give

$$
\begin{aligned}
\|y(t) - x(t)\| &\leqslant Me^{(\omega+ML_f)(t-s_i)}\Big(\epsilon c_\varphi(\psi + \varphi(t)) \\
&\quad + L_g\Big(\frac{M\epsilon c_\varphi(\psi + \varphi(t_i^-))}{1-\beta} + \epsilon M c_\varphi \varphi(t_i^-)\Big)\Big) \\
&\leqslant M\epsilon c_\varphi\Big(\Big(1 + \frac{L_g}{1-\beta}\Big)\psi + \Big(2 + \frac{L_g}{1-\beta}\Big)\varphi(t)\Big).
\end{aligned}
\tag{3.32}
$$

Using (3.30), (3.31) and (3.32) we have

$$\|y(t) - x(t)\| \leqslant c_{f,\,g,\,\varphi}\epsilon(\psi + \varphi(t)),$$

for any $t \geqslant 0$, where

$$c_{f,\,g,\,\varphi} := Mc_\varphi\Big(2 + \frac{L_g}{1-\beta}\Big).$$

Summarizing, we see that equation (3.7) is Ulam–Hyers–Rassias stable with respect to (φ, ψ). The proof is complete. □

Remark 3.1.18. *If*

$$\gamma := \sup_{i\in\mathbb{N}_0}(t_{i+1} - s_i) < \infty,$$

then assumption [H'₄] holds for any $\varphi(t) = ce^{\omega' t}$, $c > 0$ *and* $\omega' > \omega$. *Indeed, we compute*

$$\int_{s_i}^{t} e^{\omega(s_i-s)}\varphi(s)ds = c\int_{s_i}^{t} e^{\omega(s_i-s)+\omega' s}ds \leqslant c\frac{e^{\omega(s_i-t)+\omega' t}}{\omega' - \omega} \leqslant \frac{e^{-\omega\gamma}}{\omega' - \omega}\varphi(t),$$

for any $t \in [s_i, t_{i+1}]$ *and* $i \in \mathbb{N}_0$, *so* $c_\varphi = \frac{e^{-\omega\gamma}}{\omega' - \omega}$. *In particular, the constant function* $\varphi(t) = \varphi$ *can be also used with* $\omega' = 0$.

Example 3.1.19. *Consider*

$$
\begin{cases}
\dfrac{\partial}{\partial t}x(t, y) = (\Delta_y - 3I)x(t, y), \ y \in \Omega, \ t \in (2i+1, 2(i+1)], \\
\qquad\qquad i \in \{0\} \cup \mathbb{N}, \\
\dfrac{\partial}{\partial y}x(t, y) = 0, \ y \in \partial\Omega, \ t \geqslant 0, \\
x(t, y) = \sin i \cdot x(2i^-, y), \ y \in \Omega, \ (2i, 2i+1], \ i \in \mathbb{N},
\end{cases}
\tag{3.33}
$$

where $\Omega \subset \mathbb{R}^2$ *is a bounded domain,* Δ_y *is the Laplace operator in* $\mathbb{R}^2$ *and* $\partial\Omega \in C^2$.

Note now that $s_i = 2i + 1$ *and* $t_i = 2i$, $i \in \{0\} \cup \mathbb{N}$. *Here we consider infinitely many impulses on an infinite time interval* $\mathbb{R}^+$.

Let $X = L_2(\Omega)$, $D(A) = H^2(\Omega) \cap H_0^1(\Omega)$. *Define* $Ax = (\Delta_y - 3I)x$, $x \in D(A)$. *From theorem 2.5 of [8], A is just the infinitesimal generator of a contraction* C_0-*semigroup in* $L_2(\Omega)$, *that is,* $\|T(t)\| \leqslant e^{-3t}$ *for* $t \geqslant 0$, *so* $m = 1$ *and* $\omega = -3 < 0$.

Denote $x(\cdot)(y) = x(\cdot, y)$, $f(\cdot, x)(y) = 0$ *and* $g_i(\cdot, x)(y) = (\sin i) \cdot x(y)$, *then problem (3.33) can be abstracted into*

$$\begin{cases} x'(t) = Ax(t), \quad t \in (2i + 1, 2(i + 1)), \quad i \in \{0\} \cup \mathbb{N}, \\ x(2i^+) = x(2i^-), \quad i \in \mathbb{N}, \\ x(t) = (\sin i)x(2i^-), \quad t \in (2i, 2i + 1], \quad i \in \mathbb{N}. \end{cases} \tag{3.34}$$

Clearly, $[H'_0]$–$[H'_3]$ *are satisfied, and* $L_g = 1$, $\alpha = \gamma = 1$, $L_f = 0$, $\omega + ML_f = -3$, $\omega = e^{-3} < 1$ *and from remark 3.1.18 we can take* $\varphi(t) = e^{\omega' t}$, $\omega' > -3$ *and* $\psi = 1$. *Then* $[H'_4]$ *holds for* $c_\varphi = \frac{e^3}{\omega' + 3}$. *Thus, applying theorem 3.1.17, equation (3.33) is Ulam–Hyers–Rassias stable with respect to* $(e^{\omega' t}, 1)$ *on* $\mathbb{R}^+$ *with* $\omega' > -3$ *and* $c_{f, \mathbb{N}, g, \varphi} = \frac{e^3}{\omega' + 3}(2 + \frac{1}{1 - e^{-3}})$.

3.1.5 Notes and remarks

In this section we modify (3.3) to (3.5), which presents more general non-instantaneous impulses and is more suitable for showing the dynamics of evolution processes in pharmacotherapy. This fact reduces many difficulties in applying analysis methods and techniques in Bielecki's normed Banach spaces and thus makes the study of existence and uniqueness theorems interesting. The other novelties of this section are new concepts of Ulam type stability and Ulam–Hyers–Rassias stability results on compact and unbounded intervals.

The results in section 3.1 are motivated by [11].

3.2 Second-order evolution equations

3.2.1 Introduction

In this section, we study the second-order nonlinear differential equation with non-instantaneous impulses and a deviated argument in a Banach space X

$$\begin{cases} x''(t) = Ax(t) + f(t, x(t), x[h(x(t), t)]), \quad t \in (s_i, t_{i+1}), \\ \qquad i = 0, 1, \ldots, m, \\ x(t) = J_i^1(t, x(t_i^-)), \quad t \in (t_i, s_i], \quad i = 1, 2, \ldots, m, \\ x'(t) = J_i^2(t, x(t_i^-)), \quad t \in (t_i, s_i], \quad i = 1, 2, \ldots, m, \\ x(0) = x_0, \quad x'(0) = y_0, \end{cases} \tag{3.35}$$

where $x(t)$ is a state function, $0 = s_0 = t_0 < t_1 < s_1 < t_2 < \cdots < t_m < s_m < t_{m+1} = T < \infty$. We consider in (3.35) that $x \in C((t_i, t_{i+1}], X)$, $i = 0, 1, \ldots, m$ and there exist $x(t_i^-)$ and $x(t_i^+)$, $i = 1, 2, \ldots, m$ with $x(t_i^-) = x(t_i)$. The functions $J_i^1(t, x(t_i^-))$ and $J_i^2(t, x(t_i^-))$ represent non-instantaneous impulses during the intervals $(t_i, s_i]$, $i = 1, 2, \ldots, m$, so impulses at t_i^- have some duration, namely on intervals $(t_i, s_i]$. A is the infinitesimal generator of a strongly continuous cosine family of bounded linear operators $(C(t))_{t\in\mathbb{R}}$ on X. J_i^1, J_i^2, h and f are suitable functions and they will be specified later.

Many partial differential equations that arise in several problems connected with the transverse motion of an extensible beam, the vibration of hinged bars, and many other physical phenomena can be formulated as second-order abstract differential equations in infinite-dimensional spaces. A useful tool for the study of second-order abstract differential equations is the theory of strongly continuous cosine families of operators. The existence and uniqueness of the solutions of second-order nonlinear systems and the controllability of these systems in Banach spaces have been investigated extensively by many authors [12–16].

In certain real world problems, delay depends not only on time but also on an unknown quantity. The differential equations with deviated arguments are generalizations of delay differential equations. Gal [17] considered a nonlinear abstract differential equation with deviated arguments and studied the existence and uniqueness of solutions. Recently, Muslim *et al* [18] studied the exact and trajectory controllability of second-order impulsive nonlinear systems with a deviated argument. There are only a few papers discussing second-order differential equations with deviated arguments in infinite-dimensional spaces. To our knowledge, there is no paper discussing the existence, uniqueness and stability of a mild solution of a second-order differential equation with non-instantaneous impulses and a deviated argument in Banach space. In order to fill this gap, we consider a nonlinear second-order differential equation with a deviated argument. Moreover, the study of second-order differential equations with non-instantaneous impulses has not only mathematical significance, but also it has applications such as in harmonic oscillators with impulses and forced string equations, which we present in examples.

3.2.2 Preliminaries and assumptions

We briefly review definitions and some useful properties of the theory of the cosine family.

Definition 3.2.1. *(See [19].) A one parameter family $(C(t))_{t\in\mathbb{R}}$ of bounded linear operators mapping the Banach space X into itself is called a strongly continuous cosine family if and only if*

(i) $C(s + t) + C(s - t) = 2C(s)C(t)$ *for all* $s, t \in \mathbb{R}$,

(ii) $C(0) = I$,

(iii) $C(t)x$ *is continuous in* t *on* $\mathbb{R}$ *for each fixed point* $x \in X$.

$(S(t))_{t\in\mathbb{R}}$ is the sine function associated with the strongly continuous cosine family, $(C(t))_{t\in\mathbb{R}}$ which is defined by

$$S(t)x = \int_0^t C(s)x\,ds,\ \ x\in X,\ \ t\in\mathbb{R}.$$

$D(A)$ is the domain of the operator A which is defined by

$$D(A) = \{x \in X\colon C(t)x \text{ is twice continuously differentiable in t}\}.$$

$D(A)$ is the Banach space endowed with the graph norm $\|x\|_A = \|x\| + \|Ax\|$ for all $x \in D(A)$. We define a set

$$E = \{x \in X\colon C(t)x \text{ is once continuously differentiable in t}\},$$

which is a Banach space endowed with norm $\|x\|_E = \|x\| + \sup_{0\leqslant t\leqslant 1}\|AS(t)x\|$ for all $x \in E$.

With the help of $C(t)$ and $S(t)$, we define an operator valued function

$$\bar{h}(t) = \begin{bmatrix} C(t) & S(t) \\ AS(t) & C(t) \end{bmatrix}.$$

Operator valued function $\bar{h}(t)$ is a strongly continuous group of bounded linear operators on the space $E \times X$ generated by the operator

$$\bar{A} = \begin{bmatrix} 0 & I \\ A & 0 \end{bmatrix},$$

defined on $D(A) \times E$. It follows that $AS(t)\colon E \to X$ is a bounded linear operator and that $AS(t)x \to 0$ as $t \to 0$, for each $x \in E$. If $x\colon [0, \infty) \to X$ is locally integrable function then

$$y(t) = \int_0^t S(t-s)x(s)ds,$$

defines an E valued continuous function which is a consequence of the fact that

$$\int_0^t \bar{h}(t-s)\begin{bmatrix} 0 \\ x(s) \end{bmatrix}ds = \begin{bmatrix} \int_0^t S(t-s)x(s)ds \\ \int_0^t C(t-s)x(s)ds \end{bmatrix},$$

defines an $(E \times X)$ valued continuous function.

Proposition 3.2.2. *(See [19].) Let $(C(t))_{t\in\mathbb{R}}$ be a strongly continuous cosine family in X. The following are true:*

(i) *There exist constants $K \geqslant 1$ and $\omega \geqslant 0$ such that $|C(t)| \leqslant Ke^{\omega|t|}$ for all $t \in \mathbb{R}$.*

(ii) $|S(t_2) - S(t_1)| \leqslant K\left|\int_{t_1}^{t_2} e^{\omega|s|}ds\right|$ *for all $t_1, t_2 \in \mathbb{R}$.*

For more details on cosine family theory, we refer the reader to Travis, Webb and Fattorini [19–21].

Let $PC([0, T], X)$ be the space of piecewise continuous functions.

$PC([0, T], X) = \{x: J = [0, T] \to X: x \in C((t_k, t_{k+1}], X), k = 0, 1, \dots, m$ and there exist $x(t_k^-)$ and $x(t_k^+)$, $k = 1, 2, \dots, m$ with $x(t_k^-) = x(t_k)\}$. It can be seen easily that $PC([0, T], X)$ for all $t \in [0, T]$, is a Banach space endowed with the supremum norm

$$\|\psi\|_{PCB} := \sup\{\|\psi(\eta)\|e^{-\Omega\eta}, \ 0 \leqslant \eta \leqslant t\},$$

for some $\Omega > 0$. We set, $C_L(J, X) = \{y \in PC([0, T], X): \|y(t) - y(s)\| \leqslant L|t - s|, \forall t, s \in [0, T]\}$, where L is a suitable positive constant. Clearly $C_L(J, X)$ is a Banach space endowed with the PCB-norm.

In order to prove the existence, uniqueness and stability of the solution for problem (3.35), we need the following assumptions:

[A_1] A is the infinitesimal generator of a strongly continuous cosine family, $(C(t))_{t\in\mathbb{R}}$ of bounded linear operators.

[A_2] $f: J_1 \times X \times X \to X$, $J_1 = \bigcup_{i=0}^{m}[s_i, t_{i+1}]$ is a continuous function and there exist positive constants K_1 and K_2 such that

$$\|f(t, x_1, y_1) - f(t, x_2, y_2)\| \leqslant K_1(\|x_1 - x_2\| + \|y_1 - y_2\|),$$

for every $x_1, x_2, y_1, y_2 \in X, t \in J_1$. Also there exists a positive constant N such that $\|f(t, x, y)\| \leqslant N, \ \forall t \in J_1$ and $x, y \in X$.

[A_3] $h: X \times J_1 \to J$ is continuous and there exists a positive constant L_h such that

$$|h(x_1, t) - h(x_2, t)| \leqslant L_h\|x_1 - x_2\|, \ \forall x_1, x_2 \in X, \ t \in J_1,$$

and it holds that $h(\cdot, 0) = 0$.

[A_4] $J_i^l \in C(I_i \times X, X)$, $I_i = [t_i, s_i]$ and there are positive constants $L_{J_i^l}$, $i = 1, 2, \dots, m$, $l = 1, 2$, such that

$$\|J_i^l(t, x_1) - J_i^l(s, x_2)\| \leqslant L_{J_i^l}(|t - s| + \|x_1 - x_2\|), \ \forall t, s \in I_i \text{ and } x_1, x_2 \in X.$$

[A_5] There exist positive constants $C_{J_i^1}$ and $C_{J_i^2}$, $i = 1, 2, \dots, m$ such that

$$\|J_i^1(t, x)\| \leqslant C_{J_i^1} \text{ and } \|J_i^2(t, x)\| \leqslant C_{J_i^2}, \ \forall t \in I_i \text{ and } x \in X.$$

In the following definition, we introduce the concept of a mild solution for problem (3.35).

Definition 3.2.3. *A function $x \in C_L(J, X)$ is called a mild solution of the impulsive problem (3.35) if it satisfies the following relations: $x(0) = x_0$, $x'(0) = y_0$, the non-instantaneous impulse conditions $x(t) = J_i^1(t, x(t_i^-))$, $t \in (t_i, s_i]$, $i = 1, 2, \dots, m$, $x'(t) = J_i^2(t, x(t_i^-))$, $t \in (t_i, s_i]$, $i = 1, 2, \dots, m$ and $x(t)$ is the solution of the following integral equations*

$$x(t) = C(t)x_0 + S(t)y_0 + \int_0^t S(t-s)f(s, x(s), x[h(x(s), s)])ds, \quad t \in [0, t_1],$$
$$x(t) = C(t-s_i)(J_i^1(s_i, x(t_i^-))) + S(t-s_i)(J_i^2(s_i, x(t_i^-)))$$
$$+ \int_{s_i}^t S(t-s)f(s, x(s), x[h(x(s), s)])ds, \quad t \in [s_i, t_{i+1}], \; i = 1, 2, \ldots, m.$$

3.2.3 Existence and uniqueness result

Theorem 3.2.4. *Let* $x_0 \in D(A)$, $y_0 \in E$. *If all the assumptions* $[A_1]$–$[A_5]$ *are satisfied, then the second-order problem (3.35) has a unique mild solution* $x \in C_L(J, X)$.

Proof. Since $AS(t)$ is a bounded linear operator, we set $\rho = \sup_{t\in J}\|AS(t)\|$. For more details on $\|AS(t)\|$, we refer the reader to [13, 16, 22]. By choosing

$$\delta = \max_{1\leqslant i\leqslant m}\left\{\left(Ke^{\omega T}C_{J_i^1} + KTe^{\omega T}C_{J_i^2} + \frac{NKT}{\omega}e^{\omega T}\right)e^{-\Omega s_i},\right.$$
$$\left. Ke^{\omega T}\|x_0\| + KTe^{\omega T}\|y_0\| + \frac{NKT}{\omega}e^{\omega T}e^{-\Omega t_i}C_{J_i^1}\right\},$$

we set

$$\mathcal{W} = \{x \in C_L(J, X) \colon \|x\|_{PCB} \leqslant \delta\}.$$

We define a map $\mathcal{F}\colon \mathcal{W} \to \mathcal{W}$ given by

$$(\mathcal{F}x)(t) = J_i^1\left(t, C(t_i - s_{i-1})(J_i^1(s_{i-1}, x(t_{i-1}^-))) + S(t_i - s_{i-1})(J_i^2(s_{i-1}, x(t_{i-1}^-)))\right.$$
$$\left. + \int_{s_{i-1}}^{t_i} S(t_i - s)f(s, x(s), x[h(x(s), s)])ds\right),$$
$$t \in (t_i, s_i], \; i = 1, 2, \ldots, m;$$
$$(\mathcal{F}x)(t) = C(t)x_0 + S(t)y_0 + \int_0^t S(t-s)f(s, x(s), x[h(x(s), s)])ds, \quad t \in [0, t_1];$$
$$(\mathcal{F}x)(t) = C(t-s_i)(J_i^1(s_i, x(t_i^-))) + S(t-s_i)(J_i^2(s_i, x(t_i^-)))$$
$$+ \int_{s_i}^t S(t-s)f(s, x(s), x[h(x(s), s)])ds,$$
$$t \in (s_i, t_{i+1}], \; i = 1, 2, \ldots, m.$$

First, we need to show that $\mathcal{F}x \in C_L(J, X)$ for any $x \in C_L(J, X)$ and some $L > 0$. If $t_{i+1} \geqslant \tilde{t}_2 > \tilde{t}_1 > s_i$, then we obtain

$$\begin{aligned}\|(\mathcal{F}x)(\tilde{t}_2) - (\mathcal{F}x)(\tilde{t}_1)\| &\leqslant \|(C(\tilde{t}_2 - s_i) - C(\tilde{t}_1 - s_i))(J_i^1(s_i, x(t_i^-)))\| \\ &+ \|(S(\tilde{t}_2 - s_i) - S(\tilde{t}_1 - s_i))(J_i^2(s_i, x(t_i^-)))\| \\ &+ N\int_{s_i}^{\tilde{t}_1} \|(S(\tilde{t}_2 - s) - S(\tilde{t}_1 - s))\| ds \\ &+ N\int_{\tilde{t}_1}^{\tilde{t}_2} \|S(\tilde{t}_2 - s)\| ds \\ &\leqslant I_1 + I_2 + I_3 + I_4. \end{aligned} \tag{3.36}$$

We have,

$$\begin{aligned} I_1 &= \|(C(\tilde{t}_2 - s_i) - C(\tilde{t}_1 - s_i))(J_i^1(s_i, x(t_i^-)))\| \\ &= \|\int_{\tilde{t}_1 - s_i}^{\tilde{t}_2 - s_i} AS(\tau)(J_i^1(s_i, x(t_i^-)))d\tau\| \\ &\leqslant C_1(\tilde{t}_2 - \tilde{t}_1), \end{aligned} \tag{3.37}$$

where $C_1 = \rho C_{J_i^1}$.

Similarly, we have

$$\begin{aligned} I_2 &= \|(S(\tilde{t}_2 - s_i) - S(\tilde{t}_1 - s_i))(J_i^2(s_i, x(t_i^-)))\| \\ &= \|K\int_{\tilde{t}_1 - s_i}^{\tilde{t}_2 - s_i} e^{\omega\tau}(J_i^2(s_i, x(t_i^-)))d\tau\| \\ &\leqslant C_2(\tilde{t}_2 - \tilde{t}_1), \end{aligned} \tag{3.38}$$

where $C_2 = Ke^{\omega t_{i+1}}C_{J_i^2}$.

Similarly, we calculate third and fourth part of inequality (3.36) as follows

$$\begin{aligned} I_3 &= N\int_{s_i}^{\tilde{t}_1} \|(S(\tilde{t}_2 - s) - S(\tilde{t}_1 - s))\| ds \\ &\leqslant N\int_{s_i}^{\tilde{t}_1} \|K\int_{\tilde{t}_1 - s}^{\tilde{t}_2 - s} e^{\omega\tau} d\tau\| ds \\ &\leqslant C_3(\tilde{t}_2 - \tilde{t}_1), \end{aligned} \tag{3.39}$$

where $C_3 = KNt_{i+1}e^{\omega t_{i+1}}$, and

$$I_4 = N\int_{\tilde{t}_1}^{\tilde{t}_2} \|S(\tilde{t}_2 - s)\| ds \leqslant C_4(\tilde{t}_2 - \tilde{t}_1), \tag{3.40}$$

where $C_4 = KNt_{i+1}e^{\omega t_{i+1}}$.

We use inequalities (3.37)–(3.40) in inequality (3.36) and obtain the following inequality

$$\|(\mathcal{F}x)(\tilde{t}_2) - (\mathcal{F}x)(\tilde{t}_1)\| \leqslant L|\tilde{t}_2 - \tilde{t}_1|, \tag{3.41}$$

where $L \geqslant C_1 + C_2 + C_3 + C_4$.

If $t_1 \geqslant \tilde{t}_2 > \tilde{t}_1 \geqslant 0$, then we obtain

$$\begin{aligned}\|(\mathcal{F}x)(\tilde{t}_2) - (\mathcal{F}x)(\tilde{t}_1)\| &\leqslant \|(C(\tilde{t}_2) - C(\tilde{t}_1))x_0\| + \|(S(\tilde{t}_2) - S(\tilde{t}_1))y_0\| \\ &\quad + N\int_0^{\tilde{t}_1} \|(S(\tilde{t}_2-s) - S(\tilde{t}_1-s))\|ds \\ &\quad + N\int_{\tilde{t}_1}^{\tilde{t}_2} \|S(\tilde{t}_2 - s)\|ds \\ &\leqslant I_5 + I_6 + I_7 + I_8.\end{aligned} \tag{3.42}$$

We have

$$\begin{aligned}I_5 = \|(C(\tilde{t}_2) - C(\tilde{t}_1))x_0\| &= \|\int_{\tilde{t}_1}^{\tilde{t}_2} AS(\tau)x_0 d\tau\| \\ &\leqslant C_5(\tilde{t}_2-\tilde{t}_1),\end{aligned} \tag{3.43}$$

where $C_5 = \rho\|x_0\|$.

Similarly, we have

$$\begin{aligned}I_6 = \|(S(\tilde{t}_2) - S(\tilde{t}_1))y_0\| &= \|K\int_{\tilde{t}_1}^{\tilde{t}_2} e^{\omega\tau}y_0 d\tau\| \\ &\leqslant C_6(\tilde{t}_2-\tilde{t}_1),\end{aligned} \tag{3.44}$$

where $C_6 = Ke^{\omega t_1}\|y_0\|$.

Similarly, we calculate the third and fourth part of inequality (3.42) as follows

$$\begin{aligned}I_7 &= N\int_0^{\tilde{t}_1} \|(S(\tilde{t}_2-s) - S(\tilde{t}_1-s))\|ds \\ &\leqslant N\int_0^{\tilde{t}_1} \|K\int_{\tilde{t}_1-s}^{\tilde{t}_2-s} e^{\omega\tau}d\tau\|ds \\ &\leqslant C_7(\tilde{t}_2-\tilde{t}_1),\end{aligned} \tag{3.45}$$

where $C_7 = KNt_1e^{\omega t_1}$ and

$$I_8 = N\int_{\tilde{t}_1}^{\tilde{t}_2} \|S(\tilde{t}_2-s)\|ds \leqslant C_8(\tilde{t}_2-\tilde{t}_1), \tag{3.46}$$

where $C_8 = KNt_1e^{\omega t_1}$.

We use inequalities (3.43)–(3.46) in inequality (3.42) and obtain the following inequality

$$\|(\mathcal{F}x)(\tilde{t}_2) - (\mathcal{F}x)(\tilde{t}_1)\| \leqslant L|\tilde{t}_2-\tilde{t}_1|, \tag{3.47}$$

where $L \geqslant C_5 + C_6 + C_7 + C_8$.

Finally, if $s_i \geqslant \tilde{t}_2 > \tilde{t}_1 > t_i$, then we obtain

$$\|(\mathcal{F}x)(\tilde{t}_2) - (\mathcal{F}x)(\tilde{t}_1)\| \leqslant L_{J_i^1} \cdot (\tilde{t}_2-\tilde{t}_1). \tag{3.48}$$

Summarizing, we see that $\mathcal{F}x \in C_L(J, X)$ for any $x \in C_L(J, X)$ and some $L > 0$.

Next, we need to show that $\mathcal{F}: \mathcal{W} \to \mathcal{W}$. Now for $t \in (s_i, t_{i+1}]$ and $x \in \mathcal{W}$, we have

$$\begin{aligned}\|(\mathcal{F}x)(t)\| &\leqslant \|C(t-s_i)(J_i^1(s_i, x(t_i^-)))\| + \|S(t-s_i)(J_i^2(s_i, x(t_i^-)))\| \\ &\quad + \int_{s_i}^{t} \|S(t-s)f(s, x(s), x[h(x(s), s)])\|ds \\ &\leqslant Ke^{\omega(t-s_i)}C_{J_i^1} + K(t-s_i)e^{\omega(t-s_i)}C_{J_i^2} \\ &\quad + N\int_{s_i}^{t} K(t-s)e^{\omega(t-s)}ds.\end{aligned}$$

Hence,

$$\|(\mathcal{F}x)\|_{PCB} \leqslant \left(Ke^{\omega T}C_{J_i^1} + KTe^{\omega T}C_{J_i^2} + \frac{NKT}{\omega}e^{\omega T}\right)e^{-\Omega s_i}.$$

Now for $t \in [0, t_1]$ and $x \in \mathcal{W}$, we have

$$\begin{aligned}\|(\mathcal{F}x)(t)\| &\leqslant \|C(t)x_0\| + \|S(t)y_0\| + N\int_0^t \|S(t-s)\|ds \\ &\leqslant Ke^{\omega t}\|x_0\| + Kte^{\omega t}\|y_0\| + N\int_0^t K(t-s)e^{\omega(t-s)}ds.\end{aligned}$$

Hence,

$$\|(\mathcal{F}x)\|_{PCB} \leqslant Ke^{\omega T}\|x_0\| + KTe^{\omega T}\|y_0\| + \frac{NKT}{\omega}e^{\omega T}.$$

Similarly, for $t \in (t_i, s_i]$ and $x \in \mathcal{W}$ we have

$$\|(\mathcal{F}x)\|_{PCB} \leqslant e^{-\Omega t_i}C_{J_i^1}.$$

After summarizing the above inequalities, we obtain

$$\|(\mathcal{F}x)\|_{PCB} \leqslant \delta.$$

Therefore $\mathcal{F}: \mathcal{W} \to \mathcal{W}$. For any $x, y \in \mathcal{W}$, $t \in (s_i, t_{i+1}]$, $i = 1, 2, \dots, m$ we have

$$\begin{aligned}&\|(\mathcal{F}x)(t) - (\mathcal{F}y)(t)\| \\ &\leqslant Ke^{\omega(t-s_i)}L_{J_i^1}\|x(t_i^-) - y(t_i^-)\| \\ &\quad + K(t-s_i)e^{\omega(t-s_i)}L_{J_i^2}\|x(t_i^-) - y(t_i^-)\| \\ &\quad + KK_1(2 + LL_h)\int_{s_i}^{t} (t-s)e^{\omega(t-s)}\|x(s) - y(s)\|ds \\ &\leqslant Ke^{\omega(t-s_i)+\Omega t_i}L_{J_i^1}\|x - y\|_{PCB} \\ &\quad + K(t-s_i)e^{\omega(t-s_i)+\Omega t_i}L_{J_i^2}\|x - y\|_{PCB} \\ &\quad + KK_1(2 + LL_h)\int_{s_i}^{t} (t-s)e^{\omega(t-s)+\Omega s}ds\|x - y\|_{PCB} \\ &\leqslant Ke^{\omega(t-s_i)+\Omega t_i}L_{J_i^1}\|x - y\|_{PCB} \\ &\quad + K(t-s_i)e^{\omega(t-s_i)+\Omega t_i}L_{J_i^2}\|x - y\|_{PCB} \\ &\quad + \frac{KK_1(2 + LL_h)te^{\Omega t}}{(\Omega - \omega)}\|x - y\|_{PCB}.\end{aligned}$$

Hence

$$\begin{aligned}
&\|(\mathcal{F}x)(t) - (\mathcal{F}y)(t)\|e^{-\Omega t} \\
&\leqslant Ke^{\omega(t-s_i)+\Omega(t_i-t)}L_{J_i^1}\|x - y\|_{PCB} \\
&\quad + K(t - s_i)e^{\omega(t-s_i)+\Omega(t_i-t)}L_{J_i^2}\|x - y\|_{PCB} \\
&\quad + \frac{KK_1(2 + LL_h)t_{i+1}}{(\Omega - \omega)}\|x - y\|_{PCB} \\
&\leqslant \left(Ke^{\Omega(t_i-s_i)}L_{J_i^1} + Kt_{i+1}e^{\Omega(t_i-s_i)}L_{J_i^2} + \frac{KK_1(2 + LL_h)t_{i+1}}{(\Omega - \omega)}\right)\|x - y\|_{PCB}.
\end{aligned}$$

For $t \in [0, t_1]$ we obtain

$$\begin{aligned}
&\|(\mathcal{F}x)(t) - (\mathcal{F}y)(t)\| \\
&\leqslant KK_1(2 + LL_h)\int_0^t (t - s)e^{\omega(t-s)}\|x(s) - y(s)\|ds \\
&\leqslant KK_1(2 + LL_h)\int_0^t (t - s)e^{\omega(t-s)+\Omega s}ds\|x - y\|_{PCB} \\
&\leqslant \frac{KK_1(2 + LL_h)te^{\Omega t}}{(\Omega - \omega)}\|x - y\|_{PCB}.
\end{aligned}$$

Hence

$$\|(\mathcal{F}x)(t) - (\mathcal{F}y)(t)\|e^{-\Omega t} \leqslant \frac{KK_1(2 + LL_h)t_1}{(\Omega - \omega)}\|x - y\|_{PCB}.$$

Similarly, for $t \in (t_i, s_i]$, $i = 1, 2, \ldots, m$ we have

$$\begin{aligned}
&\|(\mathcal{F}x)(t) - (\mathcal{F}y)(t)\| \\
&\leqslant L_{J_i^1}\Bigg(Ke^{(t_i-s_{i-1})\omega}L_{J_i^1}\|x(t_{i-1}^-) - y(t_{i-1}^-)\| \\
&\quad + K(t_i - s_{i-1})e^{(t_i-s_{i-1})\omega}L_{J_i^2}\|x(t_{i-1}^-) - y(t_{i-1}^-)\| \\
&\quad + KK_1(2 + LL_h)\int_{s_{i-1}}^{t_i} (t_i - s)e^{(t_i-s)\omega}\|x(s) - y(s)\|ds\Bigg) \\
&\leqslant L_{J_i^1}\Bigg(Ke^{(t_i-s_{i-1})\omega+\Omega t_{i-1}}L_{J_i^1}\|x - y\|_{PCB} \\
&\quad + K(t_i - s_{i-1})e^{(t_i-s_{i-1})\omega+\Omega t_{i-1}}L_{J_i^2}\|x - y\|_{PCB} \\
&\quad + KK_1(2 + LL_h)\int_{s_{i-1}}^{t_i} (t_i - s)e^{(t_i-s)\omega+\Omega s}ds\|x - y\|_{PCB}\Bigg) \\
&\leqslant L_{J_i^1}\Bigg(Ke^{(t_i-s_{i-1})\omega+\Omega t_{i-1}}L_{J_i^1}\|x - y\|_{PCB} \\
&\quad + K(t_i - s_{i-1})e^{(t_i-s_{i-1})\omega+\Omega t_{i-1}}L_{J_i^2}\|x - y\|_{PCB} \\
&\quad + \frac{KK_1(2 + LL_h)t_ie^{\Omega t_i}}{(\Omega - \omega)}\|x - y\|_{PCB}\Bigg).
\end{aligned}$$

Therefore, we obtain

$$\begin{aligned}
&\|(\mathcal{F}x)(t) - (\mathcal{F}y)(t)\|e^{-\Omega t}\\
&\leqslant L_{J_i^1}\Bigg(Ke^{(t_i-s_{i-1})\omega+(t_{i-1}-t_i)\Omega}L_{J_i^1}\\
&\quad + K(t_i - s_{i-1})e^{(t_i-s_{i-1})\omega+(t_{i-1}-t_i)\Omega}L_{J_i^2}\\
&\quad + \frac{KK_1(2 + LL_h)t_i}{(\Omega - \omega)}\Bigg)\|x - y\|_{PCB}\\
&\leqslant L_{J_i^1}\Bigg(Ke^{(t_{i-1}-s_{i-1})\Omega}L_{J_i^1} + Ks_ie^{(t_{i-1}-s_{i-1})\Omega}L_{J_i^2}\\
&\quad + \frac{KK_1(2 + LL_h)t_i}{(\Omega - \omega)}\Bigg)\|x - y\|_{PCB}.
\end{aligned}$$

After summarizing the above inequalities, we have the following

$$\|(\mathcal{F}x) - (\mathcal{F}y)\|_{PCB} \leqslant L_F\|x - y\|_{PCB},$$

where

$$\begin{aligned}
L_F = \max_{1\leqslant i\leqslant m}\Bigg\{&\Bigg(Ke^{\Omega(t_i-s_i)}L_{J_i^1} + Kt_{i+1}e^{\Omega(t_i-s_i)}L_{J_i^2} + \frac{KK_1(2 + LL_h)t_{i+1}}{(\Omega - \omega)}\Bigg),\\
&\frac{KK_1(2 + LL_h)t_1}{(\Omega - \omega)},\ L_{J_i^1}\Bigg(Ke^{(t_{i-1}-s_{i-1})\Omega}L_{J_i^1} + Ks_ie^{(t_{i-1}-s_{i-1})\Omega}L_{J_i^2} + \frac{KK_1(2 + LL_h)t_i}{(\Omega - \omega)}\Bigg)\Bigg\}.
\end{aligned}$$

Hence, $\mathcal{F}$ is a strict contraction mapping for sufficiently large $\Omega > \omega$. Application of the Banach fixed point theorem immediately gives a unique mild solution to problem (3.35). □

3.2.4 Nonlocal problems

The nonlocal condition is a generalization of the classical initial condition. The study of nonlocal initial value problems is important because they appear in many physical systems. Byszewski [23] was the first author who studied the existence and uniqueness of mild solutions for Cauchy problems with nonlocal conditions. In this section, we investigate the existence and uniqueness of mild solution (3.35) with nonlocal conditions.

We consider the following nonlocal differential problem with a deviated argument in a Banach space X:

$$\begin{aligned}
&x''(t) = Ax(t) + f(t, x(t), x[h(x(t), t)]),\ t \in (s_i, t_{i+1}),\\
&\qquad i = 0, 1, \dots, m,\\
&x(t) = J_i^1(t, x(t_i^-)),\ t \in (t_i, s_i],\ i = 1, 2, \dots, m,\\
&x'(t) = J_i^2(t, x(t_i^-)),\ t \in (t_i, s_i],\ i = 1, 2, \dots, m,\\
&x(0) = x_0 + p(x),\ x'(0) = y_0 + q(x),
\end{aligned} \tag{3.49}$$

where $x(t)$ is a state function, $0 = s_0 < t_1 < s_1 < t_2 < \cdots < t_m < s_m < t_{m+1} = T < \infty$. The functions $J_i^1(t, x(t_i^-))$ and $J_i^2(t, x(t_i^-))$ represent non-instantaneous impulses as in system (3.35). A is the infinitesimal generator of a strongly continuous cosine family of bounded linear operators $(C(t))_{t\in\mathbb{R}}$ on X. The functions $p(x)$ and $q(x)$ will be suitably specified later.

Definition 3.2.5. *A function $x \in C_L(J, X)$ is called a mild solution of the impulsive problem (3.49) if it satisfies the following relations:*

$$x(0) = x_0 + p(x), \quad x'(0) = y_0 + q(x),$$

the non-instantaneous impulse conditions

$$x(t) = J_i^1(t, x(t_i^-)), \quad t \in (t_i, s_i], \quad i = 1, 2, \dots, m,$$
$$x'(t) = J_i^2(t, x(t_i^-)), \quad t \in (t_i, s_i], \quad i = 1, 2, \dots, m$$

and $x(t)$ is the solution of the following integral equations

$$\begin{aligned} x(t) &= C(t)(x_0 + p(x)) + S(t)(y_0 + q(x)) \\ &\quad + \int_0^t S(t-s)f(s, x(s), x[h(x(s), s)])ds, \quad t \in [0, t_1], \\ x(t) &= C(t-s_i)(J_i^1(s_i, x(t_i^-))) + S(t-s_i)(J_i^2(s_i, x(t_i^-))) \\ &\quad + \int_{s_i}^t S(t-s)f(s, x(s), x[h(x(s), s)])ds, \\ &\qquad t \in [s_i, t_{i+1}], \quad i = 1, 2, \dots, m. \end{aligned}$$

Further, we need assumptions on the functions p and q to show the existence and uniqueness of the solution for problem (3.49).

[A_6] The functions p, q: $C(J, X) \to X$ are continuous and there exist positive constants c_p and c_q such that

(i)

$$\|p(x_1) - p(x_2)\| \leqslant c_p\|x_1 - x_2\|,$$

(ii)

$$\|q(x_1) - q(x_2)\| \leqslant c_q\|x_1 - x_2\|.$$

Theorem 3.2.6. *Let $x_0 \in D(A)$, $y_0 \in E$. If all the assumptions [A_1]–[A_6] are satisfied, then the second-order nonlocal problem (3.49) has a unique mild solution $x \in C_L(J, X)$ provided that*

$$Ke^{\omega t_1}c_p + K\frac{(e^{\omega t_1} - 1)}{\omega}c_q < 1.$$

Proof. By choosing

$$\delta' = \max_{1 \leqslant i \leqslant m} \left\{ \left(Ke^{\omega T} C_{J_i^1} + KTe^{\omega T} C_{J_i^2} + \frac{NKT}{\omega} e^{\omega T} \right) e^{-\Omega s_i}, \right.$$

$$\left. Ke^{\omega T} \|x_0 + p(x)\| + KTe^{\omega T} \|y_0 + q(x)\| + \frac{NKT}{\omega} e^{\omega T},\ e^{-\Omega t_i} C_{J_i^1} \right\},$$

we set

$$\mathcal{W}' = \{x \in C_L(J, X) \colon \|x\|_{PCB} \leqslant \delta'\}.$$

We define a map $\mathcal{F}\colon \mathcal{W}' \to \mathcal{W}'$ given by

$$\begin{aligned}
(\mathcal{F}x)(t) = J_i^1 \Bigg(t,\ & C(t_i - s_{i-1})(J_i^1(s_{i-1}, x(t_{i-1}^-))) + S(t_i - s_{i-1})(J_i^2(s_{i-1}, x(t_{i-1}^-))) \\
& + \int_{s_{i-1}}^{t_i} S(t_i - s) f(s, x(s), x[h(x(s), s)]) ds \Bigg), \\
& t \in (t_i, s_i],\ i = 1, 2, \dots, m; \\
(\mathcal{F}x)(t) = {} & C(t)(x_0 + p(x)) + S(t)(y_0 + q(x)) \\
& + \int_0^t S(t - s) f(s, x(s), x[h(x(s), s)]) ds,\ t \in [0, t_1]; \\
(\mathcal{F}x)(t) = {} & C(t - s_i)(J_i^1(s_i, x(t_i^-))) + S(t - s_i)(J_i^2(s_i, x(t_i^-))) \\
& + \int_{s_i}^t S(t - s) f(s, x(s), x[h(x(s), s)]) ds, \\
& \qquad t \in (s_i, t_{i+1}],\ i = 1, 2, \dots, m.
\end{aligned}$$

We have,

$$\|(\mathcal{F}x) - (\mathcal{F}y)\|_{PCB} \leqslant L_F' \|x - y\|_{PCB},$$

where

$$L_F' = \max_{1 \leqslant i \leqslant m} \left\{ \left(Ke^{\Omega(t_i - s_i)} L_{J_i^1} + Kt_{i+1} e^{\Omega(t_i - s_i)} L_{J_i^2} + \frac{KK_1(2 + LL_h)t_{i+1}}{(\Omega - \omega)} \right), \right.$$

$$Ke^{\omega t_1} c_p + K \frac{(e^{\omega t_1} - 1)}{\omega} c_q + \frac{KK_1(2 + LL_h)t_1}{(\Omega - \omega)},$$

$$\left. L_{J_i^1} \left(Ke^{(t_{i-1} - s_{i-1})\Omega} L_{J_i^1} + Ks_i e^{(t_{i-1} - s_{i-1})\Omega} L_{J_i^2} + \frac{KK_1(2 + LL_h)t_i}{(\Omega - \omega)} \right) \right\}.$$

Thus, $\mathcal{F}$ is a strict contraction mapping for sufficiently large $\Omega > \omega$. Application of Banach fixed point theorem immediately gives a unique mild solution to problem (3.49).

The proof of this theorem is a consequence of theorem (3.2.4). □

3.2.5 Ulam's type stability

In this section, we show Ulam's type stability for the system (3.35).

Let $\epsilon > 0$, $\psi \geqslant 0$ and $\phi \in PC(J, \mathbb{R}^+)$ be nondecreasing. We consider the following inequalities

$$\begin{cases} \|y''(t) - Ay(t) - f(t, y(t), y[h(y(t), t)])\| \leqslant \epsilon, \ t \in (s_i, t_{i+1}), \\ \qquad i = 0, 1, \dots, m, \\ \|y(t) - J_i^1(t, y(t_i^-))\| \leqslant \epsilon, \ t \in (t_i, s_i], \ i = 1, 2, \dots, m, \\ \|y'(t) - J_i^2(t, y(t_i^-))\| \leqslant \epsilon, \ t \in (t_i, s_i], \ i = 1, 2, \dots, m, \end{cases} \tag{3.50}$$

and

$$\begin{cases} \|y''(t) - Ay(t) - f(t, y(t), y[h(y(t), t)])\| \leqslant \phi(t), \ t \in (s_i, t_{i+1}), \\ \qquad i = 0, 1, \dots, m, \\ \|y(t) - J_i^1(t, y(t_i^-))\| \leqslant \psi, \ t \in (t_i, s_i], \ i = 1, 2, \dots, m, \\ \|y'(t) - J_i^2(t, y(t_i^-))\| \leqslant \psi, \ t \in (t_i, s_i], \ i = 1, 2, \dots, m, \end{cases} \tag{3.51}$$

and

$$\begin{cases} \|y''(t) - Ay(t) - f(t, y(t), y[h(y(t), t)])\| \leqslant \epsilon\phi(t), \ t \in (s_i, t_{i+1}), \\ \qquad i = 0, 1, \dots, m, \\ \|y(t) - J_i^1(t, y(t_i^-))\| \leqslant \epsilon\psi, \ t \in (t_i, s_i], \ i = 1, 2, \dots, m. \\ \|y'(t) - J_i^2(t, y(t_i^-))\| \leqslant \epsilon\psi, \ t \in (t_i, s_i], \ i = 1, 2, \dots, m. \end{cases} \tag{3.52}$$

Now, we take the vector space

$$Z = C_L(J, X) \bigcap_{i=0}^m C^2((s_i, t_{i+1}), X) \bigcap_{i=0}^m C((s_i, t_{i+1}), D(A)).$$

The following definitions are inspired by Wang *et al* [11].

Definition 3.2.7. *Equation (3.35) is Ulam–Hyers stable if there exists $c(K_1 L_h L_J m) > 0$ such that for each $\epsilon > 0$ and for each solution $y \in Z$ of inequality (3.50), there exists a mild solution $x \in C_L(J, X)$ of equation (3.35) with*

$$\|y(t) - x(t)\| \leqslant c(K_1 L_h L_J m)\epsilon, \ t \in J. \tag{3.53}$$

Definition 3.2.8. *Equation (3.35) is generalized Ulam–Hyers stable if there exists $\theta_{K_1, L_h, L_J, m} \in C(\mathbb{R}^+, \mathbb{R}^+)$, $\theta(0) = 0$ such that for each $\epsilon > 0$ and for each solution $y \in Z$ of inequality (3.50), there exists a mild solution $x \in C_L(J, X)$ of equation (3.35) with*

$$\|y(t) - x(t)\| \leqslant \theta_{K_1, L_h, L_J, m}\epsilon, \ t \in J. \tag{3.54}$$

Definition 3.2.9. *Equation (3.35) is Ulam–Hyers–Rassias stable with respect to* (ϕ, ψ) *if there exists* $c(K_1 L_h L_J m\phi) > 0$ *such that for each* $\epsilon > 0$ *and for each solution* $y \in Z$ *of inequality (3.52), there exists a mild solution* $x \in C_L(J, X)$ *of equation (3.35) with*

$$\|y(t) - x(t)\| \leqslant c(K_1 L_h L_J m\phi)\epsilon(\psi + \phi(t)), \quad t \in J. \tag{3.55}$$

Definition 3.2.10. *Equation (3.35) is generalized Ulam–Hyers–Rassias stable with respect to* (ϕ, ψ) *if there exists* $c(K_1 L_h L_J m\phi) > 0$ *such that for each solution* $y \in Z$ *of inequality (3.51), there exists a mild solution* $x \in C_L(J, X)$ *of equation (3.35) with*

$$\|y(t) - x(t)\| \leqslant c(K_1 L_h L_J m\phi)(\psi + \phi(t)), \quad t \in J. \tag{3.56}$$

Remark 3.2.11. *A function* $y \in Z$ *is a solution of inequality (3.52) if and only if there is* $G \in \bigcap_{i=0}^{m} C^2((s_i, t_{i+1}), X) \bigcap_{i=0}^{m} C((s_i, t_{i+1}), D(A))$, $g_1 \in \bigcap_{i=1}^{m} C([t_i, s_i], X)$ *and* $g_2 \in \bigcap_{i=1}^{m} C^1([t_i, s_i], X)$ *such that:*

(a) $\|G(t)\| \leqslant \epsilon\phi(t), \ t \in \bigcap_{i=0}^{m}(s_i, t_{i+1})$, $\|g_1(t)\| \leqslant \epsilon\psi$ and $\|g_2(t)\| \leqslant \epsilon\psi, \ t \in \bigcap_{i=0}^{m}[t_i, s_i]$.

(b) $y''(t) = Ay(t) + f(t, y(t), y[h(y(t), t)]) + G(t), \ t \in (s_i, t_{i+1}), \ i = 0, 1, \ldots, m.$

(c) $y(t) = J_i^1(t, y(t_i^-)) + g_1(t), \ t \in (t_i, s_i], \ i = 1, 2, \ldots, m.$

(d) $y'(t) = J_i^2(t, y(t_i^-)) + g_2(t), \ t \in (t_i, s_i], \ i = 1, 2, \ldots, m.$

Easily, we can have similar remarks for inequalities (3.50) and (3.51).

Remark 3.2.12. *A function* $y \in Z$ *is a solution of inequality (3.52) then* y *is a solution of the following integral inequality*

$$\begin{cases} \|y(t) - J_i^1(t, y(t_i^-))\| \leqslant \epsilon\psi, \ t \in (t_i, s_i], \ i = 1, 2, \ldots, m, \\ \|y'(t) - J_i^2(t, y(t_i^-))\| \leqslant \epsilon\psi, \ t \in (t_i, s_i], \ i = 1, 2, \ldots, m, \\ \|y(t) - C(t)x_0 - S(t)y_0 - \int_0^t S(t-s)f(s, y(s), y[h(y(s), s)])ds\| \\ \qquad \leqslant \dfrac{\epsilon K}{\omega}\int_0^t [e^{\omega(t-s)} - 1]\phi(s)ds, \ t \in [0, t_1], \\ \|y(t) - C(t-s_i)(J_i^1(s_i, y(t_i^-))) - S(t-s_i)(J_i^2(s_i, y(t_i^-))) \\ \qquad - \int_{s_i}^t S(t-s)f(s, y(s), y[h(y(s), s)])ds\| \\ \qquad \leqslant \epsilon\psi K e^{\omega(t-s_i)} + \dfrac{\epsilon\psi K}{\omega}[e^{\omega(t-s_i)} - 1] + \dfrac{\epsilon K}{\omega}\int_{s_i}^t [e^{\omega(t-s)} - 1] \\ \qquad \phi(s)ds, \\ \qquad t \in [s_i, t_{i+1}], \ i = 1, 2, \ldots, m. \end{cases} \tag{3.57}$$

From remark 3.2.11, we have

$$\begin{cases} y''(t) = Ay(t) + f(t, y(t), y[h(y(t), t)]) + G(t), \\ \qquad t \in (s_i, t_{i+1}), \ i = 0, 1, \dots, m, \\ y(t) = J_i^1(t, y(t_i^-)) + g_1(t), \ t \in (t_i, s_i], \ i = 1, 2, \dots, m, \\ y'(t) = J_i^2(t, y(t_i^-)) + g_2(t), \ t \in (t_i, s_i], \ i = 1, 2, \dots, m. \end{cases} \tag{3.58}$$

The solution $y \in Z$ with $y(0) = x_0$ and $y'(0) = y_0$ of equation (3.58) is given by

$$\begin{cases} y(t) = J_i^1(t, y(t_i^-)) + g_1(t), \ t \in (t_i, s_i], \ i = 1, 2, \dots, m, \\ y'(t) = J_i^2(t, y(t_i^-)) + g_2(t), \ t \in (t_i, s_i], \ i = 1, 2, \dots, m, \\ y(t) = C(t)x_0 + S(t)y_0 + \int_0^t S(t-s)[f(s, y(s), y[h(y(s), s)]) \\ \qquad + G(s)]ds, \ t \in [0, t_1], \\ y(t) = C(t - s_i)((J_i^1(s_i, y(t_i^-))) \\ \qquad + g_1(s_i)) + S(t - s_i)((J_i^2(s_i, y(t_i^-))) + g_2(s_i)) \\ \qquad + \int_{s_i}^t S(t-s)[f(s, y(s), y[h(y(s), s)]) + G(s)]ds, \\ \qquad t \in [s_i, t_{i+1}], i = 1, 2, \dots, m. \end{cases} \tag{3.59}$$

We can easily have similar remarks for the solution of inequalities (3.50) and (3.51). In order to discuss the stability of problem (3.35), we need the following additional assumption:

[A_7] (H_4) *Let $\phi \in C(J, \mathbb{R}^+)$ be a nondecreasing function. There exists $c_\phi > 0$ such that*

$$\int_0^t \phi(s)ds \leqslant c_\phi \phi(t), \ \forall \ t \in J.$$

Lemma 3.2.13. *(See [24, theorem 16.4].). Let $\mathbb{M}_0 = \mathbb{M} \cup \{0\}$, where $\mathbb{M} = \{1, \dots, m\}$ and the following inequality holds*

$$u(t) \leqslant a(t) + \int_0^t b(s)u(s)ds + \sum_{0<t_k<t} \beta_k u(t_k^-), \ t \geqslant 0, \tag{3.60}$$

where $u, a, b \in PC(\mathbb{R}^+, \mathbb{R}^+)$, a is nondecreasing and $b(t) > 0$, $\beta_k > 0$, $k \in \mathbb{M}$. Then for $t \in \mathbb{R}^+$,

$$u(t) \leqslant a(t)(1+\beta)^k \exp\left(\int_0^t b(s)ds\right), \ t \in (t_k, t_{k+1}], \ k \in \mathbb{M}_0, \tag{3.61}$$

where $\beta = \sup_{k\in\mathbb{M}}\{\beta_k\}$ and $t_0 = 0$.

Theorem 3.2.14. *Let $x_0 \in D(A)$, $y_0 \in E$. If the assumptions $[A_1]$–$[A_4]$ and $[A_7]$ are satisfied. Then, equation (3.35) is Ulam–Hyers–Rassias stable with respect to (ϕ, ψ).*

Proof. Let $y \in C_L(J, D(A)) \cap C^2((s_i, t_{i+1}], X)$ be a solution of inequality (3.52) and x is the unique mild solution of problem (3.35) which is given by

$$\begin{cases} x(t) = J_i^1(t, x(t_i^-)), \ t \in (t_i, s_i], \ i = 1, 2, \dots, m, \\ x'(t) = J_i^2(t, x(t_i^-)), \ t \in (t_i, s_i], \ i = 1, 2, \dots, m, \\ x(t) = C(t)x_0 + S(t)y_0 \\ \qquad + \int_0^t S(t-s)f(s, x(s), x[h(x(s), s)])ds, \ t \in [0, t_1], \\ x(t) = C(t-s_i)(J_i^1(s_i, x(t_i^-))) + S(t-s_i)(J_i^2(s_i, x(t_i^-))) \\ \qquad + \int_{s_i}^t S(t-s)f(s, x(s), x[h(x(s), s)])ds, \\ \qquad\qquad t \in [s_i, t_{i+1}], \ i = 1, 2, \dots, m. \end{cases} \tag{3.62}$$

For $t \in [s_i, t_{i+1}]$, $i = 1, 2, \dots, m$. From inequality (3.57) we have

$$\begin{aligned} &\|y(t) - C(t-s_i)(J_i^1(s_i, y(t_i^-))) - S(t-s_i)(J_i^2(s_i, y(t_i^-))) \\ &- \int_{s_i}^t S(t-s)f(s, y(s), y[h(y(s), s)])ds\| \\ &\leqslant \epsilon\psi K e^{\omega(t-s_i)} + \frac{\epsilon\psi K}{\omega}[e^{\omega(t-s_i)} - 1] + \frac{\epsilon K}{\omega}\int_{s_i}^t [e^{\omega(t-s)} - 1]\phi(s)ds \\ &\leqslant \epsilon\psi K e^{\omega T} + \frac{\epsilon\psi K}{\omega}e^{\omega T} + \frac{\epsilon K}{\omega}e^{\omega T}\int_0^t \phi(s)ds \\ &\leqslant \epsilon\psi K e^{\omega T} + \frac{\epsilon K}{\omega}e^{\omega T}(\psi + c_\phi\phi(t)). \end{aligned}$$

For $t \in (t_i, s_i]$, $i = 1, 2, \dots, m$, we have

$$\|y(t) - J_i^1(t, y(t_i^-))\| \leqslant \epsilon\psi.$$

For $t \in [0, t_1]$, we have

$$\begin{aligned} &\|y(t) - C(t)x_0 - S(t)y_0 - \int_0^t S(t-s)f(s, y(s), y[h(y(s), s)])ds\| \\ &\leqslant \frac{\epsilon K}{\omega}\int_0^t [e^{\omega(t-s)} - 1]\phi(s)ds \\ &\leqslant \frac{\epsilon K}{\omega}e^{\omega T}c_\phi\phi(t). \end{aligned}$$

Hence, for $t \in [s_i, t_{i+1}]$, $i = 1, 2, \dots, m$, we have

$$
\begin{aligned}
&\|y(t) - x(t)\| \\
&\leqslant \|y(t) - C(t - s_i)(J_i^1(s_i, y(t_i^-))) - S(t - s_i)(J_i^2(s_i, y(t_i^-))) \\
&\quad - \int_{s_i}^{t} S(t - s)f(s, y(s), y[h(y(s), s)])ds\| \\
&\quad + Ke^{\omega T}\|J_i^1(s_i, y(t_i^-)) - J_i^1(s_i, x(t_i^-))\| \\
&\quad + KTe^{\omega T}\|J_i^2(s_i, y(t_i^-)) - J_i^2(s_i, x(t_i^-))\| \\
&\quad + KTe^{\omega T}\int_{s_i}^{t} \|f(s, y(s), y[h(y(s), s)]) - f(s, x(s), x[h(x(s), s)])ds\| \\
&\leqslant \epsilon\psi Ke^{\omega T} + \frac{\epsilon K}{\omega}e^{\omega T}(\psi + c_\phi\phi(t)) \\
&\quad + (Ke^{\omega T}L_{J_i^1} + KTe^{\omega T}L_{J_i^2})\|(y(t_i^-) - (x(t_i^-)\| \\
&\quad + K_1(2 + LL_h)KTe^{\omega T}\int_0^t \|y(s) - x(s)\|ds \\
&\leqslant \frac{\epsilon K}{\omega}e^{\omega T}[(2 + c_\phi)(\psi + \phi(t))] + K_1(2 + LL_h)KTe^{\omega T}\int_0^t \|y(s) - x(s)\|ds \\
&\quad + \sum_{j=1}^{i}(Ke^{\omega T}L_{J_j^1} + KTe^{\omega T}L_{J_j^2})\|(y(t_j^-) - (x(t_j^-)\|.
\end{aligned} \tag{3.63}
$$

For $t \in (t_i, s_i]$, $i = 1, 2, \dots, m$ we have

$$
\begin{aligned}
\|y(t) - x(t)\| &\leqslant \|y(t) - J_i^1(t, y(t_i^-))\| + \|J_i^1(t, y(t_i^-)) - J_i^1(t, x(t_i^-))\| \\
&\leqslant \epsilon\psi + \sum_{j=1}^{i} L_{J_j^1}\|(y(t_j^-) - (x(t_j^-)\| \\
&\leqslant \frac{\epsilon K}{\omega}e^{\omega T}[(2 + c_\phi)(\psi + \phi(t))] \\
&\quad + K_1(2 + LL_h)KTe^{\omega T}\int_0^t \|y(s) - x(s)\|ds \\
&\quad + \sum_{j=1}^{i}(Ke^{\omega T}L_{J_j^1} + KTe^{\omega T}L_{J_j^2})\|(y(t_j^-) - (x(t_j^-)\|.
\end{aligned} \tag{3.64}
$$

Now, for $t \in [0, t_1]$, we have

$$\begin{aligned}\|y(t)-x(t)\| &\leqslant \frac{\epsilon K}{\omega}e^{\omega T}c_{\phi}\phi(t)+K_1(2+LL_h)KTe^{\omega T}\int_0^t \\ &\quad \|y(s)-x(s)\|ds \\ &\leqslant \frac{\epsilon K}{\omega}e^{\omega T}[(2+c_{\phi})(\psi+\phi(t))] \\ &\quad +K_1(2+LL_h)KTe^{\omega T}\int_0^t \|y(s)-x(s)\|ds \\ &\quad +\sum_{j=1}^{i}(Ke^{\omega T}L_{J_j^1}+KTe^{\omega T}L_{J_j^2})\|(y(t_j^-)-(x(t_j^-)\|.\end{aligned} \tag{3.65}$$

We observe that inequalities (3.63), (3.64) and (3.65) give together an impulsive Grönwall inequality of the form of (3.60) on J. Therefore, we can apply impulsive Grönwall inequality (3.61) for $t\in J$, since $t\in(t_i, t_{i+1}]$ for some $i\in\mathbb{M}_0$. Consequently, we have

$$\begin{aligned}\|y(t)-x(t)\| &\leqslant \frac{K}{\omega}e^{\omega T}(2+c_{\phi})(1+Ke^{\omega T}L_J)^i e^{K_1(2+LL_h)KTe^{\omega T}t}\epsilon(\psi+\phi(t)) \\ &\leqslant \frac{K}{\omega}e^{\omega T}(2+c_{\phi})(1+Ke^{\omega T}L_J)^m e^{K_1(2+LL_h)KTe^{\omega T}T}\epsilon(\psi+\phi(t)) \\ &\leqslant c(K_1L_hL_Jm\phi)\epsilon(\psi+\phi(t)),\end{aligned}$$

for any $t\in J$, where $L_J=\sup_{i\in\mathbb{M}}\{L_{J_i^1}+TL_{J_i^2}\}$ and $c(K_1L_hL_Jm\phi)$ is a constant depending on K_1, L_h, L_J, m, ϕ. Hence, equation (3.35) is Ulam–Hyers–Rassias stable with respect to (ϕ, ψ). □

Theorem 3.2.15. *If the assumptions [A_1]–[A_4] and [A_7] are satisfied. Then, equation (3.35) is generalized Ulam–Hyers–Rassias stable with respect to (ϕ, ψ).*

Proof. This can easily be proved by applying the same procedure as in the theorem 3.2.14 and taking inequality (3.51). □

Theorem 3.2.16. *If the assumptions [A_1]–[A_4] and [A_7] are satisfied. Then, equation (3.35) is Ulam–Hyers stable.*

Proof. It can be easily proved by applying same procedure of theorem 3.2.14 and taking inequality (3.50). □

Example 3.2.17. *Let $X=L^2(0,\pi)$. We consider the following partial differential equations with a deviated argument*

$$\begin{cases} \partial_{tt}Z(t, y) = \partial_{yy}Z(t, y) + f_2(y, Z(h(t), y)) + f_3(t, y, Z(t, y)), \ y \in (0, \pi), \\ \qquad\qquad t \in (2i, 2i+1], \ i \in \{0\} \bigcup \mathbb{N}, \\ Z(t, 0) = Z(t, \pi) = 0, \ t \in [0, T], \ 0 < T < \infty, \\ Z(0, y) = x_0, \ y \in (0, \pi), \\ \partial_t Z(0, y) = y_0, \ y \in (0, \pi), \\ Z(t)(y) = (\sin it)Z((2i-1)^-, y), \ y \in (0, \pi), \ t \in (2i-1, 2i], \ i \in \mathbb{N}, \\ \partial_t Z(t)(y) = (i \cos it)Z((2i-1)^-, y), \ y \in (0, \pi), \ t \in (2i-1, 2i], \ i \in \mathbb{N}, \end{cases} \tag{3.66}$$

where $0 = s_0 < t_1 < s_1 < t_2 < \cdots < t_m < s_m < t_{m+1} = T < \infty$ *with* $t_i = 2i - 1$, $s_i = 2i$ *and*

$$f_3(t, y, Z(t, y)) = \int_0^y \bar{K}(y, s)(a_1|Z(t, s)| + b_1|Z(t, s)|)ds.$$

We assume that $a_1, b_1 \geqslant 0$, $(a_1, b_1) \neq (0, 0)$, $h\colon J_1 \to [0, T]$ *is locally Hölder continuous in* t *with* $h(0) = 0$ *and* $\bar{K}\colon [0, \pi] \times [0, \pi] \to \mathbb{R}$.

We define an operator A *as follows,*

$$Ax = x'' \text{ with } D(A) = \{x \in X\colon \ x'' \in X \text{ and } x(0) = x(\pi) = 0\}. \tag{3.67}$$

Here, clearly the operator A *is the infinitesimal generator of a strongly continuous cosine family of operators on* X. A *has infinite series representation*

$$Ax = \sum_{n=1}^{\infty} -n^2(x, x_n)x_n, \ x \in D(A),$$

where $x_n(s) = \sqrt{2/\pi}\sin ns$, $n = 1, 2, 3, \ldots$ *is the orthonormal set of eigenfunctions of* A. *Moreover, the operator* A *is the infinitesimal generator of a strongly continuous cosine family* $C(t)_{t\in\mathbb{R}}$ *on* X *which is given by*

$$C(t)x = \sum_{n=1}^{\infty} \cos nt(x, x_n)x_n, \ x \in X,$$

and the associated sine family $S(t)_{t\in\mathbb{R}}$ *on* X *which is given by*

$$S(t)x = \sum_{n=1}^{\infty} \frac{1}{n}\sin nt(x, x_n)x_n, \ x \in X.$$

Equation (3.66) can be reformulated as the following abstract differential equation in X:

$$\begin{cases} x''(t) = Ax(t) + f(t, x(t), x[h(x(t), t)]), \ t \in (s_i, t_{i+1}), \ i \in \{0\} \bigcup \mathbb{N}, \\ x(t) = J_i^1(t, x(t_i^-)), \ t \in (t_i, s_i], \ i \in \mathbb{N}, \\ x'(t) = J_i^2(t, x(t_i^-)), \ t \in (t_i, s_i], \ i \in \mathbb{N}, \\ x(0) = x_0, \ x'(0) = y_0, \end{cases} \tag{3.68}$$

where $x(t) = Z(t, \cdot)$, *that is* $x(t)(y) = Z(t, y)$, $y \in (0, \pi)$. *Functions* $J_i^1(t, x(t_i^-)) = (\sin it)Z((2i-1)^-, y)$ *and* $J_i^2(t, x(t_i^-)) = i(\cos it)Z((2i-1)^-, y)$ *represent non-instantaneous impulses during intervals* $(t_i, s_i]$. *The operator* A *is the same as in equation (3.67).*

The function $f: J_1 \times X \times X \to X$ *is given by*

$$f(t, \psi, \xi)(y) = f_2(y, \xi) + f_3(t, y, \psi),$$

where $f_2 : [0, \pi] \times X \to H_0^1(0, \pi)$ *is given by*

$$f_2(y, \xi) = \int_0^y \bar{K}(y, x)\xi(x)dx,$$

and

$$\|f_3(t, y, \psi)\| \leqslant V(y, t)(1 + \|\psi\|_{H^2(0, \pi)})$$

with $V(\cdot, t) \in X$ *and* V *is continuous in its second argument. For more details see* [16, 17]. *Thus, theorem 3.2.4 can be applied to problem (3.66). We can choose the functions* $p(x)$ *and* $q(x)$ *as given below*

$$p(x) = \sum_k^n \alpha_k x(t_k), \quad t_k \in J \text{ for all } k = 1, 2, 3, \dots, n,$$

$$q(x) = \sum_k^n \beta_k x(t_k), \quad t_k \in J \text{ for all } k = 1, 2, 3, \dots, n,$$

where α_k *and* β_k *are constants.*

Example 3.2.18. *We consider a particular linear case of the abstract differential equation (3.68) in the space* $X = \mathbb{R}$. *A forced string equation*

$$\begin{cases} x''(t) + a_1x(t) + a_2 \sin x(c_1t) = g(t), & t \in (s_i, t_{i+1}), \ i = 0, 1, \dots, m, \\ x(t) = a_3 \tanh(x(t_i^-))r(t), & t \in (t_i, s_i], \ i = 1, 2, \dots, m, \\ x'(t) = a_3 \tanh(x(t_i^-))r'(t), & t \in (t_i, s_i], \ i = 1, 2, \dots, m, \\ x(0) = x_0, \ x'(0) = y_0, \end{cases} \tag{3.69}$$

where $a_1 \in \mathbb{R}^+$, $a_2, a_3 \in \mathbb{R}$, $c_1 \in (0, 1]$, $g \in C(J_1, \mathbb{R})$ *and* $r \in C^1(J_2, \mathbb{R})$ *for* $J_2 = \cup_{i=1}^m I_i$. *We define* A *as follows*

$$Ax = -a_1x \text{ with } D(A) = \mathbb{R}.$$

Here, clearly the value $-a_1$ *behaves like an infinitesimal generator of a strongly continuous cosine family* $C(t) = \cos\sqrt{a_1}\,t$. *The associated sine family is given by* $S(t) = \frac{1}{\sqrt{a_1}} \sin\sqrt{a_1}\,t$. *The deviated argument in the abstract differential equation (3.68) is represented by the term* $a_2x(c_1t)$ *of the differential equation (3.69). Non-*

instantaneous impulses $a_3 \tanh(x(t_i^-))r(t)$ *and* $a_3 \tanh(x(t_i^-))r'(t)$ *are created when the string is extremely pushed on each interval* $(t_i, s_i]$.

Example 3.2.19. *We generalize the above example to consider a coupled system of strings or pendulums*

$$\begin{aligned} &x_n''(t) + a_{n1} \sin x_n(t) + a_{n2} \sin x_n(c_n t) = b_{n1} x_{n-1}(t) + b_{n2} x_{n+1}(t) \\ &\quad + g_n(t), \ \ t \in (s_i, t_{i+1}), \ i = 0, 1, \dots, m, \ \ n \in \mathbb{Z}; \\ &x_n(t) = a_{n3} \tanh(x_n(t_i^-)) r_n(t), \ \ t \in (t_i, s_i], \ \ i = 1, 2, \dots, m; \\ &x_n'(t) = a_{n3} \tanh(x_n(t_i^-)) r_n'(t), \ \ t \in (t_i, s_i], \ \ i = 1, 2, \dots, m; \\ &x_n(0) = x_{n0}, \ \ x_n'(0) = y_{n0}, \end{aligned} \tag{3.70}$$

where $a_{n1}, a_{n2}, a_{n3}, b_{n1}, b_{n2} \in \mathbb{R}$, $c_n \in (0, 1]$, $g_n \in C(J_1, \mathbb{R})$ *and* $r_n \in C^1(J_2, \mathbb{R})$. *Moreover we suppose* $\sup_n |a_{nk}| < \infty$, $k = 1, 2, 3$, $\sup_n(|b_{n1}| + |b_{n2}|) < \infty$ *and* $\sup_n(\|g_n\| + \|r_n\| + \|r_n'\|) < \infty$. *Then we consider (3.70) on* ℓ_∞ *and use exercise 1 on p 39 of [20]. The lattice ODE (3.70) is a generalization of the discrete sine-Gordon equation [25] and* $x_n(c_n t)$ *represent pantograph-like terms [26].*

3.2.6 Notes and remarks

The research presented in this chapter focuses on the existence, uniqueness and stability of solutions to the impulsive systems represented by second-order nonlinear differential equations with non-instantaneous impulses and a deviated argument. We used a strongly continuous cosine family of bounded linear operators and Banach's fixed point theorem to obtain the existence and uniqueness of the solutions. In addition, Ulam's type stability is established by using an impulsive Grönwall inequality.

The results in section 3.2 are motivated by [27].

References

[1] Hernández E and O'Regan D 2013 On a new class of abstract impulsive differential equations *Proc. Amer. Math. Soc.* **141** 1641–9

[2] Pierri M, O'Regan D and Rolnik V 2013 Existence of solutions for semi-linear abstract differential equations with not instantaneous impulses *Appl. Math. Comput.* **219** 6743–9

[3] Cădariu L 2007 *Stabilitatea Ulam–Hyers–Bourgin Pentru Ecuatii Functionale* (Timişara: Ed. Univ. Vest Timişoara)

[4] Hyers D H 1941 On the stability of the linear functional equation *Proc. Natl Acad. Sci.* **27** 222–4

[5] Hyers D H, Isac G and Rassias T M 1998 *Stability of Functional Equations in Several Variables* (Basel: Birkhäuser)

[6] Jung S M 2001 *Hyers-Ulam-Rassias Stability of Functional Equations in Mathematical Analysis* (Palm Harbor, FL: Hadronic)

[7] Rassias T M 1978 On the stability of linear mappings in Banach spaces *Proc. Amer. Math. Soc.* **72** 297–300

[8] Pazy A 1983 *Semigroup of Linear Operators and Applications to Partial Differential Equations* (Berlin: Springer)

[9] Wang J, Fečkan M and Zhou Y 2012 Ulam's type stability of impulsive ordinary differential equations *J. Math. Anal. Appl.* **395** 258–64

[10] Bainov D D and Simeonov P S 1992 *Integral Inequalities and Applications* (Dordrecht: Kluwer)

[11] Wang J and Fečkan M 2015 A general class of impulsive evolution equations *Topol. Methods Nonlinear Anal.* **46** 915–33

[12] Chalishajar D N 2009 Controllability of damped second-order initial value problem for a class of differential inclusions with nonlocal conditions on noncompact intervals *Applied Analysis and Differential Equations* (Singapore: World Scientific) 55–68

[13] Pandey D N, Das S and Sukavanam N 2014 Existence of solution for a second-order neutral differential equation with state dependent delay and non-instantaneous impulses *Int. J. Nonlinear Sci.* **18** 145–55

[14] Acharya F S 2013 Controllability of second-order semilinear impulsive partial neutral functional differential equations with infinite delay *Int. J. Math. Sci. Appl.* **3** 207–18

[15] Arthi G and Balachandran K 2014 Controllability of second order impulsive evolution systems with infinite delay *Nonlinear Anal. : Hybrid Systems* **11** 139–53

[16] Sakthivel R, Mahmudov N I and Kim J H 2009 On controllability of second order nonlinear impulsive differential systems *Nonlinear Anal.* **71** 45–52

[17] Gal C G 2007 Nonlinear abstract differential equations with deviated argument *J. Math. Anal and Appl.* **333** 971–83

[18] Muslim M, Kumar A and Agarwal R P 2016 Exact and trajectory controllability of second order nonlinear impulsive systems with deviated argument *Funct. Differ. Equ.* **23** 27–41

[19] Travis C C and Webb G F 1978 Cosine families and abstract nonlinear second order differential equations *Acta Math. Acad. Sci. Hungaricae* **32** 76–96

[20] Fattorini H O 1985 *Second Order Linear Differential Equations in Banach SpacesNorth Holland Mathematics Studies* vol 108 (Amsterdam: North Holland)

[21] Travis C C and Webb G F 1977 Compactness, regularity and uniform continuity properties of strongly continuous cosine family *Houston J. Math.* **3** 555–67

[22] Hernández E and McKibben M A 2005 Some comments on: Existence of solutions of abstract nonlinear second-order neutral functional integrodifferential equations *Comput. Math. Appl.* **50** 1–15

[23] Byszewski L 1991 Theorems about the existence and uniqueness of solutions of a semilinear nonlocal Cauchy problem *J. Math. Anal. Appl.* **162** 494–505

[24] Bainov D D and Simeonov P S 1992 *Integral Inequalities and Applications* (Dordrecht: Kluwer)

[25] Scott A 2003 *Nonlinear Sciences: Emergence and Dynamics of Coherent Structures* 2nd edn (Oxford: Oxford University Press)

[26] Derfel C and Iserles A 1997 The pantograph equation in the complex plane *J. Math. Anal. Appl.* **213** 117–32

[27] Muslim M, Kumar A and Fečkan M 2018 Existence, uniqueness and stability of solutions to second order nonlinear differential equations with non-instantaneous impulses *J. King Saud Univ. Sci.* **30** 204–13

IOP Publishing

Non-Instantaneous Impulsive Differential Equations
Basic theory and computation
JinRong Wang and Michal Fečkan

Chapter 4

Periodic solutions

In this chapter, we establish a unified framework to investigate periodic mild solutions for first-order autonomous and non-autonomous evolution equations and second-order evolution equations. In section 4.1 we discuss the existence of the periodic mild solutions. Section 4.2 is devoted to studying periodic solutions and subharmonic solutions of non-autonomous periodic evolution equations, which extend some results of section 4.1. In section 4.3 we study the existence of periodic mild solutions of second-order evolution equations.

4.1 First-order autonomous evolution equations

4.1.1 Introduction

Let $J = [0, T]$ and A: $D(A) \subseteq X \to X$ be the generator of a C_0-semigroup $\{S(t), t \geqslant 0\}$ on a Banach space X with a norm $\|\cdot\|$. Denote $M := \sup_{t\in J}\|S(t)\|$ and $C(J, X)$ by the Banach space of all continuous functions from J into X with the norm $\|u\|_C := \sup\{\|u(t)\|: t \in J\}$ for $u \in C(J, X)$. We consider the Banach space $PC(J, X) := \{u: J \to X: u \in C((t_k, t_{k+1}], X),\ k = 0, 1, \dots, m$ and there exist $u(t_k^-)$ and $u(t_k^+)$, $k = 1, \dots, m$, with $u(t_k^-) = u(t_k)\}$ endowed with the Chebyshev PC-norm $\|x\|_{PC} := \sup\{\|x(t)\|: t \in J\}$. Similarly, one can define the Banach space $PC([0, \infty), X)$ endowed with the PC-norm $\|x\|_{PC} := \sup\{\|x(t)\|: t \in [0, \infty)\}$.

Recently, Hernández and O'Regan [1] and Pierri *et al* [2] provided the first studies on Cauchy problems for a new type of first-order evolution equation with non-instantaneous impulses of the form:

$$\begin{cases} u'(t) = Au(t) + f(t, u(t)), & t \in (s_i, t_{i+1}],\ i = 0, 1, 2, \dots, m, \\ u(t) = g_i(t, u(t)), & t \in (t_i, s_i],\ i = 1, 2, \dots, m, \\ u(0) = \bar{x} \in X, \end{cases} \tag{4.1}$$

where the fixed points s_i and t_i satisfy $0 = s_0 < t_1 \leqslant s_1 \leqslant t_2 < \cdots < t_m \leqslant s_m \leqslant t_{m+1} = T$ and are pre-fixed numbers, f: $[0, T] \times X \to X$ is continuous and g_i: $[t_i, s_i] \times X \to X$ is continuous for all $i = 1, 2, \ldots, m$.

The concepts of the mild solutions of Cauchy problem (4.1) are introduced as follows.

Definition 4.1.1. *(See [1, definition 2.1].) A function $u \in PC([0, T], X)$ is called a mild solution of Cauchy problem (4.1) if u satisfies*

$$u(t) = S(t)\bar{x} + \int_0^t S(t-s)f(s, u(s))ds, \quad t \in [0, t_1];$$
$$u(t) = g_i(t, u(t)), \quad t \in (t_i, s_i], \quad i = 1, 2, \ldots, m;$$
$$u(t) = S(t - s_i)g_i(s_i, u(s_i)) + \int_{s_i}^t S(t-s)f(s, u(s))ds, \quad t \in (s_i, t_{i+1}],$$
$$i = 1, 2, \ldots, m.$$

For the Cauchy problem (4.1), the existence and uniqueness results are derived using the theory of C_0-semigroups and compact semigroups on Banach spaces or the theory of analytic compact semigroups in fractional power spaces.

Periodic motion is a very important and special phenomenon, not only in the natural sciences, but also in social sciences and, for example, the study of climate, food supplements, insect populations and sustainable development. In particular, periodic solutions of dynamic equations have attracted many researchers. We refer the reader to the monographs [3–5] and papers [6–11] (and references therein) for evolution equations.

We consider periodic solutions of nonlinear evolution equations with non-instantaneous impulses of the form:

$$\begin{cases} u'(t) = Au(t) + f(t, u(t)), & t \in (s_i, t_{i+1}), \ i = 0, 1, 2, \ldots, \infty, \\ u(t_i^+) = g_i(t_i, u(t_i^-)), & i = 1, 2, \ldots, \infty, \\ u(t) = g_i(t, u(t_i^-)), & t \in (t_i, s_i], \ i = 1, 2, \ldots, \infty, \end{cases} \tag{4.2}$$

where the fixed points s_i and t_i satisfy $0 = s_0 < t_1 \leqslant s_1 \leqslant t_2 < \cdots < t_m \leqslant s_m \leqslant t_{m+1} < \cdots$ with $\lim_{i\to\infty} t_i = \infty$ and $t_{i+m} = t_i + T$, $s_{i+m} = s_i + T$, $m \in \mathbb{N}$ denoting the number of impulsive points between 0 and T. Moreover, f: $[0, \infty) \times X \to X$ is a T-periodic (with respect to $t \in [0, \infty)$) Carathéodory function and g_i: $[t_i, s_i] \times X \to X$ is a continuous function for all $i = 1, 2, \ldots, \infty$ with $g_{i+m} = g_i$.

Definition 4.1.2. *We say that a function $u \in PC([0, T], X)$ is called a mild solution of Cauchy problem*

$$\begin{cases} u'(t) = Au(t) + f(t, u(t)), & t \in (s_i, t_{i+1}), \ i = 0, 1, 2, \ldots, m, \\ u(t_i^+) = g_i(t_i, u(t_i^-)), & i = 1, 2, \ldots, m, \\ u(t) = g_i(t, u(t_i^-)), & t \in (t_i, s_i], \ i = 1, 2, \ldots, m, \\ u(0) = \bar{x}, \end{cases} \tag{4.3}$$

if u satisfies

$$
\begin{aligned}
u(t) &= g_i(t, u(t_i^-)), \ t \in (t_i, s_i], \ i = 1, 2, \ldots, m; \\
u(t_i^+) &= g_i(t_i, u(t_i^-)), \ i = 1, 2, \ldots, m; \\
u(t) &= S(t)\overline{x} + \int_0^t S(t-s)f(s, u(s))ds, \ t \in [0, t_1]; \\
u(t) &= S(t - s_i)g_i(s_i, u(t_i^-)) + \int_{s_i}^t S(t-s)f(s, u(s))ds, \\
&\quad t \in [s_i, t_{i+1}], \ i = 1, 2, \ldots, m.
\end{aligned}
\tag{4.4}
$$

Definition 4.1.3. *A function $u \in PC([0, \infty), X)$ is said to be a T-periodic PC-mild solution of equation (4.2) if it is a PC-mild solution of Cauchy problem (4.3) corresponding to some $\overline{x}$ and $u(t + T) = u(t)$ for $t \geqslant 0$.*

4.1.2 The existence of periodic solutions

In this section, we will consider the existence of periodic solutions of equation (4.2) under the conditions below:

$[A]$ A: $D(A) \subseteq X \to X$ is the generator of a C_0-semigroup $\{S(t), t \geqslant 0\}$ on a Banach space X with a norm $\|\cdot\|$.

$[F_1]$ f: $[0, \infty) \times X \to X$ is strongly measurable for $t \geqslant 0$ and for any $u, v \in X$ satisfying $\|u\|, \|v\| \leqslant \rho$ there exists a positive constant $L_f(\rho) > 0$ such that

$$\|f(t, u) - f(t, v)\| \leqslant L_f(\rho)\|u - v\|.$$

$[F_2]$ There exists a positive constant M_f such that

$$\|f(t, u)\| \leqslant M_f(1 + \|u\|) \text{ for all } u \in X.$$

$[F_3]$ $f(t, u)$ is T-periodic in t, i.e. $f(t + T, u) = f(t, u)$, $t \geqslant 0$.

$[G_1]$ g_i: $[t_i, s_i] \times X \to X$ is continuous for all $i = 1, 2, \ldots, \infty$ and for any $u, v \in X$ satisfying $\|u\|, \|v\| \leqslant \rho$ there exists a positive constant $L_{gi}(\rho) > 0$ such that

$$\|g_i(t, u) - g_i(t, v)\| \leqslant L_{g_i}(\rho)\|u - v\|.$$

$[G_2]$ There exists a positive constant M_{gi}, $i = 1, 2, \ldots, m$ such that

$$\|g_i(t, u)\| \leqslant M_{g_i}(1 + \|u\|) \text{ for each } t \in [t_i, s_i] \text{ and all } u \in X,$$

$[G_3]$ $g_i(t, u)$ is T-periodic in t, i.e. $g_i(t + T, u) = g_i(t, u)$, $t \geqslant 0$ with $g_{i+m} = g_i$.

First, we establish the following existence results via the Banach contraction principle.

Theorem 4.1.4. *Assume the conditions $[A]$, $[F_1]$, $[F_2]$, $[G_1]$ and $[G_2]$ are satisfied. Then Cauchy problem (4.3) has a unique mild solution.*

Proof. We divide our proof into three cases.

Case 1. For $t \in [0, t_1]$. Note our assumptions and using [12, theorem 5.3.3, p 169], we known that Cauchy problem

$$\begin{cases} u'(t) = Au(t) + f(t, u(t)), & t \in [0, t_1], \\ u(0) = \bar{x} \in X, \end{cases}$$

has a unique mild solution

$$u(t) = S(t)\bar{x} + \int_0^t S(t-s)f(s, u(s))ds.$$

Case 2. For $t \in (s_i, t_{i+1})$, $i = 1, 2, \dots, m$. By using a similar method to [12, theorem 5.3.3, p 169], we can obtain that the Cauchy problem

$$\begin{cases} u'(t) = Au(t) + f(t, u(t)), & t \in (s_i, t_{i+1}), \ i = 0, 1, 2, \dots, m, \\ u(t) = g_i(t, u(t_i^-)), & t \in (t_i, s_i], \ i = 1, 2, \dots, m, \end{cases}$$

has a unique mild solution

$$u(t) = S(t - s_i)g_i(s_i, u(t_i^-)) + \int_{s_i}^t S(t-s)f(s, u(s))ds, \quad t \in (s_i, t_{i+1}],$$
$$i = 1, 2, \dots, m.$$

Case 3. For $t \in (t_i, s_i]$, $i = 1, 2, \dots, m$. One can check that

$$u(t) = g_i\left(t, S(t_i - s_{i-1})g_{i-1}(s_{i-1}, u(t_{i-1}^-)) + \int_{s_{i-1}}^{t_i} S(t_i - s)f(s, u(s))ds\right)$$

has a unique mild solution.

The proof is complete. □

Concerning periodic solutions, we define a Poincaré operator P: $X \to X$ given by

$$\begin{aligned} P(\bar{x}) &= u(T, \bar{x}) \\ &= g_m(T, u(t_m^-, \bar{x})) \\ &= g_m(T, S(t_m - s_{m-1})g_{m-1}(s_{m-1}, u(t_{m-1}^-, \bar{x})) \\ &\quad + \int_{s_{m-1}}^{t_m} S(t_m - s)f(s, u(s, \bar{x}))ds\Bigg), \end{aligned} \tag{4.5}$$

where $u(t_i, \bar{x})$ $i = 1, 2, \dots, m$ is given by (4.4).

Now we are ready to show that the fixed point P defined in (4.5) gives rise to a periodic solution by using periodicity conditions $[F_3]$ and $[G_3]$.

Lemma 4.1.5. *Equation (4.2) has a T-periodic PC-mild solution if and only if P has a fixed point.*

Proof. '$\Longrightarrow$' If $u(\cdot) = u(\cdot + T)$ then $u(0) = u(T) = P(u(0))$. So $u(0)$ is a fixed point of P.

'$\Longleftarrow$' Suppose that P has a fixed point $x_0 \in X$, i.e. $Px_0 = x_0$. For the PC-mild solution $u(\cdot, x_0)$ of Cauchy problem (4.1) corresponding to the initial value $u(0) = x_0$, we define $y(\cdot) = u(\cdot + T, x_0)$. Clearly, $y(0) = u(T, x_0) = Px_0 = x_0$.

Case 1. For $t \in [0, t_1]$. This implies $t + T \in [T, t_1 + T] = [s_m, t_{m+1}]$.

By elementary computation via $[F_3]$ and $[G_3]$, we have

$$\begin{aligned} y(t) &= u(t + T, x_0) \\ &= S(t + T - s_m)g_m(s_m, u(t_m^-, x_0)) + \int_{s_m}^{t+T} S(t + T - s)f(s, u(s, x_0))ds \\ &= S(t)g_m(T, u(t_m^-, x_0)) + \int_{T}^{t+T} S(t + T - s)f(s, u(s, x_0))ds \\ &\overset{s=\theta+T}{=} S(t)y(0) + \int_0^t S(t - \theta)f(\theta + T, u(\theta + T, x_0))d\theta \\ &= S(t)y(0) + \int_0^t S(t - \theta)f(\theta, y(\theta, y(0)))d\theta. \end{aligned}$$

Case 2. For $t \in (t_i, s_i]$, $i = 1, 2, \ldots, m$. This implies $t + T \in (t_i + T, s_i + T] = (t_{i+m}, s_{i+m}]$.

By elementary computation via $[G_3]$, we have

$$\begin{aligned} y(t) &= u(t + T, x_0) \\ &= g_{i+m}(t + T, u(t_{i+m}^-, x_0)) \\ &= g_i(t + T, u(t_i^- + T, x_0)) \\ &= g_i(t, u(t_i^- + T, x_0)) \\ &= g_i(t, y(t_i^-)). \end{aligned}$$

Case 3. For $t \in [s_i, t_{i+1}]$, $i = 1, 2, \ldots, m$. This implies $t + T \in [s_i + T, t_i + T] = [s_{i+m}, t_{i+m}]$.

By elementary computation via $[F_3]$ and $[G_3]$, we have

$$\begin{aligned} y(t) &= u(t + T, x_0) \\ &= S(t + T - s_{i+m})g_{i+m}(s_{i+m}, u(t_{i+m}^-, x_0)) + \int_{s_{i+m}}^{t+T} S(t + T - s)f(s, u(s, x_0))ds \\ &\overset{s=\theta+T}{=} S(t - s_i)g_i(s_i + T, u(t_i^- + T, x_0)) + \int_{s_i}^{t} S(t - \theta)f(\theta + T, u(\theta + T, x_0))ds \\ &= S(t - s_i)g_i(s_i, y(t_i^-)) + \int_{s_i}^{t} S(t - \theta)f(\theta, y(\theta))ds. \end{aligned}$$

From the above, we derive that $y(\cdot, y(0))$ is a PC-mild solution of Cauchy problem (4.3) with initial value $y(0) = x_0$. Thus the uniqueness implies that $u(\cdot, x_0) = y(\cdot, y(0)) = u(\cdot + T, x_0)$, so that $u(\cdot, x_0)$ is T-periodic. $\square$

Note that P is a composition of the maps given by

$$P_0(x) = u(t_1, x),\ u(t, x) = S(t)x + \int_0^t S(t-s)f(s, u(s, x))ds,\ t \in [0, t_1];$$

$$G_i(x) = g_i(s_i, x),\ i = 1, 2, \dots, m;$$

$$P_i(x) = u(t_{i+1}, x),\ u(t, x) = S(t - s_i)x + \int_{s_i}^t S(t-s)f(s, u(s, x))ds,$$

$$t \in [s_i, t_{i+1}],\ i = 1, 2, \dots, m.$$

In other words,

$$P = G_m \circ P_{m-1} \circ \cdots \circ G_1 \circ P_0. \tag{4.6}$$

Now we impose that:

$[A']$ $S(t)$ is dissipative, i.e. $\|S(t)\| \leqslant Ke^{-\omega t}$ for some $\omega > 0$ and $K \geqslant 1$.

Next, using $[G_2]$, we have

$$\|G_i(x)\| \leqslant M_{g_i}(1 + \|x\|),\ x \in X, i = 1, 2, \dots, m, \tag{4.7}$$

while using $[F_2]$, we derive

$$\begin{aligned}\|u(t, x)\| \leqslant Ke^{-\omega(t-s_i)}\|x\| + KM_f \int_{s_i}^t e^{-\omega(t-s)}(1 + \|u(s, x)\|)ds,\ t \in [s_i, t_{i+1}],\\ i = 0, 2, \dots, m-1, x \in X.\end{aligned} \tag{4.8}$$

Let $v(t, x) = e^{\omega(t-s_i)}\|u(t, x)\|$. Obviously, (4.8) reduces to

$$\begin{aligned}v(t, x) &\leqslant K\|x\| + KM_f \int_{s_i}^t (e^{\omega(s-s_i)} + v(s, x))ds\\ &= K\|x\| + KM_f \frac{e^{\omega(t-s_i)} - 1}{\omega} + KM_f \int_{s_i}^t v(s, x)ds.\end{aligned}$$

It follows from the Grönwall inequality that

$$v(t, x) \leqslant K\left(\|x\| + M_f \frac{e^{\omega(t-s_i)} - 1}{\omega}\right)e^{KM_f(t-s_i)},$$

hence

$$\|P_i(x)\| = \|u(t_{i+1}, x)\| \leqslant \frac{KM_f e^{KM_f(t_{i+1}-s_i)}}{\omega} + Ke^{(KM_f-\omega)(t_{i+1}-s_i)}\|x\|. \tag{4.9}$$

Now linking (4.7) with (4.9) we have

$$\|(G_i \circ P_{i-1})(x)\| \leqslant a + M_{g_i} K e^{(KM_f-\omega)(t_i-s_{i-1})}\|x\|,\ x \in X, i = 1, 2, \dots, m,$$

where

$$a = \max_{i=1,2,\dots,m} M_{g_i}\left(1 + \frac{KM_f e^{KM_f(t_i-s_{i-1})}}{\omega}\right).$$

Setting

$$\begin{aligned}
b &= K \max_{i=1,2,\ldots,m} M_{g_i} e^{(KM_f-\omega)(t_i-s_{i-1})};\\
A &= a\sum_{i=1}^{m-1} b^i; \qquad\qquad (4.10)\\
\Theta &= K\prod_{i=1}^{m} M_{g_i} e^{(KM_f-\omega)\sum_{i=1}^{m}(t_i-s_{i-1})}.
\end{aligned}$$

By elementary computation, we arrive at

$$\begin{aligned}
\|P(x)\| &= \|(G_m \circ P_{m-1} \circ \cdots \circ G_1 \circ P_0)(x)\|\\
&\leqslant a(1 + b + \cdots + b^{m-1}) + K\prod_{i=1}^{m} M_{g_i} e^{(KM_f-\omega)\sum_{i=1}^{m}(t_i-s_{i-1})}\|x\| \qquad (4.11)\\
&= A + \Theta\|x\|.
\end{aligned}$$

From (4.11) we immediately obtain the following result.

Lemma 4.1.6. *Every PC-mild solution of Cauchy problem (4.3) corresponding to a given initial value in a bounded subset of X is locally bounded.*

Next, we verify that P defined by (4.5) is a continuous and compact operator under our conditions. If we can show $P_j, j = 0, 1, \ldots, m$ are continuous and compact, then one can use the continuity of $G_j, j = 1, \ldots, m$ to derive that P is continuous and compact.

Lemma 4.1.7. *If $\{S(t), t \geqslant 0\}$ is a compact semigroup, then P is a continuous and compact operator.*

Proof. Let $\bar{x}, \bar{y} \in \Upsilon \subset X$, where Υ is a bounded subset of X. Suppose $u(\cdot, \bar{x})$ and $v(\cdot, \bar{y})$ are the PC-mild solutions of Cauchy problem (4.3) corresponding to the initial value $\bar{x}, \bar{y} \in X$, respectively. From lemma 4.1.6 there exists $\rho > 0$ such that $\|u(\cdot, \bar{x})\| \leqslant \rho$ and $\|v(\cdot, \bar{y})\| \leqslant \rho$.

Step 1. We show that $P_j, j = 0, 1, \ldots, m$ are continuous.

Now using $[F_1]$ we obtain

$$\|u(t, \bar{x}) - v(t, \bar{y})\| \leqslant M\|\bar{x} - \bar{y}\| + ML_f(\rho) \int_0^t \|u(s, \bar{x}) - v(s, \bar{y})\|ds, \quad t \in [0, t_1] \cup [s_i, t_{i+1}],$$

for $i = 1, 2, \ldots, m$.

By virtue of the Grönwall inequality, one can verify that there exists a constant $L > 0$ such that

$$\|u(t, \bar{x}) - v(t, \bar{y})\| \leqslant L\|\bar{x} - \bar{y}\|, \text{ for all } t \in [0, t_1] \bigcup [s_i, t_{i+1}],\ i = 1, 2, \ldots, m,$$

which implies that

$$\|P_j(\bar{x}) - P_j(\bar{y})\| \leqslant L\|\bar{x} - \bar{y}\|.$$

Hence, P_j, $j = 0, 1, \ldots, m$ are continuous operators on X.

Step 2. We show that P_j, $j = 0, 1, \ldots, m$ are compact.

Let Γ be a bounded subset of X. Define $K_j = P_j\Gamma = \{P_j(\bar{x}) \in X: \bar{x} \in \Gamma\}$, $j = 0, 1, \ldots, m$.

For $s_i < \varepsilon \leqslant t_{i+1}$, $i = 0, 1, 2, \ldots, m$, we define

$$K_{i,\,\varepsilon} = P_{i,\,\varepsilon}\Gamma = S(\varepsilon)\{u(t_{i+1} - \varepsilon, \bar{x}): \bar{x} \in \Gamma\}.$$

Next, we show that $K_{j,\varepsilon}$, $j = 0, 1, \ldots, m$ are precompact in X. From lemma 4.1.6 again, for $\bar{x} \in \Gamma$ fixed, we have $\|u(t_{i+1} - \varepsilon, \bar{x})\| \leqslant \rho$ for $t \in [s_i, t_{i+1}]$. Thus we find that the sets $\{u(t_{i+1} - \varepsilon, \bar{x}): \bar{x} \in \Gamma\}$ are totally bounded. Thus, $K_{j,\varepsilon}$, $j = 0, 1, \ldots, m$ is precompact in X by using the compactness of $S(\cdot)$.

On the other hand, for arbitrary $\bar{x} \in \Gamma$,

$$\begin{aligned} P_{i,\,\varepsilon}(\bar{x}) = S(\varepsilon)u(t_{i+1} - \varepsilon, \bar{x}) &= S(t_{i+1} - s_i)\bar{x} \\ &+ \int_{s_i}^{t_{i+1}-\varepsilon} S(t_{i+1} - s)f(s, u(s, \bar{x}))ds. \end{aligned}$$

Obviously,

$$\begin{aligned} \left\| P_{j,\,\varepsilon}(\bar{x}) - P_j(\bar{x})\right\| &\leqslant \int_{t_{j+1}-\varepsilon}^{t_{j+1}} \|S(t_{j+1} - s)f(s, u(s, \bar{x}))\|ds \\ &\leqslant MM_f(1 + \rho)\varepsilon, \end{aligned}$$

for $j = 0, 1, \ldots, m$.

This yields that the set K_j, $j = 0, 1, \ldots, m$ can be approximated with an arbitrary precision by a precompact set $K_{j,\,\varepsilon}$. Hence K_j, $j = 0, 1, \ldots, m$ are a precompact set in X. That is, P_j, $j = 0, 1, \ldots, m$ take bounded sets into precompact sets in X. Thus P_j, $j = 0, 1, \ldots, m$ are compact operators.

Finally, one can use the continuity of G_j, $j = 1, \ldots, m$ to derive that P is continuous and compact. □

Now we present the main results.

Theorem 4.1.8. *Assume that [A], [A'], [F_1], [F_2], [F_3], [G_1], [G_2], [G_3] and [G_4] are satisfied. If $\{S(t), t \geqslant 0\}$ is compact and Θ is given in (4.10)*

$$\Theta < 1, \tag{4.12}$$

then equation (4.2) has at least a T-periodic PC-mild solution.

Proof. From theorem 4.1.4, for a fixed $\overline{x} \in X$, the Cauchy problem (4.3) corresponding to the initial value $u(0) = \overline{x}$ has the PC-mild solution $u(\cdot, \overline{x})$. From lemma 4.1.7 the operator P defined by (4.5) is compact.

According to the Leray–Schauder fixed point theory, it suffices to show that the set $\{\overline{x} \in X: \overline{x} = \sigma P(\overline{x}), \sigma \in [0, 1]\}$ is a bounded subset of X. Note that from (4.12) and using (4.11) we see that P is dissipative. In particular, we obtain that if $\overline{x} = \sigma P(\overline{x})$, $\sigma \in [0, 1]$, then

$$\|\overline{x}\| \leqslant \|P(\overline{x})\| \leqslant A + \Theta\|\overline{x}\|,$$

which gives

$$\|\overline{x}\| \leqslant \frac{A}{1 - \Theta}.$$

Thus there exists $x_0 \in X$ such that $Px_0 = x_0$. Now using lemma 4.1.5 we can derive that the PC-mild solution $u(\cdot, x_0)$ of Cauchy problem (4.3) corresponding to the initial value $u(0) = x_0$ is just T-periodic. Thus, $u(\cdot, x_0)$ is a T-periodic PC-mild solution of equation (4.2). The proof is complete. □

Furthermore, (4.11) implies

$$\|P^n(x)\| \leqslant A(1 + \Theta + \cdots + \Theta^{n-1}) + \Theta^n\|x\|,$$

for any $x \in X$ and $n \in \mathbb{N}$. So if $\Theta < 1$ and $\|x\| \leqslant r$, then

$$\|P^n(x)\| \leqslant \frac{A}{1 - \Theta} + \Theta^n r,$$

so the ball $\left\{x \in X: \|x\| \leqslant \frac{A}{1-\Theta} + 1\right\}$ is an absorbing set for P (see [13, p 22]). Since P is compact and continuous under the assumptions of theorem 4.1.8, from [13, theorem 5.8] and [14, theorem 2.4.7], we know that there is a global compact connected attractor $\mathcal{A}$ of P. Moreover from [14, theorem 2.8.1], the limit capacity $c(\mathcal{A})$, the Hausdorff dimension $\dim_H \mathcal{A}$ and the dimension $\dim \mathcal{A}$ of $\mathcal{A}$ are all finite with $\dim \mathcal{A} \leqslant \dim_H \mathcal{A} \leqslant c(\mathcal{A})$. So we also know something on the dynamics of (4.2). Clearly any fixed point x_0 of P belongs to $\mathcal{A}$.

4.1.3 An example

Set $X = L^2(0,1)$, $0 = t_0 = s_0$, $t_1 = \pi$, $s_1 = T = 2\pi$ and $J = [0,2\pi]$. Define $Ax = \frac{\partial^2}{\partial y^2}x$ for $x \in D(A)$ with $D(A) = \left\{x \in X: \frac{\partial x}{\partial y}, \frac{\partial^2 x}{\partial y^2} \in X,\ x(0) = x(1) = 0\right\}$. Then, A is the infinitesimal generator of a C_0-semigroup $\{S(t), t \geqslant 0\}$ in X. Moreover, $S(\cdot)$ is dissipative and compact with $\|S(t)\| \leqslant e^{-t}$ for all $t \geqslant 0$. Obviously, $K = 1$ and $\omega = 1$.

Consider the following periodic problem

$$\begin{cases} \dfrac{\partial}{\partial t}x(t,y) = \dfrac{\partial^2}{\partial y^2}x(t,y) + \dfrac{|x(t,y)|}{1+|x(t,y)|} + y\sin t, \ y \in (0,1), \ t \in (0,\pi), \\ \dfrac{\partial}{\partial y}x(t,0) = \dfrac{\partial}{\partial y}x(t,1) = 0, \ t \in [0,2\pi], \\ x(0,y) = x(2\pi,y), \ y \in (0,1), \\ x(\pi^+,y) = -\dfrac{y|x(\pi^-,y)|}{2(e^\pi + |x(\pi^-,y)|)}, \\ x(t,y) = \dfrac{y\cos t|x(\pi^-,y)|}{2(e^\pi + |x(\pi^-,y)|)}, \ t \in (\pi,2\pi], \ y \in (0,1). \end{cases} \tag{4.13}$$

Denote $f(t,x)(y) = \frac{|x(y)|}{1+|x(y)|} + y\sin t$ and $g_1(t,x)(y) = \frac{y\cos t|x(y)|}{2(e^\pi + |x(y)|)}$. So

$$\|f(t,x)\| \leqslant \|y\sin t\| + \left\| \frac{|x(y)|}{1+|x(y)|} \right\| \leqslant 1 + \|x\|,$$

$$\|g_1(t,x)\| \leqslant \frac{1}{2}.$$

Hence, we can set $M_f = 1$ and $M_{g_1} = \frac{1}{2}$. Problem (4.13) can be abstracted into

$$\begin{cases} x' = Ax + f(t,x), \ t \in (0,\pi), \\ x(\pi^+) = g_1(\pi, x(\pi^-)), \\ x(t) = g_1(t, x(\pi^-)), \ t \in (\pi, 2\pi], \\ x(0) = x(2\pi). \end{cases}$$

Obviously, Θ given by (4.10) reduces to

$$\Theta = KM_{g_1}e^{(KM_f - \omega)(t_1 - s_0)} = \frac{1}{2} < 1.$$

Thus our above results can be used to derive the existence of 2π-periodic *PC*-mild solutions to problem (4.13). Moreover, there is a global compact connected attractor $\mathcal{A}$ of (4.13) with a finite dimension.

4.1.4 Notes and remarks

This section is devoted to studying the existence of periodic solutions for a class of nonlinear evolution equations with non-instantaneous impulses on Banach spaces. By constructing a Poincaré operator, which is a composition of the maps, and using the techniques of *a priori* estimates, we avoid assuming that the periodic solution is bounded, as it is in [15–18], and try to present new sufficient conditions on the existence of periodic mild solutions for such problems by utilizing semigroup theory and Leray–Schauder's fixed point theorem. Furthermore, the existence of a global compact connected attractor for the Poincaré operator is derived.

The results in section 4.1 are motivated by [19].

4.2 First-order non-autonomous evolution equations

4.2.1 Introduction

It has been recognized that periodically varying environments play an important role in many dynamic biological and ecological systems. The periodic evolutionary process of many non-autonomous biological and ecological dynamic models whose motions depend on abrupt changes in their states are best described using differential equations with instantaneous periodic impulses.

The hemodynamic equilibrium of a person in a random, periodically varying environment, the injection of drugs into the bloodstream and their consequent absorption in the body are periodic, gradual and continuous processes. So this situation should be regarded as continuous impulsive and periodic action, which begins at an arbitrary fixed point and stays active on one periodic time interval. Thus, this motivated us to study a model described by non-autonomous periodic evolution equations with either deterministic or random non-instantaneous periodic impulses.

The main objective of this section is to look for periodic solutions and subharmonic solutions of the following non-autonomous periodic evolution equations with random non-instantaneous impulses

$$\begin{cases} u'(t) + A(t)u(t) = q_i(t, u(t)), \ t \in (s_i, t_{i+1}), \ i = 0, 1, 2, \dots, \infty, \\ u(t_i^+) = g_i^j(t_i, u(t_i^-)), \ i = 1, 2, \dots, \infty, \ j = 1, 2, \dots, n_i, \\ u(t) = g_i^j(t, u(t_i^-)), \ t \in (t_i, s_i], \ i = 1, 2, \dots, \infty, \ j = 1, 2, \dots, n_i, \end{cases} \tag{4.14}$$

where $A(t)$: $D(A(t)) \to X$, $t \geqslant 0$, is a family of T-periodic, linear unbounded operators on a Banach space X, which can generate the strongly continuous evolutionary process $\{U(t, s), t \geqslant s \geqslant 0\}$. The fixed points s_i and t_i satisfy $0 = s_0 < t_1 \leqslant s_1 \leqslant t_2 < \cdots < t_m \leqslant s_m \leqslant t_{m+1} < \cdots$ with $\lim_{i\to\infty} t_i = \infty$ and $t_{i+m} = t_i + T$, $s_{i+m} = s_i + T$, $m \in \mathbb{N}$ denoting the number of impulsive points between 0 and T. Moreover, g_i^j: $[t_i, s_i] \times X \to X$ are T-periodic continuous functions for all $i = 1, 2, \dots, \infty$ with $g_{i+m}^j = g_i^j$, $j = 1, 2, \dots, n_i$, $n_{i+m} = n_i$, and g_i^j appearing in (4.14) with a given probability $p_{ij} > 0$. Hence $\sum_{j=1}^{n_i} p_{ij} = 1$ and $p_{(i+m)j} = p_{ij}$ for all i. We establish a lower bound for the probability on the existence of periodic and subharmonic solutions of (4.14) under addition conditions (see theorems 4.2.16 and 4.2.17). In the final theoretical section, we also study in more detail the dynamic properties of a random system (4.14) (see theorem 4.2.18). In particular, we study

$$\begin{cases} u'(t) - \mu(t)\mathbb{A}u(t) = 0, \ t \in (s_i, t_{i+1}), \ i = 0, 1, 2, \dots, \infty, \\ u(t_i^+) = \zeta_j a(t_i^+) + b(t_i^+)Bu(t_i^-), \ i = 1, 2, \dots, \infty, \ j = 1, 2, \\ u(t) = \zeta_j a(t) + b(t)Bu(t_i^-), \ t \in (t_i, s_i], \ i = 1, 2, \dots, \infty, \ j = 1, 2, \end{cases} \tag{4.15}$$

where $\mathbb{A}$ is the infinitesimal generator of a C_0-semigroup $\{S(t), t \geqslant 0\}$ in X, a: $\mathbb{R} \to X$, μ: $\mathbb{R} \to (0, \infty)$ and b: $\mathbb{R} \to \mathbb{R}$ are T-periodic continuous functions, $s_{i+1} = s_i + T$, $t_{i+1} = t_i + T$, so $m = 1$, and $\zeta_j \in \mathbb{R}$ appears in (4.15) with given probabilities $p_j > 0$. Hence $p_1 + p_2 = 1$. Furthermore, $B \in L(X)$ and μ is Lipschitz continuous. We present a method in theorem 4.2.19 such that (4.15) can have a random attractor/fractal either in the random Cantor set in $\mathbb{R}$ or random Sierpinski triangle in $\mathbb{R}^2$ [20, p 335], for example, or in any finite-dimensional space. Then we derive a lower bound for the probability that (4.14) has a globally asymptotically stable periodic mild solution (see theorem 4.2.22). Note that the results of section 4.2.2 and 4.2.3 are preparatory steps for obtaining our main results for this part of the monograph, presented in sections 4.2.4 and 4.2.5. In particular, we start this section by looking for periodic solutions of the following non-autonomous periodic evolution equations with deterministic non-instantaneous impulses:

$$\begin{cases} u'(t) + A(t)u(t) = q_i(t, u(t)), \ t \in (s_i, t_{i+1}), \ i = 0, 1, 2, \dots, \infty, \\ u(t_i^+) = g_i(t_i, u(t_i^-)), \ i = 1, 2, \dots, \infty, \\ u(t) = g_i(t, u(t_i^-)), \ t \in (t_i, s_i], \ i = 1, 2, \dots, \infty, \end{cases} \tag{4.16}$$

where q_i: $[s_i, t_{i+1}] \times X \to X$ is a T-periodic continuous function for all $i = 0, 1, 2, \dots, \infty$ with $q_{i+m} = q_i$ and g_i: $[t_i, s_i] \times X \to X$ is another T-periodic continuous function for all $i = 1, 2, \dots, \infty$ with $g_{i+m} = g_i$. To achieve our aim, we have to show the continuity and compactness of the corresponding Poincaré map P: $X \to X$ of (4.16) given by (4.20). Then, we establish new sufficient conditions for the existence of periodic mild solutions when PC-mild solutions are ultimately bounded (see theorems 4.2.11 and 4.2.12). Furthermore, a global asymptotic stability result of periodic solutions is presented in theorem 4.2.15. These results are applied in sections 4.2.4 and 4.2.5 for the random cases (4.14) and (4.15).

The final section 4.2.6 is devoted to concrete examples to illustrate the theory.

This section is a continuation of our recent related papers [19, 21]. However, we note that it seems that we are the first to study the above evolution equations with random non-instantaneous impulses, which is the main novelty of this section. It is interesting that fractals are studied similarly in [22] for an economic, random, discrete-time, two-sector optimal growth model in which the production of a homogeneous consumption good uses a Cobb–Douglas technology.

4.2.2 Preliminaries

Let $J = [0, T]$. Denote $C(J, X)$ by the Banach space of all continuous functions from J into X with the norm $\|x\|_C := \sup\{\|x(t)\|: t \in J\}$ for $u \in C(J, X)$. We consider the Banach space $PC(J, X) := \{x: J \to X: x \in C((t_k, t_{k+1}], X), \ k = 0, 1, \dots, m$ and there exist $x(t_k^-)$ and $x(t_k^+)$, $k = 1, \dots, m$, with $x(t_k^-) = x(t_k)\}$ endowed with the Chebyshev PC-norm $\|x\|_{PC} := \sup\{\|x(t)\|: t \in J\}$. Similarly, we can define the Banach space $PC([0, \infty), X)$ endowed with the PC-norm $\|x\|_{PC} := \sup\{\|x(t)\|: t \in [0, \infty)\}$.

Let $\{A(t), T \geqslant t \geqslant 0\}$ be a family of closed densely defined linear unbounded operators acting on X and assume that this family has the following three standard conditions [12]:

$[A_1]$ The domain $D(A(t)) := D$ is independent of t and is dense in X.

$[A_2]$ For $t \geqslant 0$, the resolvent $R(\lambda, A(t)) = (\lambda I - A(t))^{-1}$ exists for all λ with $\Re\lambda \leqslant 0$, and there is a constant M independent of λ and t such that

$$\| R(\lambda, A(t)) \| \leqslant M(1 + |\lambda|)^{-1} \text{ for } \Re\lambda \leqslant 0.$$

$[A_3]$ There exist constants $L > 0$ and $0 < \alpha \leqslant 1$ such that

$$\| (A(t) - A(s))A^{-1}(\tau)\| \leqslant L|t - s|^{\alpha} \text{ for } t, s, \tau \in J.$$

Lemma 4.2.1. *(See [12, p 159].) Assume that $[A_1]$–$[A_3]$ are satisfied. Then*

$$\begin{cases} x'(t) + A(t)x(t) = 0, & t \in (0, T], \\ x(0) = \bar{x}, \end{cases} \tag{4.17}$$

has a unique evolution system $\{U(t, s): 0 \leqslant s \leqslant t \leqslant T\}$ in X satisfying the following properties:

(i) There exists $M > 0$ such that $\sup_{0\leqslant s\leqslant t\leqslant T}\|U(t, s)\| \leqslant M$.

(ii) $U(t, r)U(r, s) = U(t, s)$ for $0 \leqslant s \leqslant r \leqslant t \leqslant T$.

(iii) $U(\cdot, \cdot)x \in C(\Delta, X)$ for $x \in X$, $\Delta = \{(t, \theta) \in [0, T] \times [0, T]$: $0 \leqslant s \leqslant t \leqslant T\}$.

(iv) For $0 \leqslant s < t \leqslant T$, $U(t, s): X \longrightarrow D$ and $t \longrightarrow U(t, s)$ is strongly differentiable in X. The derivative $\sup_{0\leqslant s\leqslant t\leqslant T}\|\frac{\partial}{\partial t}U(t, s)\| \leqslant M$ and it is strongly continuous on $0 \leqslant s < t \leqslant T$. Moreover,

$$\frac{\partial}{\partial t}U(t, s) = -A(t)U(t, s) \text{ for } 0 \leqslant s < t \leqslant T;$$

$$\left\| \frac{\partial}{\partial t}U(t, s)\right\| = \|A(t)U(t, s)\| \leqslant \frac{C}{t - s};$$

$$\| A(t)U(t, s)A(s)^{-1}\| \leqslant C \text{ for } 0 \leqslant s \leqslant t \leqslant T.$$

(v) For every $v \in D$ and $t \in (0, T]$, $U(t, s)v$ is differentiable with respect to s on $0 \leqslant s \leqslant t \leqslant T$

$$\frac{\partial}{\partial \theta}U(t, s)v = U(t, s)A(s)v.$$

For each $\bar{x} \in X$, equation (4.17) also has a unique classical solution $x \in C^1(J, X)$ given by $x(t) = U(t, 0)\bar{x}$, $t \in J$.

The above lemma implies that (4.17) is well-posed. For the concept of well-posedness and further details we refer the reader to [23, theorem 6.1], [24, theorem 2.6] and [25].

Moreover, we impose the following additional assumption:

$[A_4]$ $A(t)$ is T-periodic in t, i.e. $A(t + T) = A(t)$ for $t \geqslant 0$.

Lemma 4.2.2. *Assume that $[A_1]$–$[A_4]$ are satisfied. Then evolution system $\{U(t, s): 0 \leqslant s \leqslant t \leqslant T\}$ satisfies the following property:*
(vi) $U(t + T, s + T) = U(t, s)$ *for* $0 \leqslant s \leqslant t \leqslant T$.

Now we can introduce the following standard definition.

Definition 4.2.3. *A function $u \in PC(J, X)$ is called a mild solution of the Cauchy problem*

$$\begin{cases} u'(t) + A(t)u(t) = q_i(t, u(t)), & t \in (s_i, t_{i+1}), \; i = 0, 1, 2, \dots, m, \\ u(t_i^+) = g_i(t_i, u(t_i^-)), & i = 1, 2, \dots, m, \\ u(t) = g_i(t, u(t_i^-)), & t \in (t_i, s_i], \; i = 1, 2, \dots, m, \\ u(0) = \overline{x}, \end{cases} \tag{4.18}$$

if u satisfies

$$\begin{cases} u(t) = g_i(t, u(t_i^-)), \quad t \in (t_i, s_i], \; i = 1, 2, \dots, m, \\ u(t_i^+) = g_i(t_i, u(t_i^-)), \quad i = 1, 2, \dots, m, \\ u(t) = U(t, 0)\overline{x} + \displaystyle\int_0^t U(t, s)q_i(s, u(s))ds, \quad t \in [0, t_1], \\ u(t) = U(t, s_i)g_i(s_i, u(t_i^-)) + \displaystyle\int_{s_i}^t U(t, s)q_i(s, u(s))ds, \quad t \in [s_i, t_{i+1}], \\ \qquad i = 1, 2, \dots, m. \end{cases} \tag{4.19}$$

In what follows, we introduce the following assumptions and establish an existence result for (4.18).

$[Q_1]$ q_i: $[s_i, t_{i+1}] \times X \to X$ is continuous for all $i = 0, 1, 2, \dots, m$ and for any $u, v \in X$ satisfying $\|u\|, \|v\| \leqslant \rho$ there exists a positive constant $L_{qi}(\rho) > 0$ such that

$$\|q_i(t, u) - q_i(t, v)\| \leqslant L_{q_i}(\rho)\|u - v\|.$$

$[Q_2]$ There exist constants $m_{qi} \geqslant 0$ and $M_{qi} \geqslant 0$, $i = 0, 1, 2, \dots, m$ such that

$$\|q_i(t, u)\| \leqslant m_{q_i} + M_{q_i}\|u\| \text{ for all } u \in X.$$

$[G_1]$ g_i: $[t_i, s_i] \times X \to X$ is continuous for all $i = 1, 2, \dots, m$ and for any $u, v \in X$ satisfying $\|u\|, \|v\| \leqslant \rho$ there exists a positive constant $L_{gi}(\rho) > 0$ such that

$$\|g_i(t, u) - g_i(t, v)\| \leqslant L_{g_i}(\rho)\|u - v\|.$$

$[G_2]$ There exist constants $m_{gi} \geqslant 0$ and $M_{gi} \geqslant 0$, $i = 0, 1, 2, \dots, m$ such that

$$\|g_i(t, u)\| \leqslant m_{g_i} + M_{g_i}\|u\| \text{ for each } t \in [t_i, s_i] \text{ and all } u \in X.$$

By adopting a similar procedure to [19, theorem 2.1], [21, theorem 2.2] and using the standard method [12, theorem 5.3.3, p 169] via the Banach contraction principle, we can obtain the following existence and uniqueness results.

Theorem 4.2.4. *Let the assumptions $[A_1]$–$[A_3]$, $[Q_1]$–$[Q_2]$ and $[G_1]$–$[G_2]$ be satisfied. Then (4.18) has an unique mild solution.*

4.2.3 The existence of periodic solutions for a determined case

In this section, we present the existence of periodic solutions of equation (4.16). In addition to $[A_1]$–$[A_3]$, $[Q_1]$–$[Q_2]$ and $[G_1]$–$[G_2]$, we need the following periodicity conditions on q_i and g_i:

$[Q_3]$ $q_i(t, u)$ is a T-periodic in t, i.e. $q_i(t + T, u) = q_i(t, u), t \in [s_i, t_{i+1}]$; hence $q_{i+m} = q_i$.

$[G_3]$ $g_i(t, u)$ is a T-periodic in t, i.e. $g_i(t + T, u) = g_i(t, u), t \in [s_i, t_{i+1}]$; hence $g_{i+m} = g_i$.

Definition 4.2.5. *A function $u \in PC([0, \infty), X)$ is said to be a T-periodic PC-mild solution of equation (4.16) if it is a PC-mild solution of (4.18) corresponding to some $\bar{x}$ and $u(t + T) = u(t)$ for $t \geqslant 0$.*

By adopting a similar procedure as in Fečkan *et al* [19, lemma 2.2] and using lemma 4.2.2 (iv) and $[Q_3]$ and $[G_3]$, we have the following result.

Lemma 4.2.6. *Equation (4.16) has a T-periodic PC-mild solution if and only if Poincaré operator P has a fixed point where $P: X \to X$ is given by*

$$\begin{aligned} P(\bar{x}) &= u(T, \bar{x}) \\ &= g_m(T, u(t_m^-, \bar{x})) \\ &= g_m\left(T, U(t, s_{m-1})g_{m-1}(s_{m-1}, u(t_{m-1}^-, \bar{x})) + \int_{s_{m-1}}^{t} U(t, s)q_i(s, u(s, \bar{x}))ds\right). \end{aligned} \tag{4.20}$$

Now we show the basic properties of P.

Lemma 4.2.7. *$P: X \to X$ defined in (4.20) is a continuous and compact operator.*

Proof. Note that P is a composition of the maps given by

$$\begin{aligned} &P_0(x) = u(t_1, x),\ \ u(t, x) = U(t, 0)x + \int_0^t U(t, s)q_i(s, u(s, x))ds,\ \ t \in [0, t_1]; \\ &G_i(x) = g_i(s_i, x),\ \ i = 1, 2, \dots, m; \\ &P_i(x) = u(t_{i+1}, x),\ \ u(t, x) = U(t, s_i)x \\ &\qquad + \int_{s_i}^t U(t, s)q_i(s, u(s, x))ds,\ \ t \in [s_i, t_{i+1}],\ \ i = 1, 2, \dots, m. \end{aligned}$$

Thus, we can rewrite

$$P = G_m \circ P_{m-1} \circ \cdots \circ G_1 \circ P_0. \tag{4.21}$$

By following the same procedure as in Fečkan *et al* [19], we know that each P_i and G_i are locally Lipschitz, so they are continuous, thus P is continuous as well. Next, by following the proof of [26, theorem 3.1] and using the fact that the embedding $X_\eta \to X$, $\eta \in (0, 1)$ is compact, we can prove that each P_i is compact, so P is compact as well. The proof is complete. □

To proceed, we introduce the following definitions [26, 27].

Definition 4.2.8. *We say that PC-mild solutions of (4.18) are locally bounded if for each $B_1 > 0$ and $k_0 > 0$, there is a $B_2 > 0$ such that $\|\overline{x}\| \leqslant B_1$ implies $\|u(t, \overline{x})\| \leqslant B_2$ for $0 \leqslant t \leqslant k_0$.*

Definition 4.2.9. *We say that PC-mild solutions of (4.18) are ultimate bounded if there is a bound $B > 0$, such that for each $B_3 > 0$ there is a $k > 0$ such that $\|\overline{x}\| \leqslant B_3$ and $t \geqslant k$ imply $\|u(t, \overline{x})\| \leqslant B$.*

Now we show the local boundedness of the solutions.

Lemma 4.2.10. *Under the above assumptions, PC-mild solutions of (4.18) are locally bounded.*

Proof. Using lemma 4.2.1 (i) and $[G_2]$, we have

$$\|G_i(x)\| \leqslant m_{g_i} + M_{g_i}\|x\|, \quad x \in X, i = 1, 2, \ldots, m, \tag{4.22}$$

while using $[Q_2]$, we derive

$$\|u(t, x)\| \leqslant M\|x\| + M\int_{s_i}^{t} \left(m_{q_i} + M_{q_i}\|u(s, x)\|\right)ds, \quad t \in [s_i, t_{i+1}],$$
$$i = 1, 2, \ldots, m-1, x \in X.$$

From the Grönwall inequality we have

$$\|u(t, x)\| \leqslant M(\|x\| + m_{q_i}(t - s_i))e^{MM_{q_i}(t-s_i)},$$

which implies that

$$\|P_i(x)\| = \|u(t_{i+1}, x)\| \leqslant m_{q_i}(t_{i+1} - s_i)e^{MM_{q_i}(t_{i+1}-s_i)} + Me^{MM_{q_i}(t_{i+1}-s_i)}\|x\|. \tag{4.23}$$

Now linking (4.22) with (4.23) we have

$$\|(G_i \circ P_{i-1})(x)\| \leqslant a + b\|x\|, \quad x \in X, i = 1, 2, \ldots, m, \tag{4.24}$$

where

$$a = \max_{i=1,2,\ldots,m} \left\{ m_{g_i} + M_{g_i} m_{q_{i-1}} (t_i - s_{i-1}) e^{MM_{q_{i-1}}(t_i - s_{i-1})} \right\},$$

$$b = M \max_{i=1,2,\ldots,m} \left\{ M_{g_i} e^{MM_{q_{i-1}}(t_i - s_{i-1})} \right\}.$$

Note that using (4.24) and repeating a similar process again and again, we can derive from (4.21) that

$$\|P(x)\| = \|(G_m \circ P_{m-1} \circ \cdots \circ G_1 \circ P_0)(x)\| \leqslant B + C\|x\|, \tag{4.25}$$

where

$$C = M^m \prod_{i=1}^{m} M_{g_i} e^{M \sum_{i=1}^{m} M_{q_{i-1}}(t_i - s_{i-1})}, \quad B = a(1 + b + \cdots + b^{m-1}). \tag{4.26}$$

Clearly (4.25) gives the desired result. □

Now we are ready to present the main results of this section.

Theorem 4.2.11. *Assume that $[A_1]$–$[A_4]$, $[Q_1]$–$[Q_3]$ and $[G_1]$–$[G_3]$ are satisfied. If the solutions of equation (4.16) are ultimate bounded, then equation (4.16) has at least a T-periodic PC-mild solution.*

Proof. Using lemma 4.2.10, we can directly follow the proof of [26, theorem 3.2] to obtain our result via Horn's fixed point theorem, so we do not go into detail. □

To proceed in this section, we present another existence results via the well-known Schauder's fixed point theorem.

Theorem 4.2.12. *Assume that $[A_1]$–$[A_4]$, $[Q_1]$–$[Q_3]$, and $[G_1]$–$[G_3]$ are satisfied. If $C < 1$ is defined in (4.26), then equation (4.16) has at least a T-periodic PC-mild solution.*

Proof. Take $\rho \geqslant \frac{B}{1-C}$, and define

$$W := \{x \in X: \|x\| \leqslant \rho\} \subset X.$$

From lemma 4.2.7, P: $W \to W$ is continuous and compact. Hence Schauder fixed point theorem gives the result. □

Remark 4.2.13. *As a matter of fact, the condition $C < 1$ implies that the solutions of equation (4.16) are ultimate bounded. So theorem 4.2.12 also follows from theorem 4.2.11.*

Remark 4.2.14. *The results of [21] can be directly applied for Ulam's type stability problems for equation (4.16) so we refer the reader for more details to the paper [21].*

To end this section, we suppose that the functions L_{qi} in $[Q_1]$ and L_{gi} in $[G_1]$ are constants, so q_i and g_i are globally Lipschitz continuous with constants L_{qi} and L_{gi}, respectively. Moreover, we suppose the asymptotic stability of $U(\cdot, \cdot)$:

$[A_5]$ $\|U(t, s)\| \leqslant Me^{-\omega(t-s)}$ for any $t \geqslant s \geqslant 0$ and some positive constants $M \geqslant 1$ and $\omega > 0$.

Then following the above procedure, we see that each P_i is globally Lipschitz continuous with a constant $Me^{(ML_{q_i}-\omega)(t_{i+1}-s_i)}$. In fact, using our assumptions, we can obtain

$$\|u(t, \bar{x}) - v(t, \bar{y})\| \leqslant Me^{-\omega(t-s_i)}\|\bar{x} - \bar{y}\| + ML_{q_i} \int_{s_i}^{t} e^{-\omega(t-s)}\|u(s, \bar{x}) - v(s, \bar{y})\|ds,$$

for $t \in [s_i, t_{i+1}]$, $i = 0, 1, \ldots, m$. By virtue of the Grönwall inequality, we obtain

$$\|u(t, \bar{x}) - v(t, \bar{y})\| \leqslant Me^{(ML_{q_i}-\omega)(t-s_i)}\|\bar{x} - \bar{y}\| \text{ for all } t \in [s_i, t_{i+1}], \; i = 0, 1, \ldots, m,$$

which implies that

$$\|P_j(\bar{x}) - P_j(\bar{y})\| \leqslant Me^{(ML_{q_i}-\omega)(t_{i+1}-s_i)}\|\bar{x} - \bar{y}\|.$$

Then each $G_i \circ P_{i-1}$ is globally Lipschitz continuous with a constant $ML_{g_i}e^{(ML_{q_{i-1}}-\omega)(t_i-s_{i-1})}$. Consequently, P is globally Lipschitz continuous with a constant

$$L_p := M^m \prod_{i=1}^{m} L_{g_i}e^{(ML_{q_{i-1}}-\omega)(t_i-s_{i-1})}. \tag{4.27}$$

Thanks to the well-known Banach fixed point theorem, we have the following result.

Theorem 4.2.15. *Assume that $[A_1]$–$[A_5]$, $[Q_1]$, $[Q_3]$, and $[G_1]$, $[G_3]$ are satisfied. If $L_p < 1$ is defined in (4.27), then equation (4.16) has a unique T-periodic PC-mild solution which is globally asymptotically stable.*

Proof. From our assumptions, P is globally contractive. From the Banach fixed point theorem, it has a unique fixed point which is a global attractor. The proof is complete. □

4.2.4 Existence of periodic and subharmonic solutions for a random case

In this section, we seek periodic solutions of (4.14). First, we set

$$\mathscr{P} := \prod_{i=1}^{m} \{1, 2, \dots, n_i\},$$

and for any $\chi \in \mathscr{P}$, $\chi = (j_1, j_2, \dots, j_m)$, $j_i \in \{1, 2, \dots, n_i\}$, we define its probability

$$\eta(\chi) := \prod_{i=1}^{m} p_{ij_i}.$$

Note $\sum_{\chi \in \mathscr{P}} \eta(\chi) = 1$, i.e. $\eta: \mathscr{P} \to [0, \infty)$ is really a discrete probability measure on $\mathscr{P}$.

Furthermore, for any $\chi \in \mathscr{P}$, $\chi = (j_1, j_2, \dots, j_m)$, $j_i \in \{1, 2, \dots, n_i\}$, we consider

$$\begin{cases} u'(t) + A(t)u(t) = q_i(t, u(t)), \ t \in (s_i, t_{i+1}), \\ \qquad\qquad i = 0, 1, 2, \dots, m-1, \\ u(t_i^+) = g_i^{j_i}(t_i, u(t_i^-)), \ i = 1, 2, \dots, m, \\ u(t) = g_i^{j_i}(t, u(t_i^-)), \ t \in (t_i, s_i], \ i = 1, 2, \dots, m. \end{cases} \tag{4.28}$$

Now we extend $[G_1]$ and $[G_2]$, so there are corresponding constants $M_{g_i^j}$ with appropriate properties. Then as in (4.20), we can give the corresponding Poincaré map P_χ of (4.28). Furthermore, as in (4.26), we have the constant

$$C(\chi) = \prod_{i=1}^{m} M_{g_i^{j_i}} e^{\sum_{i=1}^{m} M_{g_{i-1}}(t_i - s_{i-1})}. \tag{4.29}$$

Theorem 4.2.12 implies that if $C(\chi) < 1$ then equation (4.28) has at least a T-periodic PC-mild solution. Note that the probability of P_χ in (4.14) is just $\eta(\chi)$. Summarizing, we arrive at the following result.

Theorem 4.2.16. *The probability that (4.14) has a T-periodic mild solution is greater or equal to*

$$\sum_{\prod_{i=1}^{m} M_{g_i^{j_i}} e^{\sum_{i=1}^{m} M_{g_{i-1}}(t_i - s_{i-1})} < 1}^{m} \prod_{i=1}^{m} p_{ij_i}.$$

Proof. Indeed, the probability that (4.14) has a T-periodic mild solution is greater or equal to

$$\eta(C^{-1}([0, 1))) = \sum_{C(\chi)<1} \eta(\chi) = \sum_{\prod_{i=1}^{m} M_{g_i^{j_i}} e^{\sum_{i=1}^{m} M_{g_{i-1}}(t_i - s_{i-1})} < 1}^{m} \prod_{i=1}^{m} p_{ij_i},$$

where C is considered as $C: \mathscr{P} \to [0, \infty)$. □

Certainly, we can extend the above method for studying nT-subharmonic mild solutions of (4.14) as follows. We take $\mathscr{P}^n$ with the corresponding probability measure η_n and for any $\chi = (\chi_1, \dots, \chi_n) \in \mathscr{P}^n$, we set $P_\chi = P_{\chi_n} \circ \cdots \circ P_{\chi_1}$. Note $\eta_n(\chi) = \eta(\chi_1)\cdots\eta(\chi_n)$. The fixed points of P_χ determine nT-subharmonic mild solutions of (4.14). Now we have constants $C_n(\chi) = C(\chi_1)\cdots C(\chi_n)$, so $C_n: \mathscr{P}^n \to [0, \infty)$. So we have the following result.

Theorem 4.2.17. *The probability that (4.14) has a nT-subharmonic mild solution is greater or equal to*

$$\sum_{C(\chi_1)\cdots C(\chi_n)<1} \eta(\chi_1)\cdots\eta(\chi_n).$$

Proof. Indeed, the probability that (4.14) has a nT-subharmonic mild solution is greater or equal to

$$\eta(C_n^{-1}([0, 1))) = \sum_{C_n(\chi)<1} \eta_n(\chi) = \sum_{C(\chi_1)\cdots C(\chi_n)<1} \eta(\chi_1)\cdots\eta(\chi_n).$$

The proof is complete. □

4.2.5 Dynamics of random systems

In this section, we study in more detail the dynamics of random system (4.14) by assuming

$[D_1]$ $C(\chi) < 1$ for all $\chi \in \mathscr{P}$.

Then any P_χ has a fixed point. Moreover, there are $0 < C_M < 1$ and $B_M > 0$ so that

$$\|P_\chi(x)\| \leqslant B_M + C_M\|x\|, \quad \forall\, x \in X, \forall\, \chi \in \mathscr{P}. \tag{4.30}$$

So for any $\chi \in \mathscr{P}^n$, we have

$$\begin{aligned}\|P_{\chi_n} \circ \cdots \circ P_{\chi_1}(x)\| &\leqslant B_M(1 + C_M + \cdots + C_M^{n-1}) + C_M^n\|x\| \\ &\leqslant \frac{B_M}{1 - C_M} + C_M^n\|x\|.\end{aligned}$$

This means that any random iteration $\{P_{\chi_n} \circ \cdots \circ P_{\chi_1}(x)\}_{n=1}^{\infty}$ for $(\chi_1, \chi_2, \dots) \in \mathscr{P}^\infty$ enters after a finite iteration in the ball

$$\mathfrak{B} := \left\{x \in X: \|x\| \leqslant \frac{2B_M}{1 - C_M}\right\}.$$

Consequently, the study of random iterations can be restricted on $\mathfrak{B}$. Then from $[Q_1]$, $[G_1]$ and using the Grönwall inequality as above, we see that each G_i and p_i, $i = 1, \ldots, m$ is globally Lipschitz on $\mathfrak{B}$. Thus all P_χ, $\chi \in \mathscr{P}$ are globally Lipschitz on $\mathfrak{B}$ as well with constants $L(\chi)$, respectively (see the details in (4.27)). Now we suppose

$[D_2]$ $L(\chi) < 1$ for all $\chi \in \mathscr{P}$.

Then of course $[D_2]$ implies $[D_1]$ on $\mathfrak{B}$, since $C(\chi) \leqslant L(\chi)$ on $\mathfrak{B}$. From our assumptions, the set

$$Y := \overline{\bigcup_{\chi \in \mathscr{P}} P_\chi(\mathfrak{B})}$$

is a compact subset of X. Moreover, equation (4.30) gives $Y \subset \mathfrak{B}$, so then $P_\chi(Y) \subset P_\chi(\mathfrak{B}) \subset Y$ for each $\chi \in \mathscr{P}$. Summarizing, the study of random iterations can be restricted on the compact set Y. Then we consider the compact metric space $(Y, \|\cdot\|)$ and let $(H(Y), h)$ be the corresponding space of nonempty compact subsets of Y with the Hausdorff metric h [20]. Since from $[D_2]$, all P_χ: $Y \to Y$ are contractions, they create an iterated function system (IFS) [20, definition 7.1, p 80]. Next, we define the transformation W: $H(Y) \to H(Y)$ by

$$W(Z) := \bigcup_{\chi \in \mathscr{P}} P_\chi(Z).$$

It is well known [20, theorem 7.1, p 81] that W is a contraction with a constant $L_Y = \max_{\chi \in \mathscr{P}} L(\chi) < 1$. So from the Banach fixed point theorem, W has a unique fixed point F_W, called the attractor or fractal of IFS. Furthermore, since each P_χ has its probability $\eta(\chi)$, we are dealing with IFS with probabilities (IFSP) according to [20, definition 1.1, p 330]. By following [20, p 349], we consider the set $P(Y)$ of normalized Borel measures on Y with the Hutchinson metric d_H. Then $(P(X), d_H)$ is a compact metric space. Next, our IFSP defines a Markov operator M: $P(Y) \to P(Y)$ by

$$M(\nu) := \sum_{\chi \in \mathscr{P}} \eta(\chi) \nu \circ P_\chi^{-1}.$$

We know that M is a contraction on Y with a constant L_Y, so it has a unique fixed point $\mu \in P(Y)$, i.e. $M(\mu) = \mu$. This μ is called the invariant measure of the IFSP. The relationship between F_W and μ is given by the fact that the support of μ is F_W, i.e. μ 'lives' in F_W. Furthermore, using the Carathéodry theorem [20, p 342]) η_n can be uniquely extended to η_∞ on $\mathscr{P}^\infty$. Now for any $x_0 \in Y$ and $\chi \in \mathscr{P}^\infty$, we consider the random iteration $\{x_0, x_1, \ldots\}$ for $x_n := P_{\chi_n} \circ \cdots \circ P_{\chi_1}(x_0)$. Then for a Borel set $Z \subset Y$, we define

$$N(Z, n) := \text{number of points in } \{x_0, x_1, \ldots\} \bigcap \mathfrak{B}.$$

Finally, we are ready to formulate the basic result on random iterations, a consequence of the Elton theorem [20, p 364]:

Theorem 4.2.18. *Assuming $[D_2]$, for almost all $\chi \in \mathscr{P}^\infty$ with respect to η_∞, it holds that*

$$\eta(Z) = \lim_{n\to\infty} \frac{N(Z, n)}{n + 1}$$

for any Borel set $Z \subset Y$ *and* $x_0 \in Y$.

Hence the dynamics of almost all random iterations are concentrated on F_W. This means that the dynamics of the random impulsive system (4.14) is determined by its dynamics on the compact fractal F_W. The fractal dimension of F_W can be estimated by [20, theorem 2.3, p 183] for special affine cases. Also, F_W can be approximated with respect to the Hausdorff metric h in concrete examples by using the iterations $I_n := W^n(\{0\})$ for sufficiently large n. Note I_n are finite sets and

$$h(F_W, I_n) \leqslant \frac{L_Y^n}{1 - L_Y} h(I_1, I_0) \leqslant \frac{2B_M L_Y^n}{(1 - L_Y)(1 - C_M)}.$$

To be more concrete, we consider equation (4.15) and suppose:

[B_1] The domain $D(\mathbb{A})$ is dense in X.

[B_2] The resolvent $R(\lambda, \mathbb{A}) = (\lambda I - \mathbb{A})^{-1}$ exists for all λ with $\Re\lambda \geqslant 0$, and there is a constant M independent of λ such that

$$\| R(\lambda, \mathbb{A}) \| \leqslant M(1 + |\lambda|)^{-1} \text{ for } \Re\lambda \geqslant 0.$$

Then $A(t) = -\mu(t)\mathbb{A}$ satisfies assumptions [A_1]–[A_4] and from lemma 4.2.1 we see

$$U(t, s) = S\left(\int_s^t \mu(\tau)d\tau\right), \quad t \geqslant s \geqslant 0$$

for the C_0-semigroup $\{S(t), t \geqslant 0\}$ of $\mathbb{A}$. Now $\mathscr{P} = \{1, 2\}$ and then

$$P_j = \zeta_j a(T) + b(T)BS\left(\int_0^{t_1^-} \mu(\tau)d\tau\right), \quad j = 1, 2. \tag{4.31}$$

To simplify our consideration further, we assume

[B_3] $X = X_1 \oplus X_2$, $0 < \dim X_1 < \infty$ with $D(\mathbb{A}) = D(\mathbb{A}_1) \oplus D(\mathbb{A}_2)$, $D(\mathbb{A}_i) \subset X_i$ and $\mathbb{A}$: $D(\mathbb{A}_i) \to X_i$, so $\mathbb{A}_i := \mathbb{A}/X_i$, $i = 1, 2$.

[B_4] B: $X_i \to X_i$, $i = 1, 2$.

Using the splitting $x = x_1 + x_2$, $x_i \in X_i$, $i = 1, 2$, we have $B = (B_1, B_2)$, $S = (S_1, S_2)$, B_i, S_i: $X_i \to X_i$, $i = 1, 2$. Hence (4.31) splits into

$$P_j = \left(\zeta_j a(T) + b(T)B_1S_1\left(\int_0^{t_1^-} \mu(\tau)d\tau\right), b(T)B_2S_2\left(\int_0^{t_1^-} \mu(\tau)d\tau\right)\right), \tag{4.32}$$
$$j = 1, 2.$$

Assuming

$$M_i\|b(T)\|\|B_i\|e^{\omega_i\left(\int_0^{t_1^-} \mu(\tau)d\tau\right)} < 1, \quad i = 1, 2, \tag{4.33}$$

when

$$\|S_i(t)\| \leqslant M_i e^{\omega_i t}, \quad t \geqslant 0$$

for some $M_i \geqslant 1$ and $\omega_i \in \mathbb{R}$, we can apply the above theory to (4.32). Then the corresponding $F_W \subset X_1$ is generated by the restricted IFSP

$$\left\{\zeta_1 a(T) + b(T)B_1 S_1\left(\int_0^{t_1^-} \mu(\tau)d\tau\right), \zeta_2 a(T) + b(T)B_1 S_1\left(\int_0^{t_1^-} \mu(\tau)d\tau\right)\right\} \tag{4.34}$$

on X_1. Summarizing, we have the following result.

Theorem 4.2.19. *Under assumptions [B_1]–[B_4] and (4.33), the random dynamics of equation (4.15) is restricted on finite-dimensional X_1 with an IFSP represented by (4.34), where the corresponding attractor F_W is located in X_1.*

Consequently, we can construct problem (4.15) for which F_W can be either the random Cantor set in $\mathbb{R}$ or random Sierpinski triangle in $\mathbb{R}^2$ [20, p 335], for example. Furthermore, denoting

$$E := b(T)B_1 S_1\left(\int_0^{t_1^-} \mu(\tau)d\tau\right), \quad b_i = \zeta_i a(T), \ i = 1, 2, \tag{4.35}$$

IFSP (4.34) has the form $\{w_i, w_2\}$ with $w_i(x_1) = b_i + Ex_1$, $i = 1, 2$, $x_1 \in X_1$. Let $\delta > 0$ and $x_1 \in B_\delta(b_1) := \{x_1 \in X_1: \|x_1 - b_1\| \leqslant \delta\}$, then we compute

$$\|w_1(x_1) - b_1\| = \|Ex_1\| \leqslant \|E\|(\|b_1\| + \delta). \tag{4.36}$$

From (4.33), $\|E\| < 1$, so if $\frac{\| b_1 \|\| E \|}{1 - \| E \|} \leqslant \delta$ then (4.36) gives $\|w_1(x_1) - b_1\| \leqslant \delta$, i.e. $w_1(B_\delta(b_1)) \subset B_\delta(b_1)$. Similarly, if $\frac{\| b_2 \|\| E \|}{1 - \| E \|} \leqslant \delta$ then $w_1(B_\delta(b_2)) \subset B_\delta(b_1)$. Consequently, for

$$\delta = \max\{\|b_1\|, \|b_2\|\}\frac{\|E\|}{1 - \|E\|}, \tag{4.37}$$

it holds that

$$w_i(B_\delta(b_j)) \subset B_\delta(b_i), \quad i, j = 1, 2,$$

which implies

$$w_1(B_\delta(b_1) \cup B_\delta(b_2)) \subset B_\delta(b_1) \cup B_\delta(b_2).$$

So we arrive at

$$F_W \subset B_\delta(b_1) \cup B_\delta(b_2), \quad w_i(F_W)) \subset B_\delta(b_i), \quad i = 1, 2. \tag{4.38}$$

Clearly $F_W \cap B_\delta(b_i) \neq \emptyset$, $i = 1, 2$. Next if

$$\delta < \frac{\|b_1 - b_2\|}{2}, \tag{4.39}$$

then $B_\delta(b_1) \cap B_\delta(b_2) = \emptyset$. So from (4.38), we obtain $w_1(F_W) \cap w_2(F_W) = \emptyset$, i.e. [20, theorem 2.2, p 125] gives that IFSP (4.34) is completely disconnected. Then [20, theorem 5.1, p 147, theorem 8.1, p 167] can be applied to obtain the chaotic dynamics of IFSP (4.34). Combining (4.35), (4.37) and (4.39), we obtain

$$\|E\| < \frac{\|b_1 - b_2\|}{2\max\{\|b_1\|, \|b_2\|\} + \|b_1 - b_2\|} = \frac{|\zeta_1 - \zeta_2|}{2\max\{|\zeta_1|, |\zeta_2|\} + |\zeta_1 - \zeta_2|}, \tag{4.40}$$

and summarizing we arrive at the following theorem.

Theorem 4.2.20. *Under the assumptions of theorem 4.2.19 and*

$$M_1|b(T)|\,\|B_1\|\,e^{\omega_1\left(\int_0^{t_1^-}\mu(\tau)d\tau\right)} < \frac{|\zeta_1 - \zeta_2|}{2\max\{|\zeta_1|, |\zeta_2|\} + |\zeta_1 - \zeta_2|},$$

the dynamics of IFSP on the attractor F_W is chaotic.

Now let us consider the case when E is a similitude, i.e. $\|Ex_1\| = \|E\|\|x_1\|$ for any $x_1 \in X_1$. Then [20, theorem 4.3, p 199] implies the following.

Theorem 4.2.21. *Assume $[B_1]$–$[B_4]$, (4.33) with $i = 2$. Moreover, E is a similitude satisfying (4.40). Then dynamics of IFSP on the attractor F_W is chaotic, and the Hausdorff–Besicovitch and fractal dimensions of F_W are equal to $-\frac{\ln 2}{\ln \|E\|}$.*

We end this section with the following result.

Theorem 4.2.22. *If we assume $[A_5]$ and that all mappings are globally Lipschitz continuous, then the probability that (4.14) has a globally asymptotically stable T-periodic mild solution is greater or equal to*

$$\sum_{\prod_{i=1}^m L_{g_i^{j_i}} e^{(ML_{q_{i-1}}-\omega)(t_i-s_{i-1})}<1} \prod_{i=1}^m p_{ij_i}. \tag{4.41}$$

A similar result can be stated for globally asymptotically stable subharmonic mild solutions.

Proof. Under the assumptions, each P_χ is globally Lipschitz continuous with a constant $L(\chi)$. Then, as above, the probability that (4.14) has a globally asymptotically stable T-periodic mild solution is greater or equal to

$$\eta(L^{-1}([0, 1))) = \sum_{L(\chi)<1} \eta(\chi) = \sum^{m}_{\prod_{i=1}^{m} L_{g_i}^{j_i} e^{(ML_{q_{i-1}}-\omega)(t_i-s_{i-1})}<1} \prod_{i=1}^{m} p_{ij_i},$$

where L is considered as $L: \mathscr{P} \to [0, \infty)$. The proof is complete. □

4.2.6 Examples

This section is devoted to concrete examples for illustration of the above general theory.

Example 1. First, we consider the following impulsive non-autonomous periodic problem

$$\begin{cases} \dfrac{\partial}{\partial t}x(t, y) = (1 + \sin^2 t)\dfrac{\partial^2}{\partial y^2}x(t, y) + \dfrac{2\gamma}{\pi}\arctan|x(t, y)|, \\ \qquad \gamma \in \mathbb{R},\ y \in (0, 1),\ t \in (0, \frac{\pi}{2}), \\ x(\frac{\pi}{2}^+, y) = -\dfrac{|x(\frac{\pi}{2}^-, y)|}{\zeta},\ \zeta \in \mathbb{R}, \\ x(t, y) = \cos t + \dfrac{\cos(\frac{t}{2} + \frac{3\pi}{4})|x(\frac{\pi}{2}^-, y)|}{\zeta},\ t \in (\frac{\pi}{2}, \pi],\ y \in (0, 1), \\ x(t, 0) = x(t, 1) = 0,\ t \in [0, \pi], \\ x(0, y) = x(\pi, y),\ y \in (0, 1). \end{cases} \tag{4.42}$$

In order to rewrite (4.42) in the abstract form, we set $X = L^2([0,1])$ endowed with the usual norm and chose $0 = t_0 = s_0$, $t_1 = \frac{\pi}{2}$, $s_1 = T = \pi$ and $J = [0, \pi]$. Define $\mathbb{A}x = \frac{\partial^2}{\partial y^2}x$ for $x \in D(\mathbb{A})$ with $D(\mathbb{A}) = \left\{x \in X: \frac{\partial x}{\partial y}, \frac{\partial^2 x}{\partial y^2} \in X,\ x(0) = x(1) = 0\right\}$. Set $A(t) = -\mu(t)\mathbb{A}$ where $\mu(t) := 1 + \sin^2 t$, $t \geqslant 0$. Then assumption $[A_1]$ clearly holds. Next, for all λ with $\Re\lambda \leqslant 0$, we have

$$(\lambda I - A(t))x(\xi) = \frac{2}{\pi}\sum_{n=1}^{\infty}(\lambda - n^2\mu(t))\sin n\xi\left(\int_0^{\pi} x(s)\sin nsds\right),\ \xi \in [0, 1]$$

for any $x \in D(A)$. Since $|\lambda - n^2\mu(t)|^2 \geqslant |\lambda|^2 + n^4 \geqslant 1$, we derive

$$(\lambda I - A(t))^{-1}x(\xi) = \frac{\pi}{2}\sum_{n=1}^{\infty}\frac{1}{\lambda - n^2\mu(t)}\sin n\xi\left(\int_0^{\pi} x(s)\sin nsds\right),\ \xi \in [0, 1]$$

for any $x \in X$. So $R(\lambda, A(t))$ exists. Moreover, we have

$$\|R(\lambda, A(t))x\|^2 = \sum_{n=1}^{\infty} \frac{1}{|\lambda - n^2\mu(t)|^2} \left| \int_0^{\pi} x(s)\sin nsds \right|^2$$
$$\leqslant \sum_{n=1}^{\infty} \frac{1}{|\lambda|^2 + n^4} \left| \int_0^{\pi} x(s)\sin nsds \right|^2 \leqslant \frac{\pi}{2} \sum_{n=1}^{\infty} \frac{1}{|\lambda|^2 + n^4} \|x\|^2$$
$$\leqslant \frac{\pi}{2} \|x\|^2 \int_0^{\infty} \frac{dz}{|\lambda|^2 + z^4} = \frac{\pi^2}{4\sqrt{2}\,|\lambda|^{3/2}} \|x\|^2$$
$$\leqslant \frac{4\pi^2}{3\sqrt{6}\,(|\lambda| + 1)^2} \|x\|^2$$

for $|\lambda| \geqslant 1$, since $\frac{(|\lambda|+1)^2}{|\lambda|^{3/2}} \leqslant \frac{16}{3\sqrt{3}}$ for $|\lambda| \geqslant 1$. Consequently, we have

$$\|R(\lambda, A(t))\| \leqslant \frac{2\pi}{\sqrt{3}\sqrt[4]{6}\,(|\lambda| + 1)}$$

for $|\lambda| \geqslant 1$, $\Re\lambda \leqslant 0$ and $t \geqslant 0$. Moreover, the above computation shows that

$$\|R(\lambda, A(t))\| \leqslant \sqrt{\frac{\pi}{2} \sum_{n=1}^{\infty} \frac{1}{n^4}} = \sqrt{\frac{\pi^5}{180}}$$

for $|\lambda| \leqslant 1$, $\Re\lambda \leqslant 0$ and $t \geqslant 0$. Summarizing, assumption $[A_2]$ is verified. Next, using

$$(A(t) - A(s))A^{-1}(\tau)x(\xi) = \frac{\mu(s) - \mu(t)}{\mu(\tau)} \sum_{n=1}^{\infty} \sin n\xi \left(\int_0^{\pi} x(s)\sin nsds \right)$$
$$= \frac{\pi}{2} \frac{\mu(s) - \mu(t)}{\mu(\tau)} x(\xi)$$

for any $x \in X$, we obtain

$$\|(A(t) - A(s))A^{-1}(\tau)\| = \frac{\pi}{2} \left| \frac{\mu(s) - \mu(t)}{\mu(\tau)} \right| \leqslant \frac{\pi}{2} |\mu(s) - \mu(t)|$$
$$= \frac{\pi}{2} |\sin^2 s - \sin^2 t| \leqslant \pi |t - s|.$$

Hence assumption $[A_3]$ is verified as well. So the family $\{A(t),\ t \geqslant 0\}$ is well-posed (see also [24, example 2.9b]). Moreover, $\mu(t + \pi) = \mu(t)$, so $\mathbb{A}(t + \pi) = \mathbb{A}(t)$ for $t \geqslant 0$. Thus, $\{A(t),\ t \geqslant 0\}$ is π-periodic and assumption $[A_4]$ is satisfied.

On the other hand, $\mathbb{A}$ is the infinitesimal generator of a C_0-semigroup $\{S(t),\ t \geqslant 0\}$ in X which is given by [28, example 1.3]

$$(S(t))x(\xi) = \frac{2}{\pi} \sum_{n=1}^{\infty} e^{-tn^2} \sin n\xi \left(\int_0^{\pi} x(s)\sin nsds \right), \quad \xi \in [0, 1], \quad t \geqslant 0.$$

So the associated π-periodic evolution family of $\{A(t),\ t \geqslant 0\}$ is given by

$$U(t, s)x = S\left(\int_s^t \mu(\tau)d\tau \right)x, \quad x \in X, \quad t \geqslant 0.$$

Since $\int_s^t \mu(\tau)d\tau \geqslant t - s$ for any $t \geqslant s \geqslant 0$, we have $\|U(t, s)x\| \leqslant e^{-(t-s)}\|x\| \leqslant \|x\|$ for all $t \geqslant s \geqslant 0$. Thus $M = 1$ in (i) of lemma 4.2.1.

Now denote

$$q_1(t, x)(y) = \frac{2\gamma}{\pi}\arctan|x(y)|, \quad g_1(t, x)(y) = \cos t + \frac{\cos(\frac{t}{2} + \frac{3\pi}{4})|x(y)|}{\zeta}.$$

Obviously, f and g_1 are π-periodic functions in t and satisfy global Lipschitz conditions, respectively, and growth conditions

$$\|q_1(\cdot, x) - q_1(\cdot, z)\| \leqslant |\gamma|\frac{2}{\pi}\|x - z\|, \quad \|q_1(\cdot, x)\| \leqslant |\gamma|,$$

$$\|g_1(\cdot, x) - g_1(\cdot, z)\| \leqslant \frac{1}{|\zeta|}\|x - z\|, \quad \|g_1(\cdot, x)\| \leqslant 1 + \frac{1}{|\zeta|}\|x\|.$$

- By applying theorem 4.2.11 we obtain: If PC-mild solutions of (4.42) are ultimate bounded, then problem (4.42) has at least one π-periodic PC-mild solution.
- By applying theorem 4.2.12 we obtain: If $|\zeta| > 1$ ($\Longrightarrow$ ultimately boundedness), then problem (4.42) has at least one π-periodic PC-mild solution.
- By applying theorem 4.2.15 we obtain: If $|\zeta| > e^{|\gamma| - \frac{\pi}{2}}$, then problem (4.42) has a unique globally asymptotically stable π-periodic PC-mild solution.

Example 2. Problem (4.42) can be easily extended to a random case when $\zeta \in \{\zeta_1, \dots, \zeta_n\}$ with probability $_{pi} > 0$ for given ζ_i, $i = 1, 2, \dots, n$. Then this random problem has a π-periodic PC-mild solution with a probability greater than or equal to $\sum_{|\zeta_i|>1} p_i$. More generally, this random problem has an $n\pi$-subharmonic PC-mild solution with a probability greater or equal to $\sum_{|\zeta_{i_1}\cdots\zeta_{i_n}|>1} p_{i_1}\cdots p_{i_n}$. Furthermore, this random problem has a globally asymptotically stable π-periodic PC-mild solution with a probability greater than or equal to $\sum_{|\zeta_i|>e^{|\gamma|-\frac{\pi}{2}}} p_i$.

Example 3. The final example is as follows:

$$\begin{cases} \frac{\partial}{\partial t}x(t, y) = (1 + \sin^2 t)\frac{\partial^2}{\partial y^2}x(t, y), \ y \in (0, 1), \ t \in (0, \frac{\pi}{2}), \\ x(\frac{\pi}{2}^+, y) = -x(\frac{\pi}{2}^-, y), \\ x(t, y) = \zeta_j \cos t \sin y - \frac{\sqrt{2}e^{\frac{3\pi}{4}}}{3}\cos(\frac{t}{2} + \frac{3\pi}{4})x(\frac{\pi}{2}^-, y), \\ \qquad t \in (\frac{\pi}{2}, \pi], \ y \in (0, 1), \\ x(t, 0) = x(t, 1) = 0, \ t \in [0, \pi], \\ x(0, y) = x(\pi, y), \ y \in (0, 1), \end{cases} \tag{4.43}$$

when ζ_j has a probability $p_j \in (0, 1), j = 1, 2$. Equation (4.43) has the form of (4.15) with the above X, $\mathbb{A}$ and

$$a(t) = \cos t \sin y, \quad b(t) = -\frac{\sqrt{2}e^{\frac{3\pi}{4}}}{3}\cos\left(\frac{t}{2} + \frac{3\pi}{4}\right), \quad B = I.$$

We take a splitting

$$X = X_1 \oplus X_1^{\perp}, \quad X_1 := \mathrm{span}[\sin y].$$

Then

$$\left(S_1\left(\int_0^{t_1}\mu(\tau)d\tau\right)\right)x_1(\xi) = \frac{2}{\pi}e^{-\int_0^{t_1}\mu(\tau)d\tau}\sin\xi\left(\int_0^{\pi}x_1(s)\sin s ds\right)$$
$$= c_1 e^{-\frac{3\pi}{4}}\sin\xi$$

for $x_1(\xi) = c_1 \sin\xi$, $c_1 \in \mathbb{R}$. Similarly

$$\left(S_2\left(\int_0^{t_1}\mu(\tau)d\tau\right)\right)x_2(\xi) = \frac{2}{\pi}\sum_{n=2}^{\infty}e^{-\frac{3\pi}{4}n^2}\sin n\xi\left(\int_0^{\pi}x(s)\sin nsds\right)$$

for $x_2 \in X_1^{\perp}$. Which gives

$$\left\| S_1\left(\int_0^{t_1}\mu(\tau)d\tau\right)\right\| = e^{-\frac{3\pi}{4}}, \quad \left\| S_2\left(\int_0^{t_1}\mu(\tau)d\tau\right)\right\| = e^{-3\pi}. \tag{4.44}$$

Using $b(T) = \frac{e^{\frac{3\pi}{4}}}{3}$, the inequalities of (4.33) are satisfied:

$$\frac{e^{\frac{3\pi}{4}}}{3}e^{-\frac{3\pi}{4}} = \frac{1}{3} < 1, \quad \frac{e^{\frac{3\pi}{4}}}{3}e^{-3\pi} = \frac{e^{\frac{-9\pi}{4}}}{3} < 0.0003.$$

Hence theorem 4.2.19 can be applied. Furthermore, mapping of (4.34) has the form

$$c_1 \to -\zeta_j + \frac{c_1}{3}$$

when $x_1(\xi) = c_1\sin\xi \in X_1$. Hence taking $\zeta_1 = 0$ and $\zeta_2 = -\frac{2}{3}$, mapping of (4.34) has the form

$$c_1 \to \left\{\frac{c_1}{3}, \frac{2}{3} + \frac{c_1}{3}\right\}$$

whose F_W is just the Cantor set [20, p 81]. Hence the random dynamics of (4.43) for $\zeta_1 = 0$ and $\zeta_2 = -\frac{2}{3}$ is concentrated on the Cantor set of span[$\sin y$]. We obtain something similar for general ζ_j.

4.2.7 Notes and remarks

This section is devoted to introducing a new class of impulsive models for studying the dynamics of periodic evolution processes in pharmacotherapy, which is given by random, non-instantaneous, impulsive and non-autonomous periodic evolution equations. This type of impulsive equation can describe the injection of drugs into the bloodstream, and the consequent absorption of them in the body as a random, periodic, gradual and continuous process. Sufficient conditions on the existence of periodic and subharmonic solutions are established, as are other related results such as their globally asymptotic stability. The dynamical properties are also derived for the whole system, leading to the theory of fractals

The results in section 4.2 are motivated by [29].

4.3 Second-order evolution equations

4.3.1 Introduction

Following the work in [1, 21, 30], we study the existence of periodic mild solutions of the second-order nonlinear differential equations with non-instantaneous impulses in a Banach space X of the form

$$\begin{aligned} x''(t) &= Ax(t) + f(t, x(t)), \quad t \in (s_i, t_{i+1}], \ i \in \mathbb{N}_0; \\ x(t) &= J_i^1(t, x(t_i^-)), \quad t \in (t_i, s_i], \ i \in \mathbb{N}; \\ x'(t) &= J_i^2(t, x(t_i^-)), \quad t \in (t_i, s_i], \ i \in \mathbb{N}, \end{aligned} \tag{4.45}$$

where $x(t)$ is a state function, $0 = s_0 < t_1 < s_1 < t_2, \dots, t_m < s_m < t_{m+1} < \cdots$ with $\lim_{i\to\infty} t_i = \infty$ and $t_{i+m} = t_i + T, i \in \mathbb{N}, s_{i+m} = s_i + T, i \in \mathbb{N}_0$ for some $m \in \mathbb{N}$ denoting the number of impulsive points between 0 and $T > 0$, and we set $\mathbb{N}_0 = \mathbb{N} \cup \{0\}$. We consider in (4.45) that $x \in C((t_i, t_{i+1}], X)$, $i \in \mathbb{N}$ and there exist $x(t_i^-)$ and $x(t_i^+)$, $i \in \mathbb{N}$ with $x(t_i^-) = x(t_i)$. The functions $J_i^1(t, x(t_i^-))$ and $J_i^2(t, x(t_i^-))$ represent non-instantaneous impulses during the intervals $(t_i, s_i]$, $i \in \mathbb{N}$, so impulses at t_i^- have some duration, namely on intervals $(t_i, s_i]$. A is the infinitesimal generator of a strongly continuous cosine family of bounded linear operators $(C(t))_{t\in\mathbb{R}}$ on X. J_i^1, J_i^2 and f are suitable functions and they will be specified later. We construct a Poincaré operator for (4.45) and study its fixed points and dynamics.

Second-order differential equations play a very crucial role in the modeling of physical phenomena, for example, modeling the position of the mass attached to a spring over time and modeling the motion of a simple pendulum etc. A useful tool for the study of second-order abstract differential equations in infinite-dimensional space is the theory of strongly continuous cosine families of operators. The existence and uniqueness of the solutions for second-order nonlinear systems and the controllability of these systems in Banach spaces have been studied thoroughly by many authors [31–35]. Related problems are also studied in [19, 36–39].

The stucture of this section is as follows. In section 4.3.2 we provide some important notations, definitions and assumptions which are required for the establishment of the main results of the section. In section 4.3.3 we study the periodic solutions for problem (4.45). In the final section, 4.3.4, an example is given to show the application of these abstract results.

4.3.2 Preliminaries and assumptions

First, we briefly recall some definitions from the theory of the cosine family [40, pp 32–3].

Definition 4.3.1. *A one parameter family* $(C(t))_{t\in\mathbb{R}}$ *of bounded linear operators mapping the Banach space X into itself is called a strongly continuous cosine family if and only if*

(i) $C(s+t) + C(s-t) = 2C(s)C(t)$ *for all* $s, t \in \mathbb{R}$,
(ii) $C(0)$ *is the identity operator,*
(iii) $C(t)x$ *is continuous in t on* $\mathbb{R}$ *for each fixed point* $x \in X$.

The sine function $(S(t))_{t\in\mathbb{R}}$ associated with the strongly continuous cosine family $(C(t))_{t\in\mathbb{R}}$ is defined by

$$S(t)x = \int_0^t C(s)x\,ds, \quad x \in X, \; t \in \mathbb{R}.$$

The domain $D(A)$ of the operator A is defined by

$$D(A) = \{x \in X: C(t)x \text{ is twice continuously differentiable in } t\},$$

which is a Banach space endowed with the graph norm $\|x\|_A = \|x\| + \|Ax\|$ for all $x \in D(A)$. We define a set

$$E = \{x \in X: C(t)x \text{ is once continuously differentiable in } t\},$$

which is a Banach space endowed with a suitable norm $\|\cdot\|_E$ (see [40, p 46]). We also note that if $x: I \to X$, $I = [0, \infty)$ is a locally integrable function then

$$y(t) = \int_0^t S(t-s)x(s)ds$$

defines an E valued continuous function. We refer to Fattorini and Travis, and Webb [40–42] for more details on the cosine family theory.

In order to prove the existence of the periodic solution for problem (4.45), we need the following assumptions:

[A_1] A is the infinitesimal generator of a strongly continuous cosine family $(C(t))_{t\in\mathbb{R}}$ of bounded linear operators.

[A_2] $f: I_0 \times X \to X$, $I_0 = \bigcup_{i=0}^{\infty}[s_i, t_{i+1}]$ is a continuous function and there exists a positive constant K_f such that

$$\|f(t, x_1) - f(t, x_2)\| \leqslant K_f\|x_1 - x_2\|$$

for every $x_1, x_2 \in X$, $t \in I_0$.

[A_3] There exist non-negative constants L_f and M_f such that

$$\|f(t, x)\| \leqslant L_f\|x\| + M_f, \quad x \in X, \; t \in I_0.$$

[A_4] $f(t, x)$ is T-periodic in t, i.e. $f(t+T, x) = f(t, x)$, $t \in I_0$.

[A_5] $J_i^l \in C(I_i \times X, X)$, $I_i = [t_i, s_i]$ and there are positive constants K_{Ji}, $i \in \mathbb{N}$, such that

$$\max\{\|J_i^1(t, x_1) - J_i^1(t, x_2)\|, \|J_i^2(t, x_1) - J_i^2(t, x_2)\|\} \leqslant K_{J_i}\|x_1 - x_2\|$$

for all $t \in I_i$ and $x_1, x_2 \in X$.

[A_6] There exist non-negative constants L_{Ji} and M_{Ji}, $i \in \mathbb{N}$ such that

$$\max\{\|J_i^1(t, x)\|, \|J_i^2(t, x)\|\} \leqslant L_{J_i}\|x\| + M_{J_i}, \quad t \in I_i, x \in X.$$

[A_7] The following periodicity conditions hold: $J_{i+m}^l(t + T, x) = J_i^l(t, x)$, $t \in I_i$, $x \in X$, where $i \in \mathbb{N}$ and $l = 1, 2$. Note $I_i + T = I_{i+m}$. So $K_{Ji+m} = K_{Ji}$, $L_{Ji+m} = L_{Ji}$ and $M_{Ji+m} = M_{Ji}$, $i \in \mathbb{N}$.

Let us set

$PC(I, X) = \{x: I \to X: x \in C([0, t_1], X), x \in C((t_k, t_{k+1}], X), k \in \mathbb{N}$ and there exist $x(t_k^-)$ and $x(t_k^+)$, $k \in \mathbb{N}$ with $x(t_k^-) = x(t_k)\}$.

In the following definition, we introduce the concept of the mild solution and mild dynamics for problem (4.45).

Definition 4.3.2. *A function $x \in PC(I, X)$ is called a mild solution of the impulsive problem*

$$\begin{aligned} x''(t) &= Ax(t) + f(t, x(t)), \quad t \in (s_i, t_{i+1}], \; i \in \mathbb{N}_0; \\ x(t) &= J_i^1(t, x(t_i^-)), \quad t \in (t_i, s_i], \; i \in \mathbb{N}; \\ x'(t) &= J_i^2(t, x(t_i^-)), \quad t \in (t_i, s_i], \; i \in \mathbb{N}; \\ x(0) &= x_0, \quad x'(0) = y_0, \end{aligned} \tag{4.46}$$

if it satisfies the following relations:

- *the non-instantaneous impulse conditions*

$$x(t) = J_i^1(t, x(t_i^-)), \quad x'(t) = J_i^2(t, x(t_i^-)), \quad t \in (t_i, s_i], \; i \in \mathbb{N}.$$

- *$x(t)$ is the solution of the following integral equations*

$$x(t) = C(t)x_0 + S(t)y_0 + \int_0^t S(t - s)f(s, x(s))ds, \quad t \in [0, t_1],$$

$$x(t) = C(t - s_i)(J_i^1(s_i, x(t_i^-))) + S(t - s_i)(J_i^2(s_i, x(t_i^-))) + \int_{s_i}^t S(t - s)f(s, x(s))ds, \quad t \in [s_i, t_{i+1}], \quad i \in \mathbb{N}.$$

From the mild dynamics of (4.46) we consider the iteration

$$(x_0, y_0) \cup \{(J_i^1(s_i, x(t_i^-)), J_i^2(s_i, x(t_i^-)))\}_{i \in \mathbb{N}}. \tag{4.47}$$

Definition 4.3.3. *A function $x \in PC(I, X)$ is said to be a T-periodic PC-mild solution of problem (4.45) if it is a PC-mild solution of problem (4.46) for some initial conditions x_0, y_0 and the corresponding iteration (4.47) is m-periodic, i.e. it holds that*

$$\begin{aligned}(x_0, y_0) &= (J_m^1(s_m, x(t_m^-)), J_m^2(s_m, x(t_m^-)))\\ &= (J_{i+m}^1(s_{i+m}, x(t_{i+m}^-)), J_{i+m}^2(s_{i+m}, x(t_{i+m}^-))), \quad \forall i \in \mathbb{N}.\end{aligned}$$

We recall that $s_m = T$ in the above definition.

4.3.3 The existence and uniqueness of periodic solutions

From [30, theorem 3.1], we have the following

Theorem 4.3.4. *If the assumptions $[A_1]$, $[A_2]$, $[A_3]$, $[A_5]$ and $[A_6]$ are satisfied, then the second-order problem (4.46) has a unique mild solution.*

Remark 4.3.5. *If $x \in PC(I, X)$ is a T-periodic PC-mild solution of problem (4.45) then $x(t + T) = x(t)$, $\forall\ t \geqslant 0$ by the uniqueness result of theorem 4.3.4. On the other hand, if $x_0 \in E$ then $x(t + T) = x(t)$, $\forall\ t \geqslant 0$ implies the m-periodicity of (4.47), since*

$$x(0) = x_0,\ x'(0^+) = y_0,\ x(s_i) = J_i^1(s_i, x(t_i^-)),\ x'(s_i^-) = J_i^2(s_i, x(t_i^-)), \quad i \in \mathbb{N}.$$

Moreover, if $x_0 \in E$ and $J_i^1(s_i, E) \subset E$ for any $i \in \mathbb{N}$, then $x \in C^1((t_i, t_{i+1}), X)$, $i \in \mathbb{N}$ and there exist $x'(t_i^-)$ and $x'(t_i^+)$, $i \in \mathbb{N}$, so assuming $x'(t_i^-) = x'(t_i)$, we obtain $x'(t + T) = x'(t)$, $\forall\ t \geqslant 0$.

Now we define a Poincaré operator P: $X \times X \to X \times X$, $P = (\tilde{P}_1, \tilde{P}_2)$ by

$$\begin{aligned}\tilde{P}_1(x_0, y_0) &= J_m^1(T, x(t_m^-, x_0))\\ &= J_m^1\Big(T, C(t_m - s_{m-1})(J_{m-1}^1(s_{m-1}, x(t_{m-1}^-, x_0)))\\ &\quad + S(t_m - s_{m-1})(J_{m-1}^2(s_{m-1}, x(t_{m-1}^-, x_0)))\\ &\quad + \int_{s_{m-1}}^{t_m} S(t_m - s)f(s, x(s))ds\Big),\end{aligned} \tag{4.48}$$

and

$$\begin{aligned}\tilde{P}_2(x_0, y_0) &= J_m^2(T, x(t_m^-, x_0))\\ &= J_m^2\Big(T, C(t_m - s_{m-1})(J_{m-1}^1(s_{m-1}, x(t_{m-1}^-, x_0)))\\ &\quad + S(t_m - s_{m-1})(J_{m-1}^2(s_{m-1}, x(t_{m-1}^-, x_0)))\\ &\quad + \int_{s_{m-1}}^{t_m} S(t_m - s)f(s, x(s))ds\Big).\end{aligned} \tag{4.49}$$

It is easy to show that the fixed points of P defined in (4.48) and (4.49) give rise to a periodic solution to problem (4.45), i.e. the following result holds.

Lemma 4.3.6. *Problem (4.45) has a T-periodic PC-mild solution if and only if P has a fixed point.*

We see that P is a composition of the maps:

$$P = \mathbb{J}_m \circ P_{m-1} \circ \mathbb{J}_{m-1} \circ P_{m-2} \circ \mathbb{J}_{m-2} \circ \cdots \circ \mathbb{J}_1 \circ P_0, \tag{4.50}$$

where

$$\begin{aligned} &\mathbb{J}_i: X \to X \times X, \quad \mathbb{J}_i(u) = (J_i^1(s_i, u), J_i^2(s_i, u)), \quad i = 1, 2, \dots, m; \\ &P_i(z) = x(t_{i+1}, z), \quad i = 0, 1, 2, \dots, m-1; \\ &x(t, z) = C(t - s_i)u + S(t - s_i)v + \int_{s_i}^t S(t - s)f(s, x(s, z))\,ds, \\ &\qquad t \in [s_i, t_{i+1}], \end{aligned} \tag{4.51}$$

where we set $z = (u, v) \in X \times X$ and consider a norm $\|z\| = \max\{\|u\|, \|v\|\}$.

Next, from [40, theorem 1.1], there is $K \geqslant 1$ and $\omega > 0$ such that $\|C(t)\| \leqslant Ke^{\omega t}$ for any $t \geqslant 0$. Then we derive $\|S(t)\| \leqslant \frac{K}{\omega}e^{\omega t}$. From assumption $[A_3]$, we have

$$\begin{aligned} \|x(t, z)\| \leqslant{}& Ke^{\omega(t-s_i)}\|z\| + \frac{K}{\omega}e^{\omega(t-s_i)}\|z\| \\ &+ \frac{KL_f}{\omega}\int_{s_i}^t e^{\omega(t-s)}(1 + \|x(s, z)\|)ds. \end{aligned} \tag{4.52}$$

Setting $y(t, z) = e^{-\omega(t-s_i)}\|x(t, z)\|$, equation (4.52) is reduced to

$$\begin{aligned} y(t, z) &\leqslant K\left(1 + \frac{1}{\omega}\right)\|z\| + \frac{KL_f}{\omega}\int_{s_i}^t [e^{-\omega(s-s_i)} + y(s, z)]ds \\ &\leqslant K\left(1 + \frac{1}{\omega}\right)\|z\| + \frac{KL_f}{\omega^2} + \frac{KL_f}{\omega}\int_{s_i}^t y(s, z)ds. \end{aligned}$$

Applying the Grönwall inequality, we obtain

$$y(t, u) \leqslant K\left[\left(1 + \frac{1}{\omega}\right)\|z\| + \frac{L_f}{\omega^2}\right]e^{\frac{KL_f(t-s_i)}{\omega}}.$$

Hence

$$\|P_i(z)\| = \|x(t_{i+1}, z)\| \leqslant K\left[\left(1 + \frac{1}{\omega}\right)\|z\| + \frac{L_f}{\omega^2}\right]e^{\left(\frac{KL_f}{\omega}+\omega\right)(t_{i+1}-s_i)}. \tag{4.53}$$

From assumption $[A_6]$ and inequality (4.53), we have

$$\|(\mathbb{J}_i \circ P_{i-1})(z)\| \leqslant a + L_{J_i}K\left(1 + \frac{1}{\omega}\right)e^{\left(\frac{KL_f}{\omega}+\omega\right)(t_i-s_{i-1})}\|z\|, \quad i = 1, 2, \dots, m,$$

where

$$a = \max_{i=1,2,\ldots,m}\left[M_{J_i} + \frac{L_{J_i}KL_f}{\omega^2}e^{\left(\frac{KL_f}{\omega}+\omega\right)(t_i-s_{i-1})}\right].$$

Setting

$$b = K\left(1 + \frac{1}{\omega}\right)\max_{i=1,2,\ldots,m} L_{J_i}e^{(\frac{KL_f}{\omega}+\omega)(t_i-s_{i-1})};$$

$$A = a\sum_{i=0}^{m-1} b^i;$$

$$\theta = K^m\left(1 + \frac{1}{\omega}\right)^m \prod_{i=1}^{m} L_{J_i}e^{(\frac{KL_f}{\omega}+\omega)\sum_{i=1}^{m}(t_i-s_{i-1})},$$

by an elementary computation, we have

$$\begin{aligned}\|P(z)\| &= \|(\mathbb{J}_m \circ P_{m-1} \circ \mathbb{J}_{m-1} \circ P_{m-2} \circ \mathbb{J}_{m-2} \circ \cdots \circ \mathbb{J}_1 \circ P_0)(z)\| \\ &\leqslant a(1 + b + \cdots + b^{m-1}) \\ &\quad + K^m\left(1 + \frac{1}{\omega}\right)^m \prod_{i=1}^{m} L_{J_i}\, e^{(\frac{KL_f}{\omega}+\omega)\sum_{i=1}^{m}(t_i-s_{i-1})}\|z\| \\ &\leqslant A + \theta\|z\|.\end{aligned} \tag{4.54}$$

From (4.54), we have the following result.

Lemma 4.3.7. *All PC-mild solutions of problem (4.46) corresponding to initial conditions in a bounded subset of X are uniformly bounded.*

Next, we verify that P defined by (4.50) is a continuous and compact operator under our assumptions. If we can show P_i, $i = 0, 1, \ldots, m - 1$ are continuous and compact, then we can use the continuity of $\mathbb{J}_i$, $i = 1, 2, \ldots, m$ to derive that P is continuous and compact.

Lemma 4.3.8. *If $(C(t))_{t\geqslant 0}$ is a compact cosine family, then P is a continuous and compact operator.*

Proof. Suppose $x(\cdot, z_1)$ and $z(\cdot, z_2)$ are the *PC*-mild solutions defined in (4.51) corresponding to the initial conditions $z_i = (u_i, v_i) \in X \times X$, $i = 1, 2$, respectively. From assumption $[A_2]$, we have

$$\begin{aligned}\|x(t, z_1) - x(t, z_2)\| &\leqslant Ke^{\omega(t-s_i)}\|z_1 - z_2\| + \frac{K}{\omega}e^{\omega(t-s_i)}\|z_1 - z_2\| \\ &\quad + \frac{KK_f}{\omega}\int_{s_i}^{t} e^{\omega(t-s)}\|x(s, z_1) - x(s, z_2)\|ds \\ &\leqslant Ke^{\omega T}\left(1 + \frac{1}{\omega}\right)\|z_1 - z_2\| \\ &\quad + \frac{KK_f e^{\omega T}}{\omega}\int_{s_i}^{t} \|x(s, z_1) - x(s, z_2)\|ds \\ &\quad t \in [s_i, t_{i+1}],\ i = 0, 1, \dots, m.\end{aligned}$$

By applying the Grönwall inequality, we have

$$\begin{aligned}\|x(t_{i+1}, z_1) - x(t_{i+1}, z_2)\| &\leqslant Ke^{\omega T}\left(1 + \frac{1}{\omega}\right)e^{\frac{TKK_f e^{\omega T}}{\omega}}\|z_1 - z_2\|, \\ &\quad i = 0, 1, \dots, m - 1.\end{aligned} \tag{4.55}$$

Hence, p_i, $i = 0, 1, \dots, m - 1$ are continuous operators on $X \times X$.

Now, we shall show that p_i, $i = 0, 1, \dots, m - 1$ are compact. Let Γ be a bounded subset of $X \times X$. Then $\Gamma \subset B_r \times B_r$ for a closed ball B_r in X centered at 0 with a radius $r > 0$. Recalling (4.51), we have

$$P_i(z) = C(t_{i+1} - s_i)u + S(t_{i+1} - s_i)v + \int_{s_i}^{t_{i+1}} S(t_{i+1} - s)f(s, x(s, z))ds. \tag{4.56}$$

It is easy to see that R: $[s_i, t_{i+1}] \to L(X)$ given that $R(s) = S(t_{i+1} - s)$ is continuous and $R(s)$ is compact for any $s \in [s_i, t_{i+1}]$. Then the sets $C(t_{i+1} - s_i)(B_r)$ and $S(t_{i+1} - s_i)$ (B_r) are precompact. Next, for any $n \in \mathbb{N}$, we consider continuous and linear operators Λ, Λ_k: $\Upsilon_i = C([s_i, t_{i+1}], X) \to X$, $k = 0,1, \dots, n - 1$ given by

$$\Lambda x = \int_{s_i}^{t_{i+1}} R(s)x(s)ds, \quad \Lambda_k x = \sum_{k=0}^{n-1} R(r_k)\int_{r_k}^{r_{k+1}} x(s)ds,$$

for $r_k = \frac{(n-k)s_i}{n} + \frac{kt_{i+1}}{n}$, $k = 0, 1, \dots, n - 1$. It is well known that Υ_i is a Banach space with a norm

$$\|x\|_\infty = \max_{s\in[s_i, t_{i+1}]} \|x(s)\|.$$

From

$$\|\Lambda x - \Lambda_k x\| \leqslant \|x\|_\infty \sum_{k=0}^{n-1}\int_{r_k}^{r_{k+1}} \|R(s) - R(r_k)\|ds,$$

we obtain

$$\|\Lambda - \Lambda_k\| \leqslant \sum_{k=0}^{n-1}\int_{r_k}^{r_{k+1}} \|R(s) - R(r_k)\|ds. \tag{4.57}$$

Since $[s_i, t_{i+1}]$ is compact, then $R: [s_i, t_{i+1}] \to L(X)$ is uniformly continuous and so (4.57) implies $\Lambda_k \to \Lambda$ in $L(\Upsilon_i, X)$. On the other hand, since each $x \to \int_{r_k}^{r_{k+1}} x(s)$ is linear and continuous, and so bounded, and as mapping $\Upsilon_i \to X$ and $R(r_k)$ are compact, we see that each Λ_k is compact. But then Λ is also compact. Next, the Nemytskii operator $F: \Upsilon_i \to \Upsilon_i$ defined as $(Fx)(s) = f(s, x(s))$ is continuous and bounded, since from $[A_3]$ $\|F(x)\|_\infty \leqslant K_f\|x\|_\infty + M_f$. Hence F maps bounded subsets to bounded ones. Then the mapping $G: \Upsilon_i \to X$ defined as

$$G(x) = \int_{s_i}^{t_{i+1}} S(t_{i+1} - s)f(s, x(s))ds$$

is compact and continuous, since $G = \Lambda \circ F$. According to (4.57), p_i is a sum of continuous and compact maps, so it is also continuous and compact. We just presented an alternative way of proving the continuity of p_i.

Finally, we can use the continuity of $\mathbb{J}_i$, $i = 1, 2, \ldots, m$ to derive that P is continuous and compact. □

Now we present the main theorems of this section.

Theorem 4.3.9. *Assume that $[A_1]$–$[A_7]$ are satisfied. If $(C(t))_{t\geqslant 0}$ is compact and $\theta < 1$, then equation (4.45) has at least a T-periodic PC-mild solution.*

Proof. From lemma 4.3.8, operator P is continuous and compact. Inequality (4.54) implies $P: B_{\theta_0} \times B_{\theta_0} \to B_{\theta_0} \times B_{\theta_0}$ for $\theta_0 = \frac{A}{1-\theta}$. Clearly $B_{\theta 0} \times B_{\theta 0}$ is bounded, convex and closed. So from Schauder fixed point theorem, there exist x_0, $y_0 \in X$ such that $P(z_0) = z_0$ for $z_0 = (x_0, y_0)$. By applying lemma 4.3.6, we can obtain a T-periodic PC-mild solution $x(\cdot, z_0)$ of Cauchy problem (4.46) corresponding to the initial conditions $x(0) = x_0$ and $x'(0) = y_0$. Thus, $x(\cdot, z_0)$ is a T-periodic PC-mild solution of problem (4.45). The proof is complete. □

When $(C(t))_{t\geqslant 0}$ is not compact, we can still obtain the following results.

Theorem 4.3.10. *Assume that $[A_1]$–$[A_7]$ are satisfied. If $\mathbb{J}_i$ is compact for some $i \in \{1, 2, \ldots, m\}$ and $\theta < 1$, then equation (4.45) has at least a T-periodic PC-mild solution.*

Proof. The result follows directly from the proof of theorem 4.3.9, since P is still compact. □

Theorem 4.3.11. *Assume that $[A_1]$–$[A_7]$ are satisfied. If $(S(t))_{t\geqslant 0}$ is compact and $\theta < 1$, then equation (4.45) has at least a T-periodic PC-mild solution provided that one of the following conditions holds:*

(a) $J_i^1(s_i, \cdot)$: $X \to X$ *is compact for some some* $i \in \{1, 2, \dots, m-1\}$,

(b) $J_m^1(s_m, \cdot)$: $X \to Z$ *for a finite-dimensional subspace* $Z \subset X$.

Proof. Assuming (a), from (4.56) and the arguments below it, we see that $P_{i+1} \circ \mathbb{J}_i$ is compact. Hence P is also compact. So the result follows again directly from the proof of theorem 4.3.9. Assuming (b), we have P: $Z \times X \to Z \times X$. Next, from (4.56) and the arguments below it, we see that P_0: $Z \times X \to X$ is compact. Hence P: $(Z \cap B_{\theta_0}) \times X \to (Z \cap B_{\theta_0}) \times X$ is also compact. So the result follows from the Schauder fixed point theorem. The proof is complete. □

From (4.54) we derive

$$\|P^k(z)\| \leqslant A\frac{1-\theta^k}{1-\theta} + \theta^k\|z\| \leqslant \theta_0 + \theta^k\|z\|,$$

for any $k \in \mathbb{N}_0$ and $z \in X \times X$. This implies that the mild dynamics of (4.45) is dissipative and its global compact attractor is

$$\mathcal{A} = \bigcap_{k=0}^{\infty} P^k(B_{\theta_0} \times B_{\theta_0}) \subset B_{\theta_0} \times B_{\theta_0},$$

for theorems 4.3.9, 4.3.10 and 4.3.11(a), while we have

$$\mathcal{A} = \bigcap_{k=0}^{\infty} P^k((Z \cap B_{\theta_0}) \times B_{\theta_0}) \subset (Z \cap B_{\theta_0}) \times B_{\theta_0},$$

for theorem 4.3.11(b). Finally, we can just apply the Banach fixed point theorem in the general case.

Theorem 4.3.12. *Assume that $[A_1]$–$[A_7]$ are satisfied. If*

$$\Upsilon = K^m\left(1 + \frac{1}{\omega}\right)^m \prod_{i=1}^{m} K_{J_i} e^{\left(\frac{KK_f}{\omega}+\omega\right)\sum_{i=1}^{m}(t_i - s_{i-1})} < 1,$$

then equation (4.45) has a unique T-periodic PC-mild solution, which a global attractor.

Proof. By following the above arguments to (4.54) and (4.55), we have

$$\|P(z_1) - P(z_2\| \leqslant \Upsilon\|z\|.$$

Since $\Upsilon < 1$, the result is a consequence of Banach fixed point theorem. □

4.3.4 An example

Consider

$$\begin{aligned}
\partial_{tt}x(t, y) &= \partial_{yy}x(t, y) + \frac{|x(t, y)|}{1+|x(t, y)|} + y \sin t, \quad y \in (0, 1), \\
&\quad t \in (0, \pi]; \\
x(t, 0) &= x(t, 1) = 0, \quad t \in [0, 2\pi]; \\
x(0, y) &= x_0(y), \quad y \in (0, 1); \\
\partial_t x(0, y) &= y_0(y), \quad y \in (0, 1); \\
x(t, y) &= \frac{|x(\pi^-, y)|}{1+|x(\pi^-, y)|} \sin t, \quad y \in (0, 1), \; t \in (\pi, 2\pi]; \\
\partial_t x(t, y) &= \frac{|x(\pi^-, y)|}{1+|x(\pi^-, y)|} \cos t, \quad y \in (0, 1), \; t \in (\pi, 2\pi].
\end{aligned} \tag{4.58}$$

Equation (4.58) can be reformulated as the following abstract equation in $X = L^2(0, 1)$:

$$\begin{aligned}
x''(t) &= Ax(t) + f(t, x(t)), \quad t \in (0, \pi]; \\
x(t) &= J_1^1(t, x(t_i^-)), \quad t \in (\pi, 2\pi]; \\
x'(t) &= J_1^2(t, x(t_i^-)), \quad t \in (\pi, 2\pi]; \\
x(0) &= x_0, \quad x'(0) = y_0,
\end{aligned} \tag{4.59}$$

where $0 = t_0 = s_0$, $t_1 = \pi$, $s_1 = T = 2\pi$, $x(t)(y) = x(t, y)$, $y \in (0, 1)$, and

$$\begin{aligned}
Ax &= x_{yy}, \quad D(A) = \{x \in X\colon\; x_{yy} \in X, \; x(0) = x(1) = 0\}; \\
f(t, x)(y) &= \frac{|x(\cdot, y)|}{1+|x(., y)|} + y \sin t; \\
J_1^1(t, x(t_i^-)) &= \frac{|x(\pi^-, y)|}{1+|x(\pi^-, y)|} \sin t; \\
J_1^2(t, x(t_i^-)) &= \frac{|x(\pi^-, y)|}{1+|x(\pi^-, y)|} \cos t.
\end{aligned}$$

Note that A has an infinite series representation

$$Ax = \sum_{n=1}^{\infty} - n^2(x, x_n)x_n, \quad x \in D(A),$$

where $x_n(y) = \sqrt{2} \sin n\pi y$, $n \in \mathbb{N}$ is an orthonormal set of eigenfunctions of A. Moreover, the operator A is the infinitesimal generator of a strongly continuous cosine family $C(t)_{t \in \mathbb{R}}$ on X which is given by

$$C(t)x = \sum_{n=1}^{\infty} \cos nt(x, x_n)x_n, \quad x \in X,$$

and the associated sine family $S(t)_{t \in \mathbb{R}}$ on X is given by

$$S(t)x = \sum_{n=1}^{\infty} \frac{1}{n} \sin nt(x, x_n)x_n, \quad x \in X.$$

Clearly, $C(t)_{t\in\mathbb{R}}$ is not compact, while $S(t)_{t\in\mathbb{R}}$ is compact. Next, we have $K_f = K_{J_1} = 1$, $L_f = L_{J_1} = 0$ and $M_f = M_{J_1} = 1$. Hence $\theta = 0$. Since $J_1^1(s_1, u) = 0$, we can apply theorem 4.3.11(b) with $Z = \{0\}$ to obtain a 2π-periodic PC-mild solution of (4.58).

4.3.5 Notes and remarks

In this section, we use the strongly continuous cosine family of linear operators along with the Schauder and Banach fixed point theorems to study the existence and uniqueness of the periodic solutions of a non-instantaneous impulsive system. Moreover, we construct a Poincaré operator which is a composite of the maps, and we apply the techniques of *a priori* estimates for this operator.

The results in section 4.3 are motivated by [43].

References

[1] Hernández E and O'Regan D 2013 On a new class of abstract impulsive differential equations *Proc. Amer. Math. Soc.* **141** 1641–9

[2] Pierri M, O'Regan D and Rolnik V 2013 Existence of solutions for semi-linear abstract differential equations with not instantaneous impulses *Appl. Math. Comput.* **219** 6743–9

[3] Bainov D D and Simeonov P S 1993 *Impulsive Differential Equations: Periodic Solutions and Applications* (Boca Raton, FL: CRC Press)

[4] Samoilenko A M and Perestyuk N A 1995 *Impulsive Differential Equations World Scientific Series on Nonlinear Science. Series A: Monographs and Treatises* vol 14 (Singapore: World Scientific)

[5] Benchohra M, Henderson J and Ntouyas S 2006 *Impulsive Differential Equations and Inclusions Contemporary Mathematics and Its Applications* vol 2 (New York: Hindawi)

[6] Xiang X and Ahmed N U 1992 Existence of periodic solutions of semilinear evolution equations with time lags *Nonlinear Anal.:TMA* **18** 1063–70

[7] Sattayatham P, Tangmanee S and Wei W 2002 On periodic solutions of nonlinear evolution equations in Banach spaces *J. Math. Anal. Appl.* **276** 98–108

[8] Wang J, Xiang X and Peng Y 2009 Periodic solutions of semilinear impulsive periodic system on Banach space *Nonlinear Anal.:TMA* **71** e1344–53

[9] Liu Z 2011 Anti-periodic solutions to nonlinear evolution equations *J. Funct. Anal.* **258** 2026–33

[10] Li Y 2011 Existence and asymptotic stability of periodic solution for evolution equations with delays *J. Funct. Anal.* **261** 1309–24

[11] Kokocki P 2012 Existence and asymptotic stability of periodic solution for evolution equations with delays *J. Math. Anal. Appl.* **392** 55–74

[12] Ahmed N U 1991 *Semigroup Theory with Applications to Systems and ControlPitman Research Notes in Mathematics Series* vol 246 (Harlow: Longman Scientific and Technical)

[13] Hale J K 2003 Stability and gradient dynamical systems *Rev. Mat. Comput.* **17** 7–57

[14] Hale J K 1988 *Asymptotic Behaviour of Dissipative Systems* (Providence, RI: American Mathematical Society)

[15] Liu J H 1994 Bounded and periodic solutions of differential equations in Banach space *Appl. Math. Comput.* **65** 141–50

[16] Liu J H 1995 Bounded and periodic solutions of semilinear evolution equations *Dynam. Syst. Appl.* **4** 341–50

[17] Liu J H 1998 Bounded and periodic solutions of finite delay evolution equations *Nonlinear Anal.:TMA* **34** 101–11

[18] Liu J H, Naito T and Minh N V 2003 Bounded and periodic solutions of infinite delay evolution equations *J. Math. Anal. Appl.* **286** 705–12

[19] Fečkan M, Wang J and Zhou Y 2014 Existence of periodic solutions for nonlinear evolution equations with non-instantaneous impulses *Nonauton. Dyn. Syst.* **1** 93–101

[20] Barnsley M F 1993 *Fractals Everywhere* 2nd edn (San Diego, CA: Morgan Kaufmann)

[21] Wang J and Fečkan M 2015 A general class of impulsive evolution equations *Topol. Methods Nonlinear Anal.* **46** 915–33

[22] La Torre D, Marsiglio S and Privileggi F 2011 Fractals and self-similarity in economics: the case of a stochastic two-sector growth model *Image Anal. Stereol.* **30** 143–51

[23] Pazy A 1983 *Semigroup of Linear Operators and Applications to Partial Differential Equations* (New York: Springer)

[24] Daners D and Medina K P 1992 *Abstract Evolution Equations, Periodic Problems and ApplicationsPitman Research Notes in Mathematics Series* vol 279 (Harlow: Longman Scientific and Technical)

[25] Nguyenm T L 1999 On nonautonomous functional differential equations *J. Math. Anal. Appl.* **239** 158–74

[26] Park J Y, Kwun Y C and Jeong J M 2004 Existence of periodic solutions for delay evolution integrodifferential equations *Math. Comput. Model* **40** 597–603

[27] Buşe C, Lassoued D, Nguyen T L and Saierli O 2012 Exponential stability and uniform boundedness of solutions for nonautonomous periodic abstract Cauchy problems. an evolution semigroup approach *Integr. Equ. Oper. Theory* **74** 345–62

[28] Zabczyk J 1992 *Mathematical Control Theory: An Introduction Systems and Control* (Basel: Birkhäuser)

[29] Wang J, Fečkan M and Zhou Y 2016 Random noninstantaneous impulsive models for studying periodic evolution processes in pharmacotherapy *Mathematical Modeling and Applications in Nonlinear Dynamics, Nonlinear Systems and Complexity*vol 14 ed A Luo and H Merdan (Cham: Springer)

[30] Muslim M, Kumar A and Fečkan M 2018 Existence, uniqueness and stability of solutions to second order nonlinear differential equations with non-instantaneous impulses *J. King Saud Univ. Sci.* **30** 204–13

[31] Acharya F S 2013 Controllability of second order semilinear impulsive partial neutral functional differential equations with infinite delay *Internat. J. Math. Sci. Appl.* **3** 207–18

[32] Arthi G and Balachandran K 2014 Controllability of second order impulsive evolution systems with infinite delay *Nonlinear Anal.: Hybrid Systems* **11** 139–53

[33] Chalishajar D N 2009 Controllability of damped second order initial value problem for a class of differential inclusions with nonlocal conditions on non compact intervals *Nonlinear Func. Anal. Appl. (Korea)* **14** 25–44

[34] Pandey D N, Das S and Sukavanam N 2014 Existence of solution for a second-order neutral differential equation with state dependent delay and non-instantaneous impulses *Int. J. Nonlinear Sci.* **18** 145–55

[35] Sakthivel R, Mahmudov N I and Kim J H 2009 On controllability of second order nonlinear impulsive differential systems *Nonlinear Anal.* **71** 45–52

[36] Byszewski L 1991 Theorem about the existence and uniqueness of solutions of a semilinear nonlocal Cauchy problem *J. Math. Anal. Appl.* **162** 494–505

[37] Gal C G 2007 Nonlinear abstract differential equations with deviated argument *J. Math. Anal. Appl.* **333** 971–83

[38] Hernández E and McKibben M A 2005 Existence of solutions of abstract nonlinear second-order neutral functional integrodifferential equations *Comput. Math. Appl.* **50** 655–69

[39] Muslim M and Bahuguna D 2008 Existence of solutions to neutral differential equations with deviated argument *Electron. J. Qual. Theory Differ. Equ.* **No. 27** 1–12

[40] Fattorini H O 1985 *Second Order Linear Differential Equations in Banach SpacesNorth Holland Mathematics Studies* vol 108 (Amsterdam: North Holland)

[41] Travis C C and Webb G F 1977 Compactness, regularity and uniform continuity properties of strongly continuous cosine family *Houston J. Math.* **3** 555–67

[42] Travis C C and Webb G F 1978 Cosine families and abstract nonlinear second order differential equations *Acta Math. Hungar.* **32** 76–96

[43] Muslim M, Kumar A and Fečkan M 2017 Periodic solutions to second order nonlinear differential equations with non-instantaneous impulses *Dyn. Syst. Appl.* **26** 197–210

Lightning Source UK Ltd.
Milton Keynes UK
UKHW052321151218
333888UK00004B/52/P

9 780750 317023